MEMBRANE PROTEIN TRANSPORT

A Multi-Volume Treatise

Volume 1 • 1995

MEMBRANE PROTEIN TRANSPORT

A Multi-Volume Treatise

Editor: STEPHEN S. ROTHMAN
University of California
San Francisco, California

VOLUME 1 • 1995

JAI PRESS INC.

Greenwich, Connecticut   *London, England*

CONTENTS

LIST OF CONTRIBUTORS

Twan America

Department of Molecular Cell Biology
University of Utrecht
Utrecht, The Netherlands

Robert Arkowitz

MRC Laboratory of Molecular Biology
Cambridge, England

Drusilla L. Burns

Division of Bacterial Products
Center for Biologics Evaluation
and Research
Bethesda, Maryland

Vitaly Citovsky

Department of Biochemistry and Cell
Biology
State University of New York
Stony Brook, New York

Ben de Kruijff

Department of Biochemistry of Membranes
Centre for Biomembranes and Lipid
Enzymology
University of Utrecht
Utrecht, The Netherlands

J. Fred Dice

Department of Physiology
Tufts University School of Medicine
Boston, Massachusetts

Miriam Eisenstein

Department of Structural Biology
Weizmann Institute
Rehovot, Israel

Francois Franceschi

Max Planck Institute for Molecular Genetics
Berlin, Germany

Boyd Hardesty

Department of Chemistry and Biochemistry
University of Texas at Austin
Austin, Texas

Martin Horst

Biocenter
University of Basel
Basel, Switzerland

Lois Isenman

Tufts University School of Medicine
Boston, Massachusetts

Gisela Kramer   Department of Chemistry and Biochemistry
                                    University of Texas at Austin
                                    Austin, Texas

Nafsika G. Kronidou   Biocenter
                                    University of Basel
                                    Basel, Switzerland

Wieslaw Kudlicki   Department of Chemistry and Biochemistry
                                    University of Texas at Austin
                                    Austin, Texas

Claude J. Lazdunski   Laboratoire d'Ingénierie et Dyanamique des
                                    Systèmes Membranaires
                                    Centre National de la Reserche Scientifique
                                    Marseille, France

Erwin London   Department of Biochemistry and Cell
                                    Biology
                                    State University of New York
                                    Stony Brook, New York

Randal E. Morris   Departments of Anatomy and Cell Biology
                                    University of Cincinnati
                                    Cincinnati, Ohio

O. W. Odom   Department of Chemistry and Biochemistry
                                    University of Texas at Austin
                                    Austin, Texas

Marinus Pilon   Department of Biochemistry of Membranes
                                    Centre for Biomembranes and Lipid
                                    Enzymology, and
                                    Department of Molecular Cell Biology
                                    University of Utrecht
                                    Utrecht, The Netherlands

Catharine B. Saelinger   Departments of Molecular Genetics,
                                    Biochemistry, and Microbiology
                                    University of Cincinnati
                                    Cincinnati, Ohio

Ron van't Hof   Department of Biochemistry of Membranes
                                    Centre for Biomembranes and Lipid
                                    Enzymology
                                    University of Utrecht
                                    Utrecht, The Netherlands

Peter Weisbeek   Department of Molecular Cell Biology
                                    University of Utrecht
                                    Utrecht, The Netherlands

Ada Yonath Department of Structural Biology
Weizmann Institute
Rehovot, Israel

Max Planck Institute for Molecular Genetics
Hamburg, Germany

Patricia Zambryski Department of Plant Biology
University of California
Berkeley, California

PREFACE

This treatise on *Membrane Protein Transport* is a landmark. It is the first multi-volume work published on this important subject, although most certainly it will not be the last. But its significance, in my view, goes beyond the simple fact of being first. It deals with a subject that only recently has been appreciated as being central to the organization of the cell. Membrane protein transport underlies the topologic disposition (sorting) of many proteins within cells, and it is this disposition that allows for the coordination of central cellular processes, such as metabolism, secretion, and genetic expression.

A second reason for viewing this treatise as something special is that, whatever our perception today, until recently such processes were thought impossible. That is, it was thought that the structure of both proteins and biomembranes prohibited membrane protein transport. In the mid-1970s this view began to change, albeit in a limited fashion with the proposal that proteins could pass through membranes individually during their synthesis, or cotranslationally. After synthesis, however, such processes remained prohibited. Indeed, most textbooks in biology today still describe protein transport in this fashion. However, we now know that this view is incorrect. A great variety, perhaps the lion's share, of membrane protein transport processes are posttranslational; that is, occur subsequent to protein synthesis, and recent evidence suggests that even cotranslational events may occur more in proximity (in time and place) to synthesis, than coupled to it.

Many years ago, in the early 1970s, I was giving a talk on some of our early work on this subject. I had finished presenting our evidence, and drew the conclusion that proteins appeared to be moving through the membranes that we were studying (see the paper by Goncz and Rothman in Volume 3 of this series). A scientist well known for his pioneering work on membrane structure rose and made a statement in lieu of a question. He said that our results and conclusions had to be wrong, because proteins could not pass through biological membranes. The structure of both proteins and membranes prohibited such processes; they were simply impossible, he asserted. This was the widely held view at the time, and had been so for many years. I responded by saying that if you do not believe our results, then how about those of another worker on another system. Feeling more positively inclined to this work, the questioner responded by agreeing with these observations; but added quickly that this system was an exception to generally held principles. Those general principles, he said, which applied to our work.

Well, today, I can say with some pleasure, that the exceptions have multiplied, as this treatise so aptly demonstrates. And we must ask what is the exception and what is the rule? When can it be said that there are enough exceptions to a rule, that the rule itself must be modified or discarded? I believe that membrane protein transport is more than a collection of exceptions, even interesting exceptions, to a general rule of impermeability. Rather, in my view the evidence demonstrates that such processes, although diverse and complex in nature, and as yet incompletely understood, represent a naturally occurring class of biological transport processes that can be found in all cells throughout the biological kingdoms, and that biological membranes as a quite general matter display, through a variety of mechanisms, a permeability to these most complex of organic molecules.

Our current plans are to publish three volumes, containing 30 or so papers on this and related subjects. This first volume includes papers on the secretion of various bacterial toxins and their uptake into host cells; the nature of bacterial membrane protein transport in general; the transport of proteins into chloroplasts and mitochondria: the role of membrane protein transport in protein degradation, lysosomal function and antigen presentation; the transport of protein-nucleic acid complexes in plant cells; as well as a related paper on intraribosomal folding of nascent peptide chains. Many of the papers deal with our understanding of transport mechanisms in particular cases, including a consideration of the energetics of transport, protein insertion into bilayers and structural considerations, and the role of membrane and soluble proteins in these processes.

I would like to make one technical note. It would have been nice to organize all of the papers in this series subject by subject, rather than using a journal style format in which publication order is determined by the time at which papers are received. However, given the fact that collection of these manuscripts will have occurred over a period of more than a year, requiring an additional year from receipt to publication, it seemed unfair to make authors wishing to have their data and ideas published in a timely way, whose papers were received early on in the process, to wait until a group of papers on a like subject could be compiled.

Finally, I would like to thank all of the authors for their wonderful efforts, and I hope that you, the reader, will agree with me that the results and discussion presented are fascinating. I would also like to thank JAI Press for their interest in publishing the series. Lastly, I would like to give special thanks to E. Edward Bittar for asking me to put such a series together.

Stephen S. Rothman
Editor

THE ENERGETICS OF BACTERIAL PROTEIN TRANSLOCATION

Robert Arkowitz

Membrane Protein Transport
Volume 1, pages 1–24.
Copyright © 1995 by JAI Press Inc.
All rights of reproduction in any form reserved.
ISBN: 1-55938-907-9

I. INTRODUCTION

Membrane bilayers separate and divide cells into functional compartments. The correct localization of different molecules in a cell is crucial for viability, allowing productive interactions while simultaneously preventing deleterious ones. Protein synthesis frequently occurs in a distinct location from the resident compartment of a functionally mature protein and as a result protein targeting and translocation across a membrane bilayer is essential. Within the last 10 to 20 years, there has been an explosion of research activity into the mechanisms of protein translocation across the bacterial, mitochondrial, endoplasmic reticulum, peroxisomal, and chloroplast membranes. These studies were initially directed at elucidating the protein components involved in such translocation processes. More recently, work has turned towards understanding the forms of energy driving unidirectional protein transport and the molecular mechanisms involved in catalyzing the translocation of proteins across a hydrophobic bilayer. The ability to carry out a range of genetic and biochemical manipulations in bacteria has resulted in extensive research efforts being focused on protein translocation across the cytoplasmic membrane of *Escherichia coli*. In this chapter, I will discuss recent work on the energetics of protein translocation across the *E. coli* inner membrane. The reader is directed elsewhere (Bieker et al., 1990; Mizushima and Tokuda, 1990; Schatz and Beckwith, 1990; Wickner et al., 1991; Pugsley, 1993; Arkowitz and Bassilana, 1994) for comprehensive reviews of bacterial protein translocation. This chapter will focus on the energy sources which drive *E. coli* protein translocation, specifically addressing the forms of energy involved in different stages of translocation and possible mechanisms of energy transduction. Finally, I will examine the commonalities and disparities of the energetics of bacterial protein translocation with that of yeast endoplasmic reticulum and mitochondrial protein translocation systems.

Genetic and biochemical investigations in *E. coli* have demonstrated that a complex cellular machinery comprised of SecA, SecB, SecD, SecE, SecF, SecY, and LepB is involved in protein translocation across the inner membrane. Proteins destined for secretion across the inner membrane are synthesized with an amino-terminal leader sequence (Gierasch, 1989; Randall and Hardy, 1989) which retards the folding of preproteins (Park et al., 1988). In *E. coli*, protein translation is largely uncoupled from protein translocation (Ito et al., 1980; Randall, 1983). As a result, many preproteins initially interact with the cytosolic chaperone SecB (Collier et al., 1988; Kumamoto and Gannon, 1988; Wiess et al., 1988, Lecker et al., 1989; Randall et al., 1990; Kumamoto and Francetic, 1993) which stabilizes preproteins prior to translocation and prevents aggregation (Lecker et al., 1990; Breukink et al., 1992b). Other cellular chaperones such as DnaK, DnaJ (Wild et al., 1992), and GroEL/GroES (Bochkareva et al., 1988; Kusukawa et al., 1989; Philips and Silhavy, 1990) are also involved in stabilization of preproteins. A SecB–preprotein complex is targeted to the cytoplasmic face of the inner membrane via SecA (Hartl et al., 1990).

SecA is found both in the cytosol and as a peripheral membrane protein (Oliver and Beckwith, 1982; Hartl et al., 1990; Cabelli et al., 1991). The affinity of leader sequences for the lipid bilayer (Nagaraj, 1984; Briggs et al., 1985; Nagaraj et al., 1987; Batenburg et al., 1988), particularly electrostatic interactions between positive charges in the leader and the negative charges of the acidic phospholipids, may also contribute to membrane targeting (Phoenix et al., 1993). SecA interacts with both the leader and mature domains of preproteins (Cunningham and Wickner, 1989; Lill et al., 1990). In addition, SecA binds the membrane due to its affinity for both acidic phospholipids (Hendrick and Wickner, 1991) and the integral membrane protein SecY/E (Hartl et al., 1990). SecY (Akiyama and Ito, 1987) and SecE (Schatz et al., 1989) have multiple membrane-spanning domains and these two integral membrane proteins form the stable complex SecY/E (Brundage et al., 1990, 1992; Matsuyama et al., 1990a; Nishiyama et al., 1992). In the presence of purified SecA, purified SecY/E reconstituted into liposomes catalyzes multiple cycles of preprotein translocation (Bassilana and Wickner, 1993). This multisubunit enzyme is referred to as preprotein translocase. Recent experiments suggest that SecA and SecY form a channel or groove through which preproteins may traverse the membrane (Joly and Wickner, 1993). While these studies do not exclude the possibility of the translocating preprotein contacting the lipid hydrocarbon phase, they are inconsistent with the notion that the preprotein is in an entirely lipid environment during bilayer translocation. Upon translocation across the membrane, leader peptidase (LepB) cleaves the leader sequence from the mature domain of the preprotein thereby releasing the translocated preprotein from the membrane (Dalbey and Wickner, 1985). The functions of SecD and SecF in this process are less defined, although recent experiments suggest that these proteins are involved in the later stages of translocation (Bieker-Brady and Silhavy, 1992) possibly in the release of proteins from the translocation apparatus (Matsuyama, et al., 1993; Arkowitz and Wickner, 1994).

Simplistically, translocation of a protein across a membrane can be thought of as an equi-energetic process, that is, the initial and final states of the reaction are of equal energy yet the activation barrier for the reaction is likely to be very high. This process requires energy input to drive the vectorial translocation of preproteins. Both ATP hydrolysis and the proton electrochemical gradient ($\Delta\mu H^+$) are used to drive preprotein translocation. The location and orientation of Sec proteins (including LepB) combine to make the overall translocation process irreversible.

II. CHEMICAL ENERGY: ATP HYDROLYSIS

A. *In Vivo* Studies

Energy derived from the hydrolysis of ATP is used to drive a multitude of cellular processes. Because of the ATP requirement of protein synthesis, the role of ATP hydrolysis in protein secretion has been elucidated primarily by examining post-translational translocation of preproteins into the lumen of inverted inner membrane vesicles. With this *in vitro* system, it was possible to demonstrate unequivocally that ATP is required for protein translocation (Chen and Tai, 1985; Geller et al., 1986; de Vrije et al., 1987; Yamane et al., 1987). *In vivo* studies have shown that depletion of intracellular ATP levels up to 96% did not result in an observable preprotein processing defect (Enequist et al., 1981; Geller, 1990). Although this section will focus on the hydrolysis of ATP during preprotein translocation, it should be noted that ATP hydrolysis by cytosolic chaperones such as GroEL/ES also plays a significant role in the maintenance of translocation competence of some preproteins or during specific cellular stresses (Hendrick and Hartl, 1993).

B. SecA

Initial studies with inverted inner membrane vesicles lacking F_1F_0-ATPase demonstrated that ATP hydrolysis alone (not in conjunction with the proton electrochemical gradient) was absolutely required for preprotein translocation (Chen and Tai, 1985; Geller et al., 1986; de Vrije et al., 1987; Yamane et al., 1987). It was also clear from these investigations, that the proton electrochemical gradient stimulated ATP-dependent translocation (the role of $\Delta\mu H^+$ will be discussed later). Purification of SecA, a peripheral membrane protein, showed that this essential component of the translocation apparatus catalyzes ATP hydrolysis (Cabelli et al., 1988; Cunningham, et al., 1989; Lill et al., 1989). Furthermore, ATP hydrolysis by SecA requires functional SecY, acidic phospholipids, and translocation competent preprotein (Lill et al., 1989, 1990). This ATPase activity has been termed translocation-ATPase (Lill et al., 1989), and refers to the overall ATP hydrolysis measured during *in vitro* translocation. In the absence of inner membrane vesicles and preproteins, SecA possesses a low endogenous ATPase activity (approximately 30 nmol/min·mg SecA) which is stimulated roughly three-fold upon addition of membrane vesicles, and 10- to 20-fold by addition of preproteins (Lill et al., 1989). Cross-linking studies using N_3-ATP demonstrated that SecA possesses approximately three ATP binding sites per monomer (Lill et al., 1989). More recently, photo-cross-linking studies with $[\alpha\text{-}^{32}P]$ATP suggest that there is a site of ATP binding within the amino-terminal 217 amino acyl residues of SecA (Matsuyama et al., 1990b). SecA contains several sequence motifs implicated in nucleotide binding (Oliver et al., 1990). At present the functional significance of these sites is unclear.

SecA is a large, multidomain enzyme and work from numerous laboratories has begun to identify these domains and elucidate their interactions with ATP, preprotein, and the inner membrane. Cross-linking studies of amino- and carboxy-terminal fragments of SecA in solution with a preprotein have identified a region of less than 100 amino acids (starting at amino acid residue 267) involved in SecA-preprotein interactions (Kimura et al., 1991). Interactions between SecA and the leader portion of preproteins are further supported by SecA alleles (*prlD*) which suppress leader sequence mutations *in vivo* (Fikes and Bassford, 1989). A similar region (the amino-terminal 239 amino acids) has been implicated in SecA membrane localization from studies using LacZ fusions (Cabelli et al., 1991). These results suggest that SecA–ATPase activity is regulated by its subcellular location and accessibility to preproteins. More recently, electron microscopy of thin-sectioned three-dimensional crystals of SecA at approximately 40 Å resolution, indicates large unit cell dimensions, with SecA crystallizing as an apparent tetramer (Weaver et al., 1992). SecA dominant negative mutations (Jarosik and Oliver, 1991), intragenic complementation (Cabelli et al., 1991), and reconstitution of ATP or preprotein binding sites from overlapping fragments of SecA (Matsuyama et al., 1990b; Kimura et al., 1991) all suggest SecA is multimeric. Although the functional significance of these multimeric states is unknown, multimerization may be another mechanism of regulating SecA activity.

Several membrane transport systems have components that couple ATP hydrolysis to the transport of various molecules. These ATP-dependent transporters include subunits of the bacterial periplasmic permeases, such as MalK and HisP, and exporters such as the mammalian multidrug resistance P-glycoprotein (Mdr), and the *E. coli* hemolysin exporter (HlyB). These ATPases hydrolyze ATP at stimulated rates (0.3–3 µmol/min·mg protein of transporter) (Davidson et al., 1992; Sarkadi et al., 1992; Koronakis et al., 1993) comparable to that of SecA during translocation. It is postulated that ATP hydrolysis triggers a conformational change in these transporters which opens a pore or binding site involved in membrane translocation (Ames and Joshi, 1990; Davidson et al., 1992). In contrast to the transport of small molecules, SecA, in conjunction with the other subunits of translocase, facilitates the ATP-dependent transport of polypeptides with many amino acyl residues and different side chains. One can envision three mechanistic scenarios in which translocation could be driven by SecA–ATP binding and hydrolysis for which there are ample precedents. It is conceivable that ATP binding and/or hydrolysis results in a conformational change in SecA (or change with respect membrane orientation) necessary for translocation. Another possibility would be that ATP binding and hydrolysis are involved in preprotein binding and release, similar to the mechanisms of chaperones such as GroEL/ES (Hendrick and Hartl, 1993; Jackson et al., 1993). Lastly, one can envision a combination of these two mechanisms, analogous to myosin binding, release, and movement on actin filaments. In this mechanism, ATP hydrolysis would result in changes in affinity of SecA for the preprotein and concomitant conformational change driving the preprotein across the membrane. To understand the function of ATP hydrolysis by SecA we must examine SecA's role in different stages of the translocation process.

C. Initiation of Translocation

SecA function is absolutely required for the initial stage of translocation. Although non-hydrolyzable ATP analogs such as AMP-PNP do not support preprotein translocation *in vitro*, Schiebel et al. (1991) observed that in the presence of AMP-PNP, the precursor of outer membrane protein A (proOmpA), was processed to OmpA by inverted inner membrane vesicles. Maturation of proOmpA occurred at a slower rate than the ATP-dependent translocation of proOmpA. This difference in rate could be attributed to the lower efficiency of leader peptidase cleavage. These results suggest that the binding of ATP alone is sufficient to drive the leader sequence across the membrane (presumably with the hydrophobic core of the leader domain in the membrane and the leader peptidase cleavage site exposed in the lumen). While it is unclear if this process is on the reaction pathway, it is useful to consider the implications of such a reaction.

Model studies examining SecA interactions with membranes are beginning to shed light on the process of translocation initiation. Specifically, what occurs when SecA binds ATP? Investigations of SecA in solution demonstrate that ATP binding results in a conformational change in SecA that can be detected by a loss of protease sensitivity (Shinkai et al., 1991). SecA has also been shown to directly interact with a phospholipid monolayer (Breukink et al., 1992a) and phospholipid vesicles (Hendrick and Wickner, 1991; Ulbrandt et al., 1992). Addition of SecA to the subphase of a lipid monolayer results in an increase in monolayer surface pressure indicative of insertion into the monolayer (Breukink et al., 1992a). Changes in SecA tryptophan fluorescence upon binding phospholipid vesicles also indicate deep penetration into the bilayer (Ulbrandt et al., 1992). Addition of either ATP or ADP + P_i substantially reduces the increase in monolayer surface pressure, suggesting that SecA does not insert into the monolayer when ADP + P_i are bound. In contrast, AMP-PNP has little or no effect on monolayer insertion. Recent experiments suggest that the binding of ATP may stabilize or facilitate SecA insertion into the inner membrane and that this insertion process may be responsible for the initial translocation of the leader region of proOmpA into inverted inner membrane vesicles (Economou and Wickner, 1994). Together these studies imply that hydrolysis of ATP by membrane inserted SecA to ADP + P_i results in the backing out of SecA from its membrane environment. Although this cycle of events remains to be demonstrated in phospholipid liposomes or functional inner membrane vesicles, one role of ATP hydrolysis may be the vectorial movement of SecA with respect to the membrane bilayer. This function of SecA could serve to localize the leader portion of a preprotein to the vicinity of the integral membrane proteins SecY/E, which are required for preprotein membrane transit, taking advantage of the propensity of leader sequences to bind and insert into membrane bilayers.

D. Subsequent Translocation

While ATP binding can drive the initial step of preprotein translocation, ATP hydrolysis is necessary for subsequent translocation (Schiebel et al., 1991; Arkowitz et al., 1993). Model studies of ATP interactions with SecA in solution (Shinkai et al., 1991) or bound to liposomes (Schiebel et al., 1991) in the presence of proOmpA show that the addition of ADP + P_i results in the release of proOmpA from SecA. Conversely, the addition of proOmpA results in the release of ADP from SecA (Shinkai et al., 1991). These experiments raise the possibility that during translocation, hydrolysis of ATP allows release of preproteins from SecA. Subsequent binding of preprotein upon ATP rebinding to SecA may allow repetition of this process. A mechanism consisting of repeated cycles of SecA binding and releasing preprotein, dependent on ATP hydrolysis (Schiebel et al., 1991), is reminiscent of the mechanism of chaperone-mediated protein folding. This similarity has led to the postulate that SecA functions as an ATP-dependent chaperone (Hartl et al., 1990), although experiments carried out with urea denatured preproteins suggest that this is not the sole function of SecA.

For the initiation of translocation, the binding of non-hydrolyzable ATP analogs permits the translocation of a small loop of proOmpA across the membrane. Interestingly, AMP-PNP addition to either kinetic translocation intermediates or intermediates generated by a reversible structural block (disulfide or folded dihydrofolate reductase) also results in the limited forward translocation of 20–30 amino acyl residues (Schiebel et al., 1991; Arkowitz et al., 1993). These results raise the possibility that insertion of SecA into the membrane bilayer may be a general mechanism for facilitating small domains of preproteins across the membrane or proximal to SecY/E during different stages of translocation. In order to better understand this process it will be necessary to investigate the affinity of membrane bound SecA for ATP and ADP + P_i and how these binding constants are affected by preprotein.

E. Coupling of ATP Hydrolysis to Translocation

A central question in the energetics of preprotein translocation, is whether ATP hydrolysis is strictly coupled to polypeptide chain movement across the bilayer. Strict coupling would require that ATP is not hydrolyzed in the absence of net chain movement across the membrane and should also manifest itself in a fixed coupling ratio of ATP hydrolysis to polypeptide chain (or amino acyl residue) translocated. At present, a number of different experiments all suggest ATP hydrolysis is not strictly coupled to net chain movement. Different preproteins such as proOmpA or prePhoE stimulate translocation–ATPase activity to varying extents (Cunningham and Wickner, 1989). Greater than 1000 mol of ATP are hydrolyzed per mol of proOmpA (346 amino acid residues) translocated into inverted inner membrane vesicles (Lill et al., 1989; Schiebel et al., 1991; Bassilana et al., 1992). An amino-terminal 74 residue fragment of proOmpA supports less than 2% of the translocation–ATPase activity observed with proOmpA (Bassilana et al., 1992). This 74-residue proOmpA fragment translocated at the same rate as proOmpA, resulting in less than 20 mol of ATP hydrolyzed per mol of fragment translocated. Furthermore, generation of a proton electrochemical gradient during translocation decreases the number of mol of ATP hydrolyzed per mol of preprotein translocated by approximately 10-fold (Schiebel et al., 1991; Driessen, 1993). With a preprotein that cannot complete translocation, due to the covalent attachment of bovine pancreatic trypsin inhibitor (BPTI), SecA–ATP hydrolysis continues unabated (Schiebel et al., 1991). In contrast, when this blocked intermediate completes translocation, upon cleavage of the cross-linker between BPTI and proOmpA, ATP hydrolysis ceases. These results all point to the absence of strict coupling between ATP hydrolysis and net chain movement. However, it must be kept in mind that ATP hydrolysis could have more than one function during translocation.

The process of translocation most likely involves protein targeting, protein unfolding, movement across the bilayer, and release into the periplasm. Each of these steps could have separate ATP requirements. Therefore it may be difficult or impossible to observe the coupled component of ATP hydrolysis. Indeed, the low level of ATP hydrolysis during translocation of small substrates, such as the 74-residue proOmpA fragment, may represent coupled ATP hydrolysis. Moreover the *in vitro* translocation system could result in the observed high levels of ATP hydrolysis. Reversed translocation has been observed with proOmpA translocation intermediates (Schiebel et al., 1991) and such processes could also contribute to a futile ATP hydrolysis cycle. The inability to generate a high transmembrane proton electrochemical gradient *in vitro* could highlight such a futile ATP cycle.

III. CHEMIOSMOTIC ENERGY: $\Delta\mu H^+$

A. *In Vivo* Studies

Although ATP hydrolysis is sufficient to drive the translocation of many preproteins *in vitro*, the proton electrochemical gradient is required for translocation *in vivo* and stimulates translocation *in vitro*. The role of chemiosmotic phenomena in the process of active transport across the cytoplasmic membrane of *E. coli* has been well established. As set forth by Mitchell (1968), oxidation of electron donors by the respiratory chain is accompanied by the expulsion of protons into the external media, resulting in the generation of a proton electrochemical gradient ($\Delta\mu H^+$). The electrochemical gradient of protons is comprised of two components: $\Delta\psi$, the electrical potential across the membrane (in *E. coli* inside negative), and ΔpH, the chemical gradient of protons across the membrane (in *E. coli* inside alkaline). Protein secretion is inhibited *in vivo* when the proton electrochemical gradient is dissipated (Daniels et al., 1981; Enequist et al., 1981; Zimmermann and Wickner,

1983; Geller, 1990). Dissipation of $\Delta\mu H^+$ results in the accumulation of unprocessed preproteins, largely residing in the cytoplasm (Enequist et al., 1981; Zimmermann and Wickner, 1983; Geller, 1990). This effect of $\Delta\mu H^+$ dissipation was not due to changes in intracellular ATP levels as demonstrated using cells with mutant F_1F_0-ATPase (Enequist et al., 1981). *In vivo*, different preproteins show varied sensitivities to dissipation of a proton electrochemical gradient (Daniels et al., 1981; Enequist et al., 1981; Zimmermann and Wickner, 1983). For example, pre-β-lactamase is very sensitive to changes in proton electrochemical gradient, with a drop in total $\Delta\mu H^+$ of 30 mV (190 to 160 mV) resulting in half maximal inhibition of processing (Bakker and Randall, 1984). In contrast, proOmpA processing is much less affected by the uncoupler CCCP than pre-β-lactamase (Daniels et al., 1981).

Geller (1990) has shown that treatment of cells with CCCP, resulted in the accumulation of maltose binding protein (MBP) on the periplasmic face of the cytoplasmic membrane. These *in vivo* results suggest that the proton electrochemical gradient functions at a late stage of translocation, perhaps the release of the translocated preprotein from the cytoplasmic membrane. Although $\Delta\psi$ is the main component of $\Delta\mu H^+$ *in vivo*, the total proton electrochemical gradient drives secretion, and not specifically $\Delta\psi$ or ΔpH (Bakker and Randall, 1984). Therefore, the translocation machinery must be able to utilize the energy of both $\Delta\psi$ and ΔpH. It must be noted, however, that it is difficult to interpret the effects of $\Delta\psi$ and ΔpH specific uncouplers on protein secretion *in vivo* because of the complexity of the cell.

B. *In Vitro* Studies

The role of the proton electrochemical gradient has been studied in greater detail using *in vitro* translocation systems. Although $\Delta\mu H^+$ alone is not sufficient for translocation of preproteins into inverted inner membrane vesicles, this energy increases the rate of ATP-dependent translocation *in vitro* (Chen and Tai, 1986; Geller et al., 1986; de Vrije et al., 1987; Yamane et al., 1987). Accordingly, it has been suggested that $\Delta\mu H^+$ acts later in the translocation process, rather than during initiation of translocation (Geller and Green, 1989). In agreement with *in vivo* observations, the dependency of translocation on $\Delta\mu H^+$ *in vitro* varies with each preprotein. While $\Delta\mu H^+$ stimulates to various extents the translocation of proOmpA (Geller and Green, 1989; Yamada et al., 1989a; Driessen and Wickner, 1991), preMBP (Arkowitz and Wickner, 1994), and a mutant of proOmpF-Lpp with an uncleavable leader sequence (Yamada et al., 1989a), wild-type proOmpF-Lpp strictly requires $\Delta\mu H^+$ (Yamada et al., 1989a). These results suggest that at least part of the effect of $\Delta\mu H^+$ is preprotein-specific, perhaps with the proton electrochemical gradient directly affecting the preprotein. *In vitro* both components of $\Delta\mu H^+$, $\Delta\psi$ and ΔpH, can independently stimulate preprotein translocation (Yamane et al., 1987; Driessen and Wickner, 1991). The $\Delta\psi$ and ΔpH could affect charged, acidic, or basic residues in the leader or mature portion of a preprotein.

Inverted inner membrane vesicles of *Vibrio alginolyticus*, which contain an active Na^+ pump, have been used to demonstrate that an electrochemical gradient of Na^+, $\Delta\mu Na^+$ (and not protons), can stimulate ATP-dependent translocation (Tokuda et al., 1990). In these experiments *E. coli* SecA was used, suggesting that this specificity for $\Delta\mu Na^+$ resides within the integral membrane component of translocase.

The most convincing demonstration that a proton electrochemical gradient is directly involved as a driving force in preprotein translocation came with the solubilization, purification, and reconstitution of the integral membrane proteins SecY/E (Brundage et al., 1990; Akimaru et al., 1991). Generation of a proton electrochemical gradient in SecY/E proteoliposomes using either a potassium acetate diffusion gradient (Driessen and Wickner, 1991) or the light driven proton pump bacteriorhodopsin (Brundage et al., 1990), results in an increase in the rate of proOmpA translocation. Therefore, in this purified system, the $\Delta\mu H^+$ must be acting on either SecA, SecY/E, the membrane bilayer, or proOmpA. More recently, Driessen (1993) showed with SecY/E proteoliposomes that both $\Delta\psi$ and ΔpH are equally effective in their stimulation of proOmpA translocation. These results suggest that $\Delta\psi$ and ΔpH may be acting on a common step during preprotein translocation.

C. Preprotein Specific Effects

Although the $\Delta\mu H^+$ effect is preprotein-specific, it seems unlikely as discussed above that $\Delta\mu H^+$ is solely acting at the level of the preprotein. Despite the large number of experiments addressing the role of $\Delta\mu H^+$ in preprotein translocation, the molecular mechanism of this driving force remains an enigma. Does this driving force act on the preprotein or on the translocase? The translocation of a derivative of the preprotein proOmpF-Lpp with no charged residues in its mature domain was still stimulated by $\Delta\mu H^+$ (Kato et al., 1992). This observation is inconsistent with $\Delta\mu H^+$ only driving the translocation of negatively charged residues of the preprotein into the positively charged lumen of the vesicle (i.e., electophoresis). Although electrophoresis appears not to be involved in wild-type proOmpA translocation, the translocation of a mutant proOmpA, with two glutamate residues following the leader sequence (proOmpA EE), is only stimulated by the electrical component of $\Delta\mu H^+$, suggestive of electrophoresis (Geller et al., 1993). One possibility is that the $\Delta\mu H^+$ affects the conformation of these different mutant preproteins, altering their $\Delta\mu H^+$ dependency. Clearly, it is difficult to generalize about the effect of $\Delta\mu H^+$ using data from different preproteins because it is likely that there are different rate limiting steps with each of these substrates.

Generation of a proton electrochemical gradient *in vitro*, allows stable secondary structural domains to be translocated. A disulfide loop containing 13 amino acyl residues within the mature portion of proOmpA blocks ATP-dependent translocation (Tani et al., 1989, 1990; Schiebel et al., 1991). In contrast, generation of a $\Delta\mu H^+$

allows completion of translocation of this secondary structural element (Tani et al., 1989, 1990). In addition, kinetic intermediates in proOmpA translocation, observable at low ATP concentrations, can complete translocation when a $\Delta\mu H^+$ is generated (Tani et al., 1989; Schiebel et al., 1991). More recently, it has been shown that a 20-amino acid polypeptide covalently cross-linked to the carboxy-terminus of a proOmpA derivative can be translocated in a $\Delta\mu H^+$-dependent fashion (Kato and Mizushima, 1993). These studies suggest that a $\Delta\mu H^+$ may be involved in "gating" or "opening" a translocation channel. However, there must be a limit to the width of this putative channel because a $\Delta\mu H^+$ does not allow the translocation of the bovine pancreatic trypsin inhibitor (BPTI) portion of a proOmpA–BPTI chimera (Bassilana and Wickner, 1993), or the translocation of a stably folded dihydrofolate reductase domain (Dhfr) fused to the carboxy-terminus of proOmpA (Arkowitz et al., 1993).

It is interesting to note that in the above three cases of $\Delta\mu H^+$-dependent completion of translocation, ATP hydrolysis by SecA was not required. Either removal of ATP (Schiebel et al., 1991), inactivation of SecA (Schiebel et al., 1991), or addition of AMP-PNP (an inhibitor of SecA-dependent translocation) (Tani et al., 1990; Kato and Mizushima, 1993) did not prevent the $\Delta\mu H^+$-dependent completion of translocation. Nonetheless, ATP-dependent initiation of translocation is prerequisite for this $\Delta\mu H^+$-dependent "gating". Hence, translocase must exist in a productive state with preprotein partially translocated in order to utilize the energy of the proton electrochemical gradient. Together, these results indicate that the proton electrochemical gradient facilitates the translocation of large, bulky groups and may be involved in allowing the transit of small amounts of folded structure.

D. Later Stages of Translocation

Studies on proOmpA (Geller and Green, 1989; Tani et al., 1989; Schiebel et al., 1991) and preMBP (Geller, 1990; Arkowitz and Wickner, 1994) translocation all point to the proton electrochemical gradient being involved in the later stages of translocation. The processing of proOmpA to OmpA *in vitro* is not affected by a $\Delta\mu H^+$, whereas $\Delta\mu H^+$ is sufficient for the completion of translocation (Geller and Green, 1989; Schiebel et al., 1991). Similar results have been observed with preMBP translocation (Geller, 1990; Arkowitz and Wickner, submitted). One possible explanation is that $\Delta\mu H^+$ is involved in the release of preproteins following translocation. If release of the preprotein into the periplasm is a rate-limiting step in translocation, generation of a $\Delta\mu H^+$ would result in an increase in the apparent rate of translocation. In this case, the $\Delta\mu H^+$ could act directly on the preprotein (perhaps changing its conformation), or more likely, via a membrane component of the translocation apparatus. Recent studies suggest that SecD and SecF may be involved in release of the preprotein (Matsuyama et al., 1993), which may be facilitated by $\Delta\mu H^+$ (Arkowitz and Wickner, 1994).

E. $\Delta\mu H^+$ and ATP

While *in vitro* studies have emphasized the separate contributions of ATP and $\Delta\mu H^+$, both of these driving forces are present *in vivo* and required for preprotein translocation. While the majority of results is consistent with $\Delta\mu H^+$ being involved in the later stages of translocation, it appears that during the translocation process SecA–ATP-dependent translocation can affect $\Delta\mu H^+$-dependent translocation. For example, increasing the concentration of SecA decreases the $\Delta\mu H^+$ stimulatory effect on proOmpF-Lpp translocation (Yamada et al., 1989b). In addition, the effect of $\Delta\mu H^+$ on preprotein translocation is more striking at low ATP concentrations. Shiozuka et al. (1990) have shown that $\Delta\mu H^+$ lowers the apparent $K_{1/2}$ of ATP for translocation by approximately 50-fold. It was proposed that $\Delta\mu H^+$ may affect the release of ADP from SecA (Shiozuka et al., 1990). A possible interpretation of these results is that SecA–ATP-dependent and $\Delta\mu H^+$-dependent translocation can compete during preprotein membrane transit (Schiebel et al., 1991). As a result, increasing the SecA concentration will favor the SecA–ATP-dependent pathway, whereas generation of a $\Delta\mu H^+$ will result in a decrease in the SecA–ATP-dependent pathway. Evidence for this competition comes from studies on kinetic translocation intermediates of proOmpA (Schiebel et al., 1991). One translocation intermediate in which approximately 16 kDa of proOmpA have translocated into the lumen of the vesicles, can complete translocation with $\Delta\mu H^+$ alone. The addition of ATPγS prevents this $\Delta\mu H^+$-dependent completion of translocation. When SecA is inactivated by treatment with anti-SecA antibodies, the $\Delta\mu H^+$-dependent translocation occurs at an increased rate. It appears that when a substantial portion of the translocating chain is on the cytoplasmic side of the membrane, competition between $\Delta\mu H^+$-dependent translocation and SecA–ATP-dependent translocation pathways can occur. In contrast, $\Delta\mu H^+$-dependent translocation can predominate when only 6–7 kDa of proOmpA remain on the cytoplasmic side of the membrane. These studies suggest that the contribution from SecA–ATP-driven translocation decreases as the amount of the polypeptide chain available to bind SecA decreases.

In summary, SecA–ATP-dependent translocation is absolutely required for the initiation of translocation. ATP hydrolysis and $\Delta\mu H^+$ can be used to drive different stages of the translocation process. To understand the roles of these energy sources in polypeptide chain movement across the membrane, it will be necessary to examine partial translocation reactions in which chain movement is rate-limiting and requires only one of these forms of energy.

IV. OTHER DRIVING FORCES

In addition to ATP and $\Delta\mu H^+$, other forces driving protein secretion can be envisioned. For example, the refolding of preproteins on the periplasmic face of the membrane could drive translocation. Little is known about the periplasmic components involved in translocation. Proteolytic cleavage of the leader peptide by LepB occurs at the periplasmic side of the cytoplasmic membrane, yet is not necessary for translocation *in vivo* (Dalbey and Wickner, 1985; Fikes and Bassford, 1987; Barkocy-Gallagher and Bassford, 1992) or *in vitro* (Yamane et al., 1988). This processing step could promote the folding of the translocated protein in the periplasm. Disulfide bond formation in preproteins as they reach the periplasm could also contribute to driving translocation across the membrane. Mutations in two genes, *dsbA* (*ppfA*) (Bardwell et al., 1991; Kamitani et al., 1992) and *dsbB* (Bardwell et al., 1993; Missiakas et al., 1993) were identified which result in the failure to form disulfide bonds in translocated periplasmic proteins. Deletion of *dsbA* however does not confer a secretion defect, suggesting that disulfide bond formation occurs subsequent to protein translocation (Bardwell et al., 1991).

It has been observed that even when ATP and $\Delta\mu H^+$ are removed *in vitro*, partial translocation (forward and reversed) of proOmpA and proOmpA–Dhfr translocation intermediates can occur (Schiebel et al., 1991; Arkowitz et al., 1993). The driving forces for such partial translocation reactions are not established, yet it is possible that folding (von Heijne and Blomberg, 1979) or transfer of residues into and out of the transmembrane environment could promote these reactions. For example, folding on the periplasmic side of the membrane could drive forward translocation, whereas folding on the cytoplasmic side of the membrane could drive reversed translocation. Irrespective of the driving force involved in these partial translocation reactions, the occurrence of partial translocation in the absence of ATP and $\Delta\mu H^+$ suggests that the interactions of the integral membrane subunits of translocase with the transiting preprotein are relatively weak and reversible. Furthermore, such bidirectional chain movements are consistent with the Brownian "ratchet" hypothesis put forth by Simon et al. (1992) in which the membrane transitting polypeptide fluctuates back and forth through a translocation channel or pore with overall bias toward forward movement. It is attractive to speculate that preprotein folding or interactions with other proteins in the periplasm provide the bias for forward movement across the bilayer. While there is little known about the refolding of preproteins on the periplasmic side of the inner membrane of *E. coli*, a role for lumenal chaperones in the translocation of proteins across the yeast endoplasmic reticulum and mitochondrial membranes is well established.

V. COMPARISON WITH YEAST ENDOPLASMIC RETICULUM AND MITOCHONDRIAL TRANSLOCATION SYSTEMS

Protein translocation into the yeast endoplasmic reticulum (ER) and mitochondria has many similarities with translocation across the cytoplasmic membrane of bacteria. Nonetheless, translocation into both mitochondria and ER is oriented in the opposite direction compared to bacterial protein secretion. Translocation across eukaryotic membranes occurs within the cell, where ATP is readily available. In contrast, due to the permeable nature of the bacterial outer membrane, it is unlikely that ATP is available as a periplasmic energy source for translocation. Proteins which are translocated into the matrix of mitochondria must cross two membrane bilayers. Posttranslational protein translocation into yeast (Hansen et al., 1986; Rothblatt and Meyer, 1986; Walters and Blobel, 1986) endoplasmic reticulum has been demonstrated, suggesting that polypeptide chain elongation during translation is not a driving force for translocation.

Translocation across both the yeast ER (Hansen et al., 1986; Rothblatt and Meyer, 1986; Walters and Blobel, 1986; Sanz and Meyer, 1989) and mitochondrial (Pfanner and Neupert, 1986; Chen and Douglas, 1987; Eilers et al., 1987; Pfanner et al., 1987) membranes requires ATP hydrolysis. A major role for ATP hydrolysis on the cytoplasmic side of these organelles appears to be in maintaining translocation competent conformations of preproteins (Chen and Douglas, 1987; Pfanner et al., 1987; Verner and Schatz, 1987; Chirico et al., 1988; Deshaies et al., 1988). Specifically cytosolic Hsp70 heat-shock proteins have been shown by both *in vivo* (Deshaies et al., 1988) and *in vitro* studies (Chirico et al., 1988; Murakami et al., 1988; Brodsky et al., 1993) to stimulate protein translocation into ER and mitochondria. This class of chaperones stabilizes preproteins prior to translocation by preventing aggregation and nonproductive interactions (Pelham, 1986; Sheffield et al., 1990). *In vitro*, the requirement for these chaperones is not absolute; translocation into both organelles occurs efficiently with urea denatured preproteins in the absence of chaperones (Eilers and Schatz, 1986; Eilers et al., 1987, 1988; Pfanner et al., 1988; Sanz and Meyer, 1988a,b; Ostermann et al., 1989; Sheffield et al., 1990; Becker et al., 1992). However, even under these conditions, translocation into ER microsomes and the mitochondrial matrix still requires ATP hydrolysis. Nevertheless, a functional homolog of SecA in ER or mitochondria has yet to be identified. It appears that part of the ATP requirement for translocation into yeast ER and mitochondrial matrix is due to factors (chaperones) requiring ATP in the lumen of these organelles.

Inhibition of ATP uptake into yeast microsomal vesicles blocks translocation (Mayinger and Meyer, 1993), indicative of a lumenal ATP requirement. Kar2, the yeast homolog of the mammalian ER lumenal chaperone BiP (Normington et al., 1989; Rose et al., 1989), is required for yeast ER translocation (Vogel et al., 1990; Nguyen et al., 1991; Sanders et al., 1992; Brodsky et al., 1993). Kar2 can be cross-linked to translocating preproteins, indicating that it is directly involved in translocation (Sanders et al., 1992). Interaction of translocating preproteins with ER lumenal proteins such as Kar2 is likely to play an important role in the energetics of translocation into the yeast ER.

Translocation across the inner membrane of yeast mitochondria also requires ATP hydrolysis (Hwang et al., 1989; Hwang and Schatz, 1989; Ostermann et al., 1989; Rassow and Pfanner, 1991). Two matrix localized chaperones, mHsp70 and mHsp60, sequentially facilitate translocation and folding of preproteins in an

ATP-dependent fashion (Cheng et al., 1989; Ostermann et al., 1989; Kang et al., 1990; Ostermann et al., 1990; Scherer et al., 1990; Manning-Krieg et al., 1991; Glick et al., 1992; Koll et al., 1992). Genetic and biochemical studies suggest that mHsp70 is involved in "pulling" preproteins into the mitochondrial matrix (Neupert et al., 1990; Manning-Krieg et al., 1991). It appears that in both mitochondria and ER translocation ATP hydrolysis in the lumen by chaperones is a major driving force.

ATP-independent movement of preproteins across the ER membrane has not been observed. In contrast, translocation of protein domains across the outer membrane of yeast mitochondria does not necessarily require ATP. For example, cytochrome *c* heme lyase translocates across the outer membrane of intact mitochondria (Lill et al., 1992) and mitochondrial outer membrane vesicles in the absence of ATP (Mayer et al., 1993). Furthermore, fusion proteins consisting of pre-cytochrome b_2 (Pfanner et al., 1990; Glick et al., 1992), pre-cytochrome c_1 (Glick et al., 1992), or pre-cytochrome oxidase IV (Hwang et al., 1991; Jascur et al., 1992) and dihydrofolate reductase have all been shown to cross the mitochondrial outer membrane in the absence of ATP. In these cases of ATP-independent translocation, preprotein folding or interaction with membranes or other proteins could drive translocation.

In contrast to the requirement for a proton electrochemical gradient in *E. coli* translocation, translocation into the yeast ER is insensitive to $\Delta\mu H^+$ uncouplers (Hansen et al., 1986; Rothblatt and Meyer, 1986; Waters and Blobel, 1986). These studies strongly suggest that neither $\Delta\psi$ or ΔpH is involved in driving ER translocation. However, $\Delta\mu H^+$ is required for translocation across the yeast mitochondrial inner membrane (Gasser et al., 1982; Schleyer et al., 1982) through only its electrical component $\Delta\psi$ (inside negative), and not ΔpH (Pfanner and Neupert, 1985; Martin et al., 1991). The $\Delta\psi$ is involved in the initiation of translocation across the inner membrane, resulting in processing of the preprotein by the matrix peptidase (Schleyer and Neupert, 1985; Chen and Douglas, 1987; Pfanner et al., 1987; Eilers et al., 1988). Even the translocation of preproteins which do not have a cleavable presequence, such as the ADP/ATP (Pfanner and Neupert, 1985) and the phosphate (Zara et al., 1992) carriers, requires a $\Delta\psi$. Studies with synthetic yeast mitochondrial presequences have shown that their translocation across mitochondria (Glaser and Cumsky, 1990; Roise, 1992) or phospholipid vesicles (Maduke and Roise, 1993) depends upon $\Delta\psi$. Different preproteins show different sensitivities to $\Delta\psi$-dependent import, not attributable to the length of the mature domain or folded state of the preprotein (Martin et al., 1991). The presequence dictates the $\Delta\psi$ dependency and therefore it appears likely that $\Delta\psi$ is exerting an electrophoretic/electrostatic effect on the positive charges of the presequence (Martin et al., 1991). Consistent with this postulate, the translocation of mammalian phosphate carrier, which has a presequence, can be driven by a reduced magnitude of $\Delta\psi$ compared to the mature mammalian preprotein (similar to the yeast phosphate carrier) (Zara et al., 1992). These studies all point to the role of the proton electrochemical gradient in translocation across yeast ER and mitochondrial membranes being significantly different.

While yeast ER, mitochondrial, and bacterial protein translocation systems all require ATP, the stage of preprotein translocation in which ATP is required appears to differ. In *E. coli*, an ATP-dependent driving force in the periplasm is unlikely, whereas preprotein folding, interaction with periplasmic proteins, or the proton electrochemical gradient could very well be involved in the later stages of translocation. Energy-independent chain movement across a bilayer is a common theme in *E. coli* and mitochondrial translocation systems. Lastly, the role of the proton electrochemical gradient in translocation appears to be vastly different in yeast ER, mitochondria, and bacteria.

VI. OUTLOOK

The simplicity and experimental accessibility of *E. coli* has made it an ideal system to study the translocation of preproteins across a membrane bilayer. Most, if not all, of the protein components required for translocation in *E. coli* have been isolated and characterized. SecA-ATP binding is required for the initiation of translocation and hydrolysis drives subsequent preprotein domains across the membrane. ATP hydrolysis results in conformational changes in SecA which may contribute to preprotein translocation across the membrane. The proton electrochemical gradient can drive the later stages of translocation most likely via a direct effect on translocase. This system is now at the stage where a number of fundamental questions can be addressed with respect to the energetics of translocation. How is the energy from ATP hydrolysis converted to polypeptide chain movement? How does the proton electrochemical gradient increase the rate of translocation? Are there energetic contributions to protein translocation from the periplasmic side of the membrane? In order to understand these and other questions it will be necessary to examine the translocation reaction at a quantitative level both with respect to thermodynamic parameters, such as binding constants; and kinetic parameters, such as analyses of rate-limiting steps and other individual rate constants. Only through detailed kinetic and thermodynamic studies will we gain an understanding of the driving forces of protein translocation at a molecular level.

ACKNOWLEDGMENTS

I would like to thank B. Wickner for many stimulating discussions. I am indebted to M. Bassilana, B. Conradt, T. Economou, and D. Epstein for critically reading the manuscript and helpful suggestions. R.A. was supported by a Damon Runyon–Walter Winchell Cancer Research Fund Fellowship (DRG-1106).

REFERENCES

Akimaru, J., Matsuyama, S.-I., Tokuda, H., & Mizushima, S. (1991). Reconstitution of a protein translocation system containing purified SecY, SecE, and SecA from *Escherichia coli*. Proc. Natl. Acad. Sci. USA 88, 6545–6549.

Akiyama, Y. & Ito, K. (1987). Topology analysis of the SecY protein, an integral membrane protein involved in protein export in *Escherichia coli*. EMBO J. 6, 3465–3470.

Ames, G.F. & Joshi, A.K. (1990). Energy coupling in bacterial periplasmic permeases. J. Bacteriol. 172, 4133–4137.

Arkowitz, R.A. & Bassilana, M. (1994). Protein translocation in *Escherichia coli*. Biochem. Biophys. Acta 1197, 311–343.

Arkowitz, R.A. & Wickner, W. (1994). SecD and SecF are required for the proton electrochemical gradient stimulation of preprotein translocation. EMBO J. 13, 954–963.

Arkowitz, R.A., Joly, J.C., & Wickner, W. (1993). Translocation can drive unfolding of a preprotein domain. EMBO J. 12, 243–253.

Bakker, E.P. & Randall, L.L. (1984). The requirement for energy during export of β-lactamase in *Escherichia coli* is fulfilled by the total protonmotive force. EMBO J. 3, 895–900.

Bardwell, J.C.A., McGovern, K., & Beckwith, J. (1991). Identification of a protein required for disulfide bond formation *in vivo*. Cell 67, 581–589.

Bardwell, J.C.A., Lee, J., Jander, G., Martin, N., Belin, D., & Beckwith, J. (1993). A pathway for disulfide bond formation *in vivo*. Proc. Natl. Acad. Sci. USA 90, 1038–1042.

Barkocy-Gallagher & Bassford, P.J. Jr. (1992). Synthesis of precursor maltose-binding protein with proline in the +1 position of the cleavage site interferes with the activity of *Escherichia coli* signal peptidase *in vivo*. J. Biol. Chem. 267, 1231–1238.

Bassilana, M., Arkowitz, R.A., & Wickner, W. (1992). The role of the mature domain of proOmpA in the translocation ATPase reaction. J. Biol. Chem. 267, 25246–25250.

Bassilana, M. & Wickner, W. (1993). Purified *Escherichia coli* preprotein translocase catalyzes multiple cycles of precursor protein translocation. Biochemistry 32, 2626–2630.

Batenburg, A.M., Demel, R.A., Verkleij, A.J., & de Kruijff, B. (1988). Penetration of signal sequence of *Escherichia coli* PhoE protein into phospholipid model membranes leads to lipid-specific changes in signal peptide structure and alterations of lipid organization. Biochemistry 27, 5678–5685.

Becker, K., Guiard, B., Rassow, J., Soellner, T., & Pfanner, N. (1992). Targeting of a chemically pure preprotein to mitochondria does not require the addition of a cytosolic signal recognition factor. J. Biol. Chem. 267, 5637–5643.

Bieker, K.L., Phillips, G.J., & Silhavy, T. (1990). The *sec* and *prl* genes of *E. coli*. J. Bioenerg. Biomembr. 22, 291–310.

Bieker-Brady, K. & Silhavy, T.J. (1992). Suppressor analysis suggests a multistep, cyclic mechanism for protein secretion in *Escherichia coli*. EMBO J. 11, 3165–3174.

Bochkareva, E.S., Lissin, N.M., & Girshovich, A.S. (1988). Transient association of newly synthesized unfolded proteins with the heat-shock GroEL protein. Nature 336, 254–257.

Breukink, E., Demel, R.A., de Korte-Kool, G., & de Kruijff, B. (1992a). SecA insertion into phospholipids is stimulated by negatively charged lipid and inhibited by ATP: a monolayer study. Biochemistry 31, 1119–1124.

Breukink, E., Kusters, R., & de Kruijff, B. (1992b). *In vitro* studies on the folding characteristics of the *Escherichia coli* precursor protein prePhoE. Eur. J. Biochem. 208, 419–425.

Briggs, M.S., Gierasch, L.M., Zlotnick, A., Lear, J.D., & DeGrado, W.F. (1985). *In vivo* function and membrane binding properties are correlated for *Escherichia coli* LamB signal peptides. Science 228, 1096–1098.

Brodsky, J.L., Hamamoto, S., Feldheim, D., & Schekman, R. (1993). Reconstitution of protein translocation from solubilized yeast membranes reveals topologically distinct roles for BiP and Hsc70. J. Cell Biol. 120, 95–102.

Brundage, L., Hendrick, J.P., Schiebel, E., Driessen, A.J.M., & Wickner, W. (1990). The purified integral membrane protein SecY/E is sufficient for reconstitution of SecA-dependent precursor protein translocation. Cell 62, 649–657.

Brundage, L., Fimmel, C.J., Mizushima, S., & Wickner, W. (1992). SecY, SecE, and Band 1 form the membrane-embedded domain of E. coli preprotein translocase. J. Biol. Chem. 267, 4166–4170.

Cabelli, R.J., Chen, L., Tai, P.C., & Oliver, D.B. (1988). SecA protein is required for secretory protein translocation into E. coli membrane vesicles. Cell 55, 683–692.

Cabelli, R.J., Dolan, K.M., Qian, L., & Oliver, D.B. (1991). Characterization of membrane-associated and soluble states of SecA protein from wild-type and SecA51(TS) mutant stains of Escherichia coli. J. Biol. Chem. 266, 24420–24427.

Chen, L. & Tai, P.C. (1985). ATP is essential for protein translocation into Escherichia coli membrane vesicles. Proc. Natl. Acad. Sci. USA 82, 4384–4388.

Chen, L. & Tai, P.C. (1986). Roles of H^+-ATPase and proton motive force in ATP-dependent protein translocation in vitro. J. Bacteriol. 167, 389–392.

Chen, W.J. & Douglas, M.G. (1987). Phosphodiester bond cleavage outside the mitochondria is required for the completion of protein import into the mitochondrial matrix. Cell 49, 651–658.

Cheng, M.Y., Hartl, F.-U., Martin, J., Pollock, R.A., Kalousek, F., Neupert, W., Hallberg, E.M., Hallberg, R.L., & Horwich A.L. (1989). Mitochondrial heat-shock protein hsp60 is essential for assembly of proteins imported into yeast mitochondria. Nature 337, 620–625.

Chirico, W.J., Waters, M.G., & Blobel, G. (1988). 70K heat shock related proteins stimulate protein translocation into microsomes. Nature 332, 805–810.

Collier, D.B., Bankaitis, V.A., Weiss, J.B., & Bassford, P.J. Jr. (1988). The antifolding activity of SecB promotes the export of the E. coli maltose binding protein. Cell 53, 273–283.

Cunningham, K., Lill, R., Crooke, E., Rice, M., Moore, K., Wickner, W., & Oliver, D. (1989). Isolation of SecA protein, a peripheral membrane protein of E. coli plasma membrane that is essential for the functional binding and translocation of proOmpA. EMBO J. 8, 955–959.

Cunningham, K. & Wickner, W. (1989). Specific recognition of the leader region of precursor proteins is required for the activation of translocation ATPase of E. coli. Proc. Natl. Acad. Sci. USA 86, 8630–8634.

Dalbey, R.E. & Wickner, W. (1985). Leader peptidase catalyzes the release of exported proteins from the outer surface of the Escherichia coli plasma membrane. J. Biol. Chem. 260, 15925–15931.

Daniels, C.J., Quay, S.C., & Oxender, D.L. (1981). Role for membrane potential in the secretion of protein into the periplasm of Escherichia coli. Proc. Natl. Acad. Sci. USA 78, 5396–5400.

Davidson, A.L., Shuman, H.A., & Nikaido, H. (1992). Mechanism of maltose transport in Escherichia coli transmembrane signalling by periplasmic binding proteins. Proc. Natl. Acad. Sci. USA 89, 2360–2364.

Deshaies, R.J., Koch, B.D., Werner-Washburne, M., Craig, E.A., & Schekman, R. (1988). A subfamily of stress proteins facilitates translocation of secretory and mitochondrial precursor polypeptides. Nature 332, 800–805.

de Vrije, T., Tommassen, J., & de Kruijff, B. (1987). Optimal posttranslational translocation of the precursor of PhoE protein across Escherichia coli membrane vesicles requires both ATP and the proton motive force. Biochim. Biophys. Acta 900, 63–72.

Driessen, A.J.M. (1993). Precursor protein translocation by the Escherichia coli translocase is directed by the protonmotive force. EMBO J. 11, 847–853.

Driessen, A.J.M. & Wickner, W. (1991). Proton transfer is rate limiting for translocation of precursor proteins by the Escherichia coli translocase. Proc. Natl. Acad. Sci. USA 88, 2471–2475.

Economov, A. & Wickner, W. (1994). SecA promotes preprotein insertion by undergoing ATP-driven cycles of membrane insertion and deinsertion. Cell 78, 835–843.

Eilers, M., Hwang, S., & Schatz, G. (1988). Unfolding and refolding of a purified precursor protein during import into isolated mitochondria. EMBO J. 7, 1139–1145.

Eilers, M., Oppliger, W., & Schatz, G. (1987). Both ATP and an energized inner membrane are required to import a purified precursor protein into mitochondria. EMBO J. 6, 1073–1077.

Eilers, M. & Schatz, G. (1986). Binding of a specific ligand inhibits import of a purified precursor protein into mitochondria. Nature 322, 228–232.

Enequist, H.G., Hirst, T.R., Harayama, S., Hardy, S.J.S., & Randall, L.L. (1981). Energy is required for maturation of exported proteins in *Escherichia coli*. Eur. J. Biochem. 116, 227–233.

Fikes, J.D. & Bassford, P.J. Jr. (1987). Export of unprocessed precursor maltose-binding protein to the periplasm of *Escherichia coli* cells. J. Bacteriol. 169, 2352–2359.

Fikes, J.D. & Bassford, P.J. Jr. (1989). Novel secA alleles improve export of maltose-binding protein synthesized with a defective signal peptide. J. Bacteriol. 171, 402–409.

Gasser, S.M., Daum, G., & Schatz, G. (1982). Import of proteins into mitochondria. Energy-dependent uptake precursors by isolated mitochondria. J. Biol. Chem. 257, 13034–13041.

Geller, B.L. (1990). Electrochemical potential releases a membrane bound secretion intermediate of maltose-binding protein in *Escherichia coli*: J. Bacteriol. 172, 4870–4876.

Geller, B.L. & Green, H.M. (1989). Translocation of pro-OmpA across inner membrane vesicles of *Escherichia coli* occurs in two consecutive energetically distinct steps. J. Biol. Chem. 264, 16465–16469.

Geller, B., Zhu, H.-Y., Cheng, S., Kuhn, A., & Dalbey, R. (1993). Charged residues render pro-OmpA potential dependent for initiation of membrane translocation. J. Biol. Chem. 268, 9442–9447.

Geller, B.L., Movva, N.R., & Wickner, W. (1986). Both ATP and the electrochemical potential are required for optimal assembly of pro-OmpA into *Escherichia coli* inner membrane vesicles. Proc. Natl. Acad. Sci. USA 83, 4219–4222.

Gierasch, L.M. (1989). Signal sequences. Biochemistry 28, 923–930.

Glaser, S.M. & Cumsky, M.G. (1990). Localization of a synthetic presequence that blocks protein import into mitochondria. J. Biol. Chem. 265, 8817–8822.

Glick, B.S., Brandt, A., Cunningham, K., Mueller, S., Hallberg, R.L., & Schatz, G. (1992). Cytochromes c_1 and b_2 are sorted to the intermembrane space of yeast mitochondria by a stop-transfer mechanism. Cell 69, 809–822.

Hansen, W., Garcia, P.D., & Walter, P. (1986). *In vitro* protein translocation across the yeast endoplasmic reticulum: ATP-dependent posttranslational translocation of the prepro-α-factor. Cell 45, 397–406.

Hartl, F.-U., Lecker, S., Schiebel, E., Hendrick, J.P., & Wickner, W. (1990). The binding cascade of SecB to SecA to SecY/E mediates preprotein targeting to the *E. coli* plasma membrane. Cell 63, 269–279.

Hendrick, J.P. & Hartl, F.-U. (1993). Molecular chaperone functions of heat-shock proteins. Ann. Rev. Biochem. 63, 349–384.

Hendrick, J.P. & Wickner, W. (1991). SecA protein needs both acidic phospholipids and SecY/E protein for functional high-affinity binding to the *Escherichia coli* plasma membrane. J. Biol. Chem. 266, 24596–24600.

Hwang, S., Jascur, T., Vestweber, D., Pon, L., & Schatz, G. (1989). Disrupted yeast mitochondria can import precursor proteins directly through their inner membrane. J. Cell Biol. 109, 487–493.

Hwang, S.T. & Schatz, G. (1989). Translocation of proteins across the mitochondrial inner membrane, but not into the outer membrane, requires nucleoside triphosphates in the matrix. Proc. Natl. Acad. Sci. USA, 86, 8432–8436.

Hwang, S.T., Wachter, C., & Schatz, G. (1991). Protein import into the yeast mitochondria matrix. J. Biol. Chem. 266, 21083–21089.

Ito, K., Date, T., & Wickner, W. (1980). Synthesis, assembly into the cytoplasmic membrane, and proteolytic processing of the precursor of coliphage M13 coat protein. J. Biol. Chem. 255, 2123–2130.

Jackson, G.S., Staniforth, R.A., Halsall, D.J., Atkinson, T., Holbrook, J.J., Clarke, A.R., & Burston, S.G. (1993). Binding and hydrolysis of nucleotides in the chaperonin catalytic cycle: implications for the mechanism of assisted protein folding. Biochemistry 32, 2554–2563.

Jarosik, G.P. & Oliver, D.B. (1991). Isolation and analysis of dominant secA mutations in *Escherichia coli*. J. Bacteriol. 173, 860–868.

Jascur, T., Goldenberg, D.P., Vestweber, D., & Schatz, G. (1992). Sequential translocation of an artificial precursor protein across the two mitochondrial membranes. J. Biol. Chem. 267, 13636–13641.

Joly, J.C. & Wickner, W. (1993). The SecA and SecY subunits of translocase are the nearest neighbors of the translocating preprotein, shielding it from phospholipids. EMBO J. 12, 255–263.

Kamitani, S., Akiyama, Y., & Ito, K. (1992). Identification and characterization of an *Escherichia coli* gene required for the formation of correctly folded alkaline phosphatase, a periplasmic enzyme. EMBO J. 11, 57–62.

Kang, P.J., Ostermann, J., Shilling, J., Neupert, W., Craig, E.A., & Pfanner N. (1990). Requirement for hsp70 in the mitochondrial matrix for translocation and folding of precursor proteins. Nature 348, 137–143.

Kato, M. & Mizushima, S. (1993). Translocation of conjugated presecretory proteins possessing an internal non-peptide domain into everted membrane vesicles in *Escherichia coli*. J. Biol. Chem. 268, 3586–3593.

Kato, M., Tokuda, H., & Mizushima, S. (1992). *In vitro* translocation of secretory proteins possessing no charges at the mature domain takes place efficiently in a protonmotive force-dependent manner. J. Biol. Chem. 267, 413–418.

Kimura, E., Akita, M., Matsuyama, S., & Mizushima, S. (1991). Determination of a region in SecA that interacts with presecretory proteins in *Escherichia coli*. J. Biol. Chem. 266, 6600–6606.

Koll, H., Guiard, B., Rassow, J., Ostermann, J., Horwich, A.L., Neupert, W., & Hartl, F.-U. (1992). Antifolding activity of hsp60 couples protein import into the mitochondrial matrix with export to the intermembrane space. Cell 68, 1163–1175.

Koronakis, V., Hughes, C., & Koronakis, E. (1993). ATPase activity and ATP/ADP-induced conformational change in the soluble domain of the bacterial protein translocator HlyB. Mol. Microbiol. 8, 1163–1175.

Kumamoto, C.A. & Francetic, O. (1993). Highly selective binding of nascent polypeptides by an *Escherichia coli* chaperone protein *in vivo*. J. Bacteriol. 175, 2184–2188.

Kumamoto, C.A. & Gannon, P.M. (1988). Effects of *Escherichia coli* secB mutations on pre-maltose binding protein conformation and export kinetics. J. Biol. Chem. 263, 11554–11558.

Kusukawa, N., Yura, T., Ueguchi, C., Akiyama, Y., & Ito, K. (1989). Effects of mutations in heat-shock genes groES and groEL on protein export in *Escherichia coli*. EMBO J. 8, 3517–3521.

Lecker, S.H., Driessen, A.J.M., & Wickner, W. (1990). ProOmpA contains secondary and tertiary structure prior to translocation and is shielded from aggregation by association with SecB protein. EMBO J. 9, 2309–2314.

Lecker, S.H., Lill, R., Ziegelhoffer, T., Georgopoulos, C., Bassford, P.J. Jr., Kumamoto, C.A., & Wickner, W. (1989). Three pure chaperone proteins of *Escherichia coli*-SecB, trigger factor and GroEL-form soluble complexes with precursor proteins *in vitro*. EMBO J. 8, 2703–2709.

Lill, R., Cunningham, K., Brundage, L., Ito, K., Oliver, D., & Wickner, W. (1989). The SecA protein hydrolyzes ATP and is an essential component of the protein translocation ATPase of *E. coli*. EMBO J. 8, 961–966.

Lill, R., Dowhan, W., & Wickner, W. (1990). The ATPase activity of SecA is regulated by acidic phospholipids, SecY, and the leader and mature domains of precursor proteins. Cell 60, 271–280.

Lill, R., Stuart, R.A., Drygas, M.E., Nargang, F.E., & Neupert, W. (1992). Import of cytochrome c heme lyase into mitochondria: a novel pathway into the intermembrane space. EMBO J. 11, 449–456.

Maduke, M. & Roise, D. (1993). Import of mitochondrial presequence into protein-free phospholipid vesicles. Science 260, 364–367.

Manning-Krieg, U.C., Scherer, P.E., & Schatz G. (1991). Sequential action of mitochondrial chaperones in protein import into the matrix. EMBO J. 10, 3273–3280.

Martin, J., Mahlke, K., & Pfanner, N. (1991). Role of an energized inner membrane in mitochondrial protein import. J. Biol. Chem. 266, 18051–18057.

Matsuyama, S., Akimaru, J., & Mizushima, S. (1990a). SecE-dependent overproduction of SecY in *Escherichia coli*. FEBS Lett. 269, 96–100.

Matsuyama, S., Fujita, Y., & Mizushima, S. (1993). SecD is involved in the release of the translocated secretory proteins from the cytoplasmic membrane of *Escherichia coli*. EMBO J. 12, 265–270.

Matsuyama, S., Kimura, E., & Mizushima, S. (1990b). Complementation of two overlapping fragments of SecA, a protein translocation ATPase of *Escherichia coli*, allows ATP binding to its amino terminal region. J. Biol. Chem. 265, 8760–8765.

Mayer, A., Lill, R., & Neupert, W. (1993). Translocation and insertion of precursor proteins into isolated outer membrane of mitochondria. J. Cell Biol. 121, 1233–1243.

Mayinger, P. & Meyer, D.I. (1993). An ATP transporter is required for protein translocation in the yeast endoplasmic reticulum. EMBO J. 12, 659–666.

Missiakas, D., Georgopoulos, C., & Raina, S. (1993). Identification and characterization of *Escherichia coli* gene *dsbB*, whose product is involved in the formation of disulfide bonds *in vivo*. Proc. Natl. Acad. Sci. USA 90, 7084–7088.

Mitchell, P. (1968). Chemiosmotic coupling in oxidative and photosynthetic phosphorylation, Glynn Research, Ltd., Bodmin, England.

Mizushima, S. & Tokuda, H. (1990). *In vitro* translocation of bacterial secretory proteins and energy requirements. J. Bioenerg. Biomembr. 22, 389–399.

Murakami, H., Pain, D., & Blobel G. (1988). 70-kD heat shock-related protein is one of at least two distinct cytosolic factors stimulating protein import into mitochondria. J. Cell Biol. 107, 2051–2057.

Nagaraj, R. (1984). Interaction of synthetic signal sequence fragments with model membranes. FEBS Lett. 165, 79–82.

Nagaraj, R., Joseph, M., & Reddy, G.L. (1987). Perturbation of the lipid bilayer of model membranes by synthetic signal peptides. Biochim. Biophys. Acta 903, 465–472.

Neupert, W., Hartl, F.-U., Craig, E.A., & Pfanner, N. (1990). How do polypeptides cross the mitochondrial membranes? Cell 63, 447–450.

Nguyen, T.H., Law, D.T.S., & Williams, D.B. (1991). Binding protein BiP is required for translocation of secretory proteins into the endoplasmic reticulum in *Saccharomyces cerevisiae*. Proc. Natl. Acad. Sci. USA 88, 1565–1569.

Nishiyama, K., Mizushima, S., & Tokuda, H. (1992). The carboxyl-terminal region of SecE interacts with SecY and is functional in the reconstitution of protein translocation activity in *Escherichia coli*. J. Biol. Chem. 267, 7170–7176.

Normington, K., Kohno, K., Kozutsumi, Y., Gething, M.J., & Sambrook, J. (1989). *S. cerevisiae* encodes an essential protein homologous in sequence and function to mammalian BiP. Cell 57, 1223–1236.

Oliver, D.B. & Beckwith, J. (1982). Regulation of a membrane component required for protein secretion in *Escherichia coli*. Cell 30, 311–319.

Oliver, D.B., Cabelli, R.J., & Jarosik, G.P. (1990). SecA protein: autoregulated initiator of secretory precursor protein translocation across the *E. coli* plasma membrane. J. Bioenerg. Biomembr. 22, 311–336.

Ostermann, J., Horwich, A.L., Neupert, W., & Hartl, F.-U. (1989). Protein folding in mitochondria with hsp60 and ATP hydrolysis. Nature 341, 125–130.

Ostermann, J., Voos, W., Kang, P.J., Craig, E.A., Neupert, W., & Pfanner, N. (1990). Precursor proteins in transit through mitochondrial contact sites interact with hsp70 in the matrix. FEBS Lett. 277, 281–284.

Park, S., Liu, G., Topping, T.B., Cover, W.H., & Randall, L.L. (1988). Modulation of folding of exported proteins by the leader sequences. Science 239, 1033–1035.

Pelham, H.R. (1986). Speculations on the functions of the major heat shock and glucose-regulated proteins. Cell 46, 959–961.

Pfanner, N. & Neupert, W. (1985). Transport of proteins into mitochondria: a potassium diffusion potential is able to drive the import of ADP/ATP carrier. EMBO J. 4, 2819–2825.

Pfanner, N. & Neupert, W. (1986). Transport of F1-ATPase subunit beta into mitochondria depends on both a membrane potential and nucleoside triphosphates. FEBS Lett. 209, 152–156.

Pfanner, N., Pfaller, R., Kleene, R., Ito, M., Tropschug, M., & Neupert, W. (1988). Role of ATP in mitochondrial protein import. Conformational alteration of a precursor protein can substitute for ATP requirement. J. Biol. Chem. 263, 4049–4051.

Pfanner, N., Rassow, J., Guiard, B., Soellner, T., Hartl, F.-U., & Neupert, W. (1990). Energy requirements for unfolding and membrane translocation of precursor proteins during import into mitochondria. J. Biol. Chem. 265, 16324–16329.

Pfanner, N., Tropschug, M., & Neupert, W. (1987). Mitochondrial protein import: nucleoside triphosphates are involved in conferring import-competence to precursors. Cell 49, 815–823.

Phillips, G.J. & Silhavy, T.J. (1990). Heat-shock proteins DnaK and GroEL facilitate export of LacZ hybrid proteins in E. coli. Nature 344, 882–884.

Phoenix, D.A., Kusters, R., Hikita, C., Mizushima, S., & de Kruijff, B. (1993). OmpF-Lpp signal sequence mutants with varying charge hydrophobicity ratios provide evidence for a phosphatidyl-glycerol-signal sequence interaction during protein translocation across the Escherichia coli inner membrane. J. Biol. Chem. 268, 17069–17073.

Pugsley, A.P. (1993). The complete general secretory pathway in gram-negative bacteria. Microbiol. Rev. 57, 50–108.

Randall, L.L. (1983). Translocation of domains of nascent periplasmic proteins across the cytoplasmic membrane is independent of elongation. Cell 33, 231–240.

Randall, L.L. & Hardy, S.J.S. (1989). Unity in function in the absence of consensus in sequence: Role of leader peptides in export. Science 243, 1156–1159.

Randall, L.L., Topping, T.B., & Hardy, S.J.S. (1990). No specific recognition of leader peptide by SecB, a chaperone involved in protein export. Science 248, 860–863.

Rassow, J. & Pfanner, N. (1991). Mitochondrial preproteins en route from the outer membrane to the inner membrane are exposed to the intermembrane space. FEBS Lett. 293, 85–88.

Roise, D. (1992). Interaction of a synthetic mitochondrial presequence with isolated yeast mitochondria: mechanism of binding and kinetics of import. Proc. Natl. Acad. Sci. USA 89, 608–612.

Rose, M.D., Misra, L.M., & Vogel, J.P. (1989). KAR2, a karyogamy gene, is the yeast homolog of the mammalian BiP/GRP78 gene. Cell 57, 1211–1221.

Rothblatt, J.A. & Meyer, D.I. (1986). Secretion in yeast: translocation and glycosylation of prepro-α-factor in vitro can occur via an ATP-dependent posttranslational mechanism. EMBO J. 5, 1031–1036.

Sanders, S.L., Whitfield, K.M., Vogel, J.P., Rose, M.D., & Schekman, R.W. (1992). Sec61p and BiP directly facilitate polypeptide translocation into ER. Cell 69, 353–365.

Sanz, P. & Meyer, D.I. (1988a). Secretion in yeast: preprotein binding to a membrane receptor and ATP-dependent translocation are sequential and separable events in vitro. J. Cell: Biol. 108, 2101–2106.

Sanz, P. & Meyer, D.I. (1988b). Signal recognition particle (SRP) stabilizes the translocation competent conformation of pre-secretory proteins. EMBO J. 7, 3553–3557.

Sanz, P. & Meyer, D.I. (1989). Secretion in yeast: preprotein binding to a membrane receptor and ATP-dependent translocation are sequential and separable events in vitro. J. Cell Biol. 108, 2101–2106.

Sarkadi, B., Price, E.M., Boucher, R.C., Germann, V.A., & Scarborough, G.A. (1992). Expression of the human multidrug resistance cDNA in insect cells generates a high drug-stimulated membrane ATPase. J. Biol. Chem. 267, 4854–4858.

Schatz, P.J. & Beckwith, J. (1990). Genetic analysis of protein export in *Escherichia coli*. Ann. Rev. Genet. 24, 215–248.

Schatz, P.J., Riggs, P.D., Jacq, A., Fath, M.J., & Beckwith, J. (1989). The secE gene encodes an integral membrane protein required for protein export in *Escherichia coli*. Genes Dev. 3, 1035–1044.

Schiebel, E., Driessen, A.J.M., Hartl, F.-U., & Wickner, W. (1991). ΔμH⁺ and ATP function at different steps of the catalytic cycle of preprotein translocase. Cell 64, 927–939.

Schleyer, M. & Neupert, W. (1985). Transport of proteins in mitochondria: translocational intermediates spanning contact sites between outer and inner membranes. Cell 43, 339–350.

Schleyer, M., Schmidt, B., & Neupert, W. (1982). Requirement of a membrane potential for the posttranslational transfer of proteins into mitochondria. Eur. J. Biochem. 125, 109–116.

Sheffield, W.P., Shore, G.C., & Randall, S.K. (1990). Mitochondrial precursor protein. Effects of 70-kilodalton heat shock protein on polypeptide folding, aggregation, and import competence. J. Biol. Chem. 265, 11069–11076.

Sherer, P.E., Krieg, U.C., Hwang, S.T., Dietmar, V., & Schatz, G. (1990). A precursor protein partly translocated into yeast mitochondria is bound to a 70 kDa mitochondrial stress protein. EMBO J. 9, 4315–4322.

Shinkai, A., Mei, L.H., Tokuda, H., & Mizushima, S. (1991). The conformation of SecA, as revealed by its protease sensitivity, is altered upon interaction with ATP, presecretory proteins, everted membrane vesicles, and phospholipids. J. Biol. Chem. 266, 5827–5833.

Shiozuka, K., Tani, K., Mizushima, S., & Tokuda, H. (1990). The proton motive force lowers the level of ATP required for the *in vitro* translocation of a secretory protein in *Escherichia coli*. J. Biol. Chem. 265, 18843–18847.

Simon, S.M., Peskin, C.S., & Oster, G.F. (1992). What drives the translocation of proteins? Proc. Natl. Acad. Sci. USA 89, 3770–3774.

Tani, K., Shiozuka, K., Tokuda, H., & Mizushima, S. (1989). *In vitro* analysis of the process of translocation of OmpA across the *Escherichia coli* cytoplasmic membrane. J. Biol. Chem. 264, 18582–18588.

Tani, K., Tokuda, H., & Mizushima, S. (1990). Translocation of proOmpA possessing an intramolecular disulfide bridge into membrane vesicles of *Escherichia coli*. J. Biol. Chem. 265, 17341–17347.

Tokuda, H., Kim, Y.J., & Mizushima, S. (1990). *In vitro* translocation into inverted membrane vesicles prepared from *Vibrio alginolyticus* is stimulated by the electrochemical potential of Na⁺ in the presence of *Escherichia coli* SecA. FEBS Lett. 264, 10–12.

Ulbrandt, N.D., London, E., & Oliver, D.B. (1992). Deep penetration of a portion of *Escherichia coli* SecA protein into model membranes is promoted by anionic phospholipids and by partial unfolding. J. Biol. Chem. 267, 15184–15192.

Verner, K. & Schatz, G. (1987). Import of an incompletely folded precursor protein into isolated mitochondria requires an energized inner membrane, but no added ATP. EMBO J. 6, 2449–2456.

Vogel, J.P., Misra, L.M., & Rose, M.D. (1990). Loss of BiP/GRP78 function blocks translocation of secretory proteins in yeast. J. Cell Biol. 110, 1885–1895.

von Heijne, G. & Blomberg, C. (1979). Trans-membrane translocation of proteins. Eur. J. Biochem. 97, 175–181.

Walters, M.G. & Blobel, G. (1986). Secretory protein translocation in a yeast cell-free system can occur posttranslationally and requires ATP hdyrolysis. J. Cell Biol. 102, 1543–1550.

Weaver, A.J., McDowall, A.W., Oliver, D.B., & Diesenhofer, D. (1992). Electron microscopy of thin-sectioned three-dimensional crystals of SecA protein from *Escherichia coli*: structure in projection at 40 A resolution. J. Struct. Biol. 109, 87–96.

Weiss, J.B., Ray, P.H., & Bassford, P.J. Jr. (1988). Purified SecB protein of *Escherichia coli* retards folding and promotes membrane translocation of the maltose-binding protein *in vitro*. Proc. Natl. Acad. Sci. USA 85, 8978–8982.

Wickner, W., Driessen, A.J.M., & Hartl, F.-U. (1991). The enzymology of protein translocation across the *Escherichia coli* plasma membrane. Ann. Rev. Biochem. 60, 101–124.

Wild, J., Altman, E., Yura, T., & Gross, C.A. (1992). DnaK and DnaJ heat-shock proteins participate in protein export in *Escherichia coli*. Genes Dev. 6, 1165–1172.

Yamada, H., Tokuda, H., & Mizushima, S. (1989). Proton motive force-dependent and -independent protein translocation revealed by an efficient *in vitro* assay system of *Escherichia coli*. J. Biol. Chem. 264, 1723–1728.

Yamada, H., Matsuyama, S.-I., Tokuda, H., & Mizushima, S. (1989). A high concentration of SecA allows proton motive force-independent translocation of a model secretory protein into *Escherichia coli* membrane vesicles. J. Biol. Chem. 264, 18577–18581.

Yamane, K., Ichihara, S., & Mizushima, S. (1987). *In vitro* translocation of protein across *Escherichia coli* membrane vesicles requires both the proton motive force and ATP. J. Biol. Chem. 262, 2358–2362.

Yamane, K., Matsuyama, S., & Mizushima, S. (1988). Efficient *in vitro* translocation into *Escherichia coli* membrane vesicles of a protein carrying an uncleavable signal peptide. J. Biol. Chem. 263, 5368–5372.

Zara, V., Palmieri, F., Mahlke, K., & Pfanner, N. (1992). The cleavable presequence is not essential for import and assembly of the phosphate carrier of mammalian mitochondria but enhances the specificity and efficiency of import. J. Biol. Chem. 267, 12077–12081.

Zimmermann, R. & Wickner, W. (1983). Energetics and intermediates of the assembly of protein OmpA into the outer membrane of *Escherichia coli*. J. Biol. Chem. 258, 3920–3925.

TRANSPORT OF PERTUSSIS TOXIN ACROSS BACTERIAL AND EUKARYOTIC MEMBRANES

Drusilla L. Burns

ABSTRACT

Pertussis toxin, a protein toxin produced by the Gram-negative bacterium *Bordetella pertussis*, the causative agent of pertussis or whooping cough, plays an important role in pathogenesis by inhibiting signal transduction in host cells. In order to gain access to host cells, the toxin must cross three membrane barriers: the inner and outer

Membrane Protein Transport
Volume 1, pages 25–38.
Copyright © 1995 by JAI Press Inc.
All rights of reproduction in any form reserved.
ISBN: 1-55938-907-9

membranes of the bacterium, and the eukaryotic cell membrane. Pertussis toxin utilizes a set of accessory proteins to transit the membranes of the bacterium. The toxin itself contains the necessary information to cross the final membrane barrier of the eukaryotic cell.

I. INTRODUCTION

Bacterial protein toxins that are exported from Gram-negative bacteria and have target substrates within eukaryotic cells must cross at least three membrane barriers before inflicting their damage on the eukaryotic cell. These barriers include the inner and outer membranes of the bacterium and the plasma membrane of the eukaryotic cell. Toxins may utilize accessory proteins to cross the bacterial membranes, but often cross the last barrier, the eukaryotic cell membrane without the help of other bacterial proteins. One such bacterial protein toxin is pertussis toxin (PT) which is secreted from the Gram-negative organism, *Bordetella pertussis*, the causative agent of the disease pertussis or whooping cough.

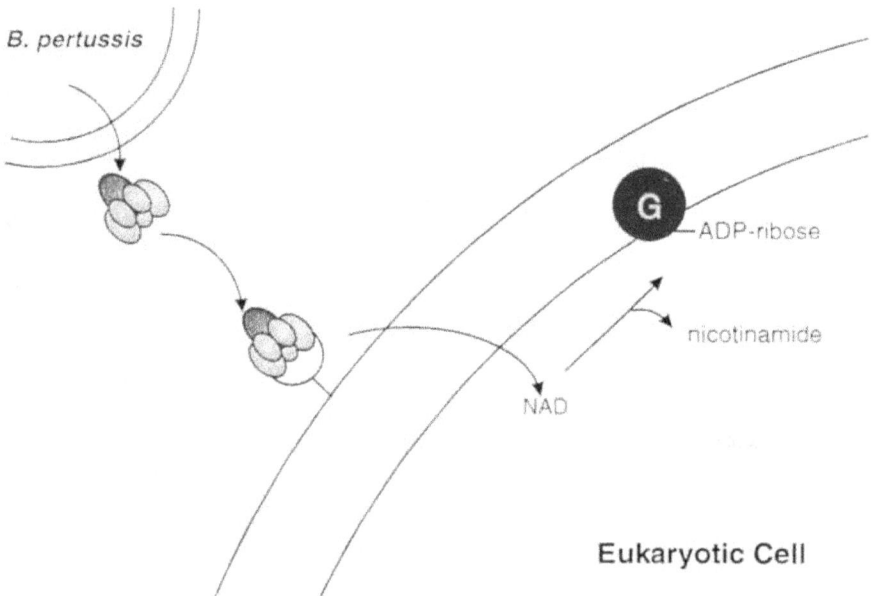

Figure 1. Mechanism of action of PT.

II. STRUCTURE AND BIOLOGICAL ACTIVITIES OF PT

PT is believed to contribute to pathogenesis, at least in part, by impairing the immune system (Kaslow and Burns, 1992). Biological activities of PT include inhibition of the recirculation of leukocytes (Spangrude et al., 1984), inhibition of the migration of macrophages (Meade et al., 1984), and enhancement in rats of insulin secretion in response to glucose load (Yajima et al., 1978). Passive administration of anti-PT immunoglobulin to children with pertussis has been shown to result in decreased duration and frequency of whooping, suggesting that the toxin might also play a role in inducing this symptom of the disease (Granström et al., 1991).

As illustrated in Figure 1, PT acts on a molecular level by catalyzing the transfer of the ADP-ribose moiety of NAD to a family of GTP-binding regulatory proteins (G proteins) which transmit signals from the outside of the cell to effector molecules on the inside of the cell (Katada and Ui, 1982a,b). When G proteins are ADP-ribosylated by PT, their function is inhibited such that the cell no longer responds to a variety of hormones (reviewed by Gilman, 1987).

As shown in Figure 2, PT is an oligomeric protein which contains two moieties, an enzymatically active S1 subunit and a B oligomer composed of one copy each of subunits S2, S3, and S5 and two copies of S4 (Tamura et al., 1982, 1983). The S1 subunit ADP-ribosylates G proteins, whereas the B oligomer contains the eukaryotic cell receptor binding sites (Tamura et al., 1982).

Before PT can inflict its damage on the eukaryotic cell, it must be properly assembled and exported across the inner and outer membranes of *B. pertussis*. It must then travel to the eukaryotic cell and gain access to the G proteins which are believed to be located on the inner leaflet of the plasma membrane as well as NAD which is believed to be in the cytoplasm of the cell.

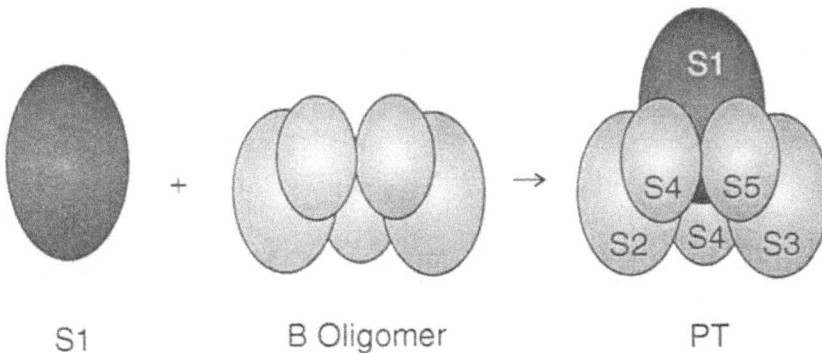

S1   B Oligomer   PT

Figure 2. Structure of PT.

III. BIOSYNTHESIS OF PT AND ITS SECRETION
FROM *B. PERTUSSIS*

The genes for the PT subunits are located within a single operon (Locht and Keith, 1986; Nicosia et al., 1986) and are expressed only in the virulent form of the organism. The structure of this operon is shown in Figure 3. Each PT subunit is synthesized with its own signal sequence (Locht and Keith, 1986; Nicosia et al., 1986) suggesting that the subunits may be individually transported through the inner membrane, perhaps utilizing a general secretion pathway homologous to the *sec* system of *Escherichia coli* (Wickner et al., 1991). While the individual subunits might cross the inner membrane in an extended, unfolded state, evidence exists which indicates that the assembled form of PT is found in the periplasmic space of the bacterium suggesting that the folded, assembled toxin must be transported across at least the outer membrane of the bacterium (Nicosia and Rappuoli, 1987; Lee et al., 1989).

Support for the idea that only the fully assembled form of PT is competent for secretion comes from studies in which secretion of mutant forms of PT were studied. Pizza et al. (1990) analyzed four mutants of *B. pertussis* which have single or multiple changes in the gene encoding the S1 subunit and which synthesize S1 subunits which are not assembled into a holotoxin molecule. These mutants were found to secrete only low levels of the B oligomer. Complete assembly of the B oligomer of PT may also be required for secretion since a mutant of *B. pertussis* containing a Tn*5* transposon insertion in the structural gene for S3 does not secrete PT (Marchitto et al., 1987; Nicosia and Rappuoli, 1987). Since the S3 gene is the last gene in the *ptx* operon, the mutant should produce S1, S2, S4, S5, and the first part of S3. While this mutant contained detectable levels of biologically active pertussis toxin (Nicosia and Rappuoli, 1987), and the presence of S1, S2, and S4 were verified by immunoblot analysis (Marchitto et al., 1987), secretion of the subunits was not detected. These results suggest that the C-terminal half of S3 is important for proper secretion of PT, perhaps because it is necessary for the stability of the holotoxin molecule.

Thus, assembled PT, having a molecular weight of 105,000 (Nicosia et al., 1986), must cross at least one membrane barrier. Conceivably, PT might be able to cross this barrier without the aid of other proteins. However, often accessory proteins are required for the secretion of macromolecules from Gram-negative bacteria. Several

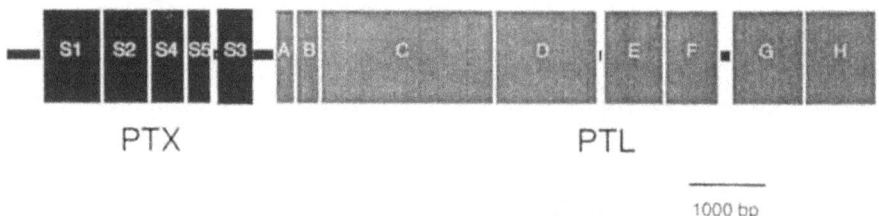

1000 bp

Figure 3. *ptx* and *ptl* regions of the *B. pertussis* chromosome. *ptx* genes encoding the structural subunits of PT are shown. Also shown are the orfs of the *ptl* region.

families of accessory proteins have been identified which transport large proteins or DNA across the outer membrane, or both the inner and outer membranes of bacteria. Such families include the Pul family of proteins which transport proteins such as the enzyme pullulanase from *Klebsiella* (Pugsley, 1993) and pili from *Pseudomonas aeruginosa*, the Pap family of proteins involved in the export of pili from various Gram-negative organisms (Hultgren et al., 1991), and the RTX family of proteins involved in the export of proteins such as hemolysin from *Escherichia coli* and adenylate cyclase toxin from *B. pertussis* (Welch, 1991). Other transport systems such as the VirB proteins of *Agrobacterium tumefaciens* are required for transport of DNA across bacterial membranes (Zambryski, 1988, 1992). Recent evidence suggests that PT may also utilize a set of accessory proteins, termed Ptl proteins, to exit the bacterial cell. The Ptl proteins display homology to the VirB system of *A. tumefaciens*.

The Ptl set of accessory proteins was discovered by mutational analysis of the *B. pertussis* chromosome (Weiss et al., 1993). Insertion of a Tn*5lac* transposon approximately 3.2 kb downstream from the pertussis toxin structural genes generated a mutant, BPM 3171, which was found to secrete less toxin than the parental strain, although the total amount of PT produced by the mutant was approximately the same as that produced by the parent. Other virulence factors exported by *B. pertussis*, such as filamentous hemagglutinin and pertactin, did not appear to be affected by this mutation. Interestingly, this mutant appeared to be less virulent than wild-type *B. pertussis* in mice when they were challenged intranasally with the organism (Weiss and Goodwin, 1989), suggesting that secretion of PT, not simply production of the toxin, is important for the pathogenesis.

Additional mutations were made in this region of the chromosome in order to define the boundaries of the region which is important for PT secretion (Weiss et al., 1993). Suicide vectors were constructed which contained a gentamicin-resistance determinant and fragments of *B. pertussis* DNA which were immediately downstream from the PT structural genes. Since these plasmids could not replicate in *B. pertussis*, drug-resistance was conferred only when the entire plasmid integrated into the bacterial chromosome by homologous recombination. The *B. pertussis* sequences therefore are present twice in the genome of these strains but are interrupted by the vector. Mutations resulted if the cloned region was internal to an operon since the vector interrupted the end of the operon in the first copy and the second copy lacked promoter sequences. Three mutants, BPPTL2, BPPTL4, and BPPTL9, were constructed such that the vector would disrupt the *B. pertussis* chromosome approximately 7.5, 8.5 and 9.5 kb, respectively, downstream from the PT structural genes. BPPTL2 and BPPTL4 were defective in the secretion of PT, whereas BPPTL9 secreted the toxin. Thus the region of the chromosome important for PT secretion might extend up to 9.5 kb downstream of the PT structural genes. These data also suggested that the region important for PT secretion might be organized in an operon structure. This region has been termed the pertussis-toxin-liberation (*ptl*) locus.

Table 1. Properties of the Predicted Ptl
Proteins

Predicted Protein	M_r	pl
PtlA	10,812	10.8
PtlB	12,312	12.2
PtlC	92,637	7.6
PtlD	48,678	11.6
PtlE	31,010	10.6
PtlF	29,471	6.8
PtlG	35,176	9.9
PtlH	37,158	7.2

Nucleotide sequence analysis of this region revealed eight open reading frames (orfs) which have been termed orfA-H (Weiss et al., 1993). These orfs are shown in Figure 3. The molecular weights and isoelectric points for the proteins predicted by these orfs are listed in Table 1.

Proteins predicted by each of these orfs were found to be homologous to eight of the VirB proteins produced by *Agrobacterium tumefaciens* (Covacci and Rappuoli, 1993; Weiss et al., 1993). These homologies (Figure 4) are intriguing because the VirB proteins are believed to play a role in the transport of a large piece of single-stranded DNA, termed T-DNA, across bacterial and perhaps plant membranes (Ward et al., 1988; Kuldau et al., 1990; Zambryski, 1992). Once T-DNA is transferred into the plant, it integrates into the plant genome and encodes proteins involved in the biosynthesis of plant hormones. Overexpression of these hormones results in loss of control of cell division and tumor growth (reviewed by Nester et al., 1984).

The homologies observed between proteins predicted to be encoded by the *ptl* genes and the VirB proteins suggests that the *ptl* locus may encode proteins that belong to a family of transport proteins involved in secretion of macromolecules from bacteria. Other operons which encode homologous proteins include the Tra2 operon of plasmid RP4 (Lessl et al., 1992) and the PilW operon of plasmid R388 (Shirasu and Kado, 1993). Both of these operons are required for conjugative transfer between bacteria. It is interesting that this family of proteins is used to transport both DNA and proteins. Since, at least in the case of the VirB system, the T-DNA is associated with proteins (Zambryski, 1992), all systems may transport proteins, some of which are associated with DNA and thus co-transport the DNA across membranes.

While sequence data predicts the existence of Ptl proteins, the proteins have yet to be identified in extracts of *B. pertussis*. Preliminary evidence for the existence of PtlE, PtlF, and PtlG has been obtained by using PCR to amplify these open reading frames, producing recombinant forms of these proteins in *E. coli*, isolating

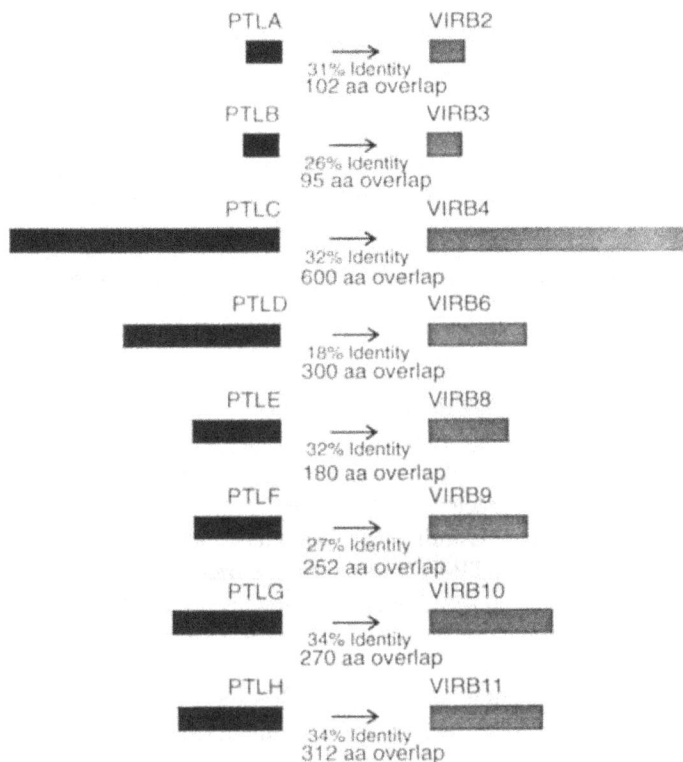

Figure 4. Homologies between *ptl* orfs and the VirB proteins of *A. tumefaciens*. The percent identity between the predicted Ptl protein and VirB homolog is indicated along with the length of the amino acid (aa) sequences over which the proteins exhibit identity.

these proteins, immunizing mice with the preparations, and using these sera to probe for the Ptl proteins. Proteins which are recognized by the PtlE, PtlF, and PtlG antisera were found in extracts of virulent *B. pertussis* but not the avirulent form of this organism (Johnson and Burns, unpublished data). Each of the predicted Ptl proteins is discussed in some detail below.

PtlA: Some controversy exists as to whether orfA encodes a protein or whether this region of DNA is actually part of the promoter region for the *ptl* operon since data exist which would suggest that the *ptl* genes are part of an operon distinct from the *ptx* operon and one of the few possible locations for the promoter region of the *ptl* operon would be in orfA (Covacci and Rappuoli, 1993; Weiss et al., 1993).

PtlB: OrfB is predicted to encode a small, basic protein. The amino-terminal half of the protein is expected to be very hydrophobic with the carboxy-terminal half of the protein being hydrophilic.

Table 2. Predicted Nucleotide Binding Sites of PtlC, PtlH, VirB4, and VirB11

Protein	Sequence	Location
PtlC	II GQS GS GKTVL	amino acids 453–464
PtlH	VA GQT GS GKTTL	amino acids 167–178
VirB4	IF GPI GRGKTTL	amino acids 427–438
VirB11	LC GPT GS GKTTM	amino acids 162–173

Note: Ptl sequences were taken from Weiss et al., 1993. VirB sequences were from Thompson et al. (1988).
        Consensus sequences are indicated.

PtlC: PtlC is predicted to be the largest of the Ptl proteins. Its VirB analog, VirB4, copurifies with total membranes from *vir*-induced *Agrobacterium* (Engström et al., 1987). Both PtlC and VirB4 (Thompson et al., 1988) contain consensus sequences which are characteristic of nucleotide binding sites (Table 2). This finding has suggested the possibility that a potential function of these proteins might be to provide energy, via ATP-hydrolysis, for the transport process (Thompson et al., 1988). Recently, the nucleotide-binding domain of VirB4 has been shown to be essential since mutations in this region are critical for VirB4 function (Berger and Christie, 1993).

PtlD: PtlD is predicted to be basic and to have a pI of 11.6. Its analog, VirB6, which has a pI of 4.4 (Ward et al., 1990a), is the most hydrophobic of the VirB proteins, and contains six potential membrane-spanning regions (Zambryski, 1992).

PtlE: PtlE has been identified in extracts of virulent *B. pertussis* (Johnson and Burns, unpublished results). Its apparent molecular weight on polyacrylamide gels containing sodium dodecyl sulfate (SDS) is 30,000 compared with its predicted molecular weight of 31,000. The VirB analog of PtlE, VirB8, has been determined to be a membrane-associated protein (Thorstenson et al., 1993).

PtlF: PtlF has been identified in extracts of virulent *B. pertussis* (Johnson and Burns, unpublished results). The predicted molecular weight of this protein is 29,500 compared to its apparent molecular weight on polyacrylamide gels containing SDS of 31,000. PtlF has an N-terminal sequence, MMAARMMAAGLAATALSAHA (Weiss et al., 1993), which has the characteristics of a signal sequence (Briggs and Gierasch, 1986) in that it contains a basic residue at its extreme N-terminal end followed by a hydrophobic core of amino acids, and finally alanines at the -3 and -1 positions. The VirB analog of PtlF, VirB9, contains a signal sequence (Zambryski, 1992). Cell fractionation studies have shown that most VirB9 is found in the total membrane fraction (Thorstenson et al., 1993).

PtlG: Like PtlE and PtlF, PtlG has been identified in extracts of *B. pertussis* (Johnson and Burns, unpublished results). Its analog, VirB10, has been isolated as large aggregates contained in the inner membrane fraction of *vir*-induced *Agrobacterium* (Ward et al., 1990b).

PtlH: PtlH exhibits homology with VirB11 as well as with a number of other bacterial proteins involved in transport. These include PulE which is necessary for the secretion of pullulanase from *Klebsiella* (Pugsley, 1993) and PilB which is required for export of pili from *Pseudomonas aeruginosa* (Lory, 1992). Thus, this protein is particularly interesting because it seems to be a necessary component of many types of secretion systems. PtlH is predicted to contain a consensus nucleotide binding domain that is conserved among members of this family of proteins (Table 2). Christie et al. (1989) have demonstrated that VirB11 has ATPase activity and exhibited autophosphorylation activity *in vitro*. These workers suggested that VirB11 may modulate its own activity or perhaps the activity of other VirB proteins by phosphorylation. ATP hydrolysis might also provide energy for the transport process. VirB11 is found in the total membrane fraction of cell lysates suggesting that this protein is either an integral membrane protein or associates tightly with membrane proteins (Thorstenson et al., 1993). Another protein belonging to this family, PulE, has been shown to be cytoplasmic (Possot et al., 1992).

Recent data have demonstrated that when the cellular locations of seven of the eleven VirB proteins—VirB1, VirB4, VirB5, VirB8, VirB9, VirB10, and VirB11— were examined, six of the seven were found to be in both the inner and outer membrane fractions (Thorstenson et al., 1993). The remaining protein, VirB5 was found in both the cytoplasm and inner membrane fractions. Since at least six of the VirB proteins are found in both the inner and outer membranes, it was postulated that they may form a complex pore structure that spans both membranes. The T-DNA would then traverse both membranes in a single step. The question of whether PT might follow a similar transport mechanism or whether this protein is transported by a two step mechanism in which it is first transported through the inner membrane via the general export pathway and then transported across the outer membrane via a gate or channel formed by the Ptl proteins remains to be answered. The presence of signal sequences on each of the PT subunits (Locht and Keith, 1986; Nicosia et al., 1986) as well as the localization of some assembled PT in the periplasm (Nicosia and Rappuoli, 1987) might argue for a periplasmic intermediate.

While it is now known that a set of accessory proteins is required for secretion of PT from *B. pertussis*, a number of questions remain unanswered including the structure of the transport complex, the mechanism by which PT associates with this complex, and the sequence of events which must occur in order for the toxin to be properly transported across bacterial membrane barriers.

IV. ENTRY OF PT INTO EUKARYOTIC CELLS

Once PT is secreted from the bacterium, it diffuses to its target eukaryotic cell where it binds to receptors on the surface of that cell. Since most mammalian cells are susceptible to PT action, receptors for PT must be present on many cell types. While the identity of the receptor(s) for PT is not known, the chemical nature of the PT receptor on Chinese hamster ovary (CHO) cells has been elucidated.

Variant CHO cells which have altered carbohydrate structures on glycoproteins and glycolipids have been used to determine the structure of the PT receptor. A variant CHO cell line (15B) which lacks the carbohydrate sequence NeuAc→GalβGlcNAc on complex type glycoproteins was found to be resistant to PT action, suggesting that the functional receptor for PT on CHO cells is a glycoprotein (Brennan et al., 1988). Other CHO cell variants were also found to be resistant to the toxin (Witvliet et al., 1989) including LEC 2 cells and LEC 8 cells which lack NeuAc residues and NeuAc-Gal residues, respectively, on glycoproteins and glycolipids (Deutscher et al., 1984; Deutscher and Hirschberg, 1986). These latter findings suggested that the functional receptor for PT on CHO cells likely contains terminal sialic acid residues.

After the PT binds to its receptor on the surface of the eukaryotic cell, the toxin must gain access to its substrates, G proteins and NAD, both of which are believed to be located on the inside of the cell. The toxin might penetrate the plasma membrane directly. Alternatively, it might be internalized via endocytosis and cross a vesicle membrane. The question of how the toxin might cross the final membrane barrier remains unanswered; however, interaction of PT subunits with lipids has been studied and information obtained from these studies may provide clues to the mechanism by which PT penetrates membranes.

The ability of PT subunits to interact with radioactive photoactive phospholipid probes which were dispersed in detergent micelles was studied (Montecucco et al., 1986). Two types of probes were used: PC I, which had a photoreactive group located at the polar head group level, and PC II, which had the photoreactive group located at the fatty acid methyl terminus. After mixture of PT and micelles of Triton X-100 containing the photoreactive lipid probes followed by illumination to induce covalent linkage of the photoreactive group to the PT subunits with which it was in contact, subunits S2, S3, S4, and possibly S5 were labeled with PC I, but only S2 and S3 were labeled with PC II. These results indicated that while most, if not all, of the B subunits were capable of interacting with the detergent micelles, only S2 and S3 penetrated the micelles deeply. When the lipid content of the micelles was altered, different results were obtained. When micelles of lysolecithin were used, S2, S3, S4, and possibly S5 were labeled with PC I, but no subunit was labeled with PC II, indicating that the B subunits did not penetrate this type of micelle as deeply they penetrated the detergent micelles. In this study, S1 was not labeled under any condition.

The interaction of PT and its isolated S1 and B oligomer moieties with lipids was also studied by examining the binding of these species to phospholipid vesicles composed of phosphatidylcholine (Hausman and Burns, 1992). Interaction of PT subunits with the vesicles was found to be dependent on the presence of ATP and a reducing agent, both of which are required for activation of PT (Katada et al.,

1983; Moss et al., 1983; Lim et al., 1985; Moss et al., 1986). While both ATP and reducing agent are necessary for activation of the toxin, they work by different mechanisms. ATP is thought to activate PT by weakening the intersubunit bonds between the S1 subunit and the B oligomer (Burns and Manclark, 1986), whereas reducing agents break a critical disulfide bond within the S1 subunit. In the presence of ATP (0.5 mM), neither PT nor B oligomer bound to the lipid vesicles; however the isolated S1 subunit exhibited some ability to bind to the vesicles. When reducing agent (200 mM dithiothreitol) was present, binding of S1 to the vesicles increased. In the presence of both ATP and reducing agent, the holotoxin molecule appeared to dissociate into the S1 subunit and B oligomer moieties. The liberated S1 subunit bound to the vesicles, whereas the B oligomer demonstrated little binding to the vesicles. The results from this study indicate that the S1 subunit, when dissociated from the B oligomer, interacts with lipid vesicles. One might therefore postulate that if the S1 subunit were to dissociate from the B oligomer after binding to the surface of the eukaryotic cell, the S1 subunit might interact directly with the membrane, perhaps even penetrating the membrane so as to gain access to its intracellular substrates. While studies with photoreactive lipid probes did not detect interaction of S1 subunit with lipids, those studies were done under conditions in which the holotoxin molecule did not dissociate into its S1 and B oligomer moieties; therefore, the B oligomer might have prevented the S1 subunit from interacting with membranes.

V. CONCLUSIONS

Bacterial protein toxins such as PT are complex molecules in that they must contain information needed to perform multiple tasks in order to inflict damage on eukaryotic cells. Bacterial toxins must cross the membrane barriers of the bacterial cell; travel to the eukaryotic cell in an intact state; bind to receptors on the eukaryotic cell and perhaps cross the membrane barrier of the eukaryotic cell; and finally catalyze the reaction which results in alteration of cellular metabolism of the eukaryotic cell. PT accomplishes the first of these tasks by utilizing a set of accessory proteins to transit the bacterial membrane barriers. These accessory proteins appear to belong to a family of transport proteins involved in the secretion of macromolecules, including both proteins and DNA, across bacterial membranes of a number of organisms. After PT exits the bacterial cell, it utilizes information encoded in the primary sequence of its subunits to bind to eukaryotic receptors, gain access to its substrates, and ADP-ribosylate G proteins of the eukaryotic cell. The possibility remains that PT may utilize eukaryotic proteins to help it carry out these latter tasks.

REFERENCES

Berger, B.R. & Christie, P.J. (1993). The *Agrobacterium tumefaciens virB4* gene product is an essential virulence protein requiring an intact nucleoside triphosphate-binding domain. J. Bacteriol. 175, 1723–1734.

Brennan, M.J., David, J.L., Kenimer, J.G., & Manclark, C.R. (1988). Lectin-like binding of pertussis toxin to a 165-kilodalton Chinese hamster cell glycoprotein. J. Biol. Chem. 263, 4895–4899.

Briggs, M.S. & Gierasch, L.M. (1986). Molecular mechanisms of protein secretion: the role of the signal sequence. Adv. Prot. Chem. 38, 109–180.

Burns, D.L. & Manclark, C.R. (1986). Adenine nucleotides promote dissociation of pertussis toxin subunits. J. Biol. Chem. 261, 4324–4327.

Christie, P.J., Ward, J.E., Gordon, M.P., & Nester, E.W. (1989). A gene required for transfer of T-DNA to plants encodes an ATPase with autophosphorylating activity. Proc. Natl. Acad. Sci. USA 86, 9677–9681.

Covacci, A. & Rappuoli, R. (1993). Pertussis toxin export requires accessory genes located downstream from the pertussis toxin operon. Mol. Microbiol. 8, 429–434.

Deutscher, S.L., Nuwayhid, N., Stanley, P., Briles, E., & Hirschberg, C. (1984). Translocation across Golgi vesicle membranes: a CHO glycosylation mutant deficient in CMP-sialic acid transport. Cell 39, 295–299.

Deutscher, S.L. & Hirschberg, C.B. (1986). Mechanism of galactosylation in the Golgi apparatus. J. Biol. Chem. 261, 96–100.

Engström, P., Zambryski, P., Van Montagu, M.X., & Stachel, S. (1987). Characterization of *Agrobacterium tumefaciens* virulence proteins induced by the plant factor acetosyringone. J. Mol. Biol. 197, 635–645.

Gilman, A.G. (1987). G proteins: transducers of receptor generated signals. Annu. Rev. Biochem. 56, 615–649.

Granström, M., Linder-Nielsen, A.M., Holmblad, P., Mark, A., & Hanngren, K. (1991). Specific immunoglobulin for treatment of whooping cough. Lancet 338, 1230–1233.

Hausman, S.Z. & Burns, D.L. (1992). Interaction of pertussis toxin with cells and model membranes. J. Biol. Chem. 267, 13735–13739.

Hultgren, S.J., Normark, S., & Abraham, S.N. (1991). Chaperone-assisted assembly and molecular architecture of adhesive pili. Annu. Rev. Microbiol. 45, 383–415.

Kaslow, H.R. & Burns, D.L. (1992). Pertussis toxin and target eukaryotic cells: binding, entry, and activation. FASEB J. 6, 2684–2690.

Katada, T., Tamura, M., & Ui, M. (1983). The A protomer of islet-activating protein, pertussis toxin, as an active peptide catalyzing ADP-ribosylation of a membrane protein. Arch. Biochem. Biophys. 224, 290–298.

Katada, T. & Ui, M. (1982a). ADP-ribosylation of the specific membrane protein of C6 cells by islet-activating protein associated with modification of adenylate cyclase activity. J. Biol. Chem. 257, 7210–7216.

Katada, T. & Ui, M. (1982b). Direct modification of the membrane adenylate cyclase system by islet-activating protein due to ADP-ribosylation of a membrane protein. Proc. Natl. Acad. Sci. USA 79, 3129–3133.

Kuldau, G.A., de Vos, G., Owen, J., McCaffrey, G., & Zambryski, P. (1990). The *virB* operon of *Agrobacterium tumefaciens* pTiC58 encodes 11 open reading frames. Mol. Gen. Genet. 221, 256–266.

Lee, C.K., Roberts, A., & Perrin, S. (1989). Expression of pertussis toxin in *Bordetella bronchiseptica* and *Bordetella parapertussis* carrying recombinant plasmids. Infect. Immun. 57, 1413–1418.

Lessl, M., Balzer, D., Pansegrau, W., & Lanka E. (1992). Sequence similarities between the RP4 Tra2 and the Ti VirB region strongly support the conjugation model for T-DNA transfer. J. Biol. Chem. 267, 20471–20480.

Lim, L.K., Sekura, R.D., & Kaslow, H.R. (1985). Adenine nucleotides directly stimulate pertussis toxin. J. Biol. Chem. 260, 2585–2588.

Locht, C. & Keith, J.M. (1986). Pertussis toxin gene: nucleotide sequence and genetic organization. Science 232, 1258–1264.

Lory, S. (1992). Determinants of extracellular protein secretion in Gram-negative bacteria. J. Bacteriol. 174, 3423–3428.

Marchitto, K.S., Munoz, J.J., & Keith, J.M. (1987). Detection of subunits of pertussis toxin in Tn5-induced *Bordetella* mutants deficient in toxin biological activity. Infect. Immun. 55, 1309–1313.

Meade, B.D., Kind, P.D., & Manclark, C.R. (1984). Lymphocytosis-promoting factor of *Bordetella pertussis* alters mononuclear phagocyte circulation and response to inflammation. Infect. Immun. 46, 733–739.

Montecucco, C., Tomasi, M., Schiavo, G., & Rappuoli, R. (1986). Hydrophobic photolabelling of pertussis toxin subunits interacting with lipids. FEBS Lett. 194, 301–304.

Moss, J., Stanley, S.J., Burns, D.L., Hsia, J.A., Yost, D.A., Myers, G.A., & Hewlett, E.L. (1983). Activation by thiol of the latent NAD glycohydrolase and ADP-ribosyltransferase activities of *Bordetella pertussis* toxin (islet-activating protein). J. Biol. Chem. 258, 11879–11882.

Moss, J., Stanley, S.J., Watkins, P.A., Burns, D.L., Manclark, C.R., Kaslow, H.R., & Hewlett, E.L. (1986). Stimulation of the thiol-dependent ADP-ribosyltransferase and NAD glycohydrolase activities of *Bordetella pertussis* toxin by adenine nucleotides, phospholipids, and detergents. Biochemistry 25, 2720–2725.

Nester, E.W., Gordon, M.P., Amasino, R.M., & Yanofsky, M.F. (1984). Crown gall: A molecular and physiological analysis. Ann. Rev. Plant Physiol. 35, 387–413.

Nicosia, A. & Rappuoli, R. (1987). Promoter of the pertussis toxin operon and production of pertussis toxin. J. Bacteriol. 169, 2843–2846.

Nicosia, A., Perugini, M., Franzini, C., Casagli, M.C., Borri, M.G., Antoni, G., Almoni, M., Neri, P, Ratti, G., & Rappuoli, R. (1986). Cloning and sequencing of the pertussis toxin genes: Operon structure and gene duplication. Proc. Natl. Acad. Sci. USA 83, 4631–4635.

Pizza, M., Bugnoli, M., Manetti, R., Covacci, A., & Rappuoli, R. (1990). The subunit S1 is important for pertussis toxin secretion. J. Biol. Chem. 265, 17759–17763.

Possot, O., d'Enfert, C., Reyss, I., & Pugsley, A.P. (1992). Pullulanase secretion in *Escherichia coli* K-12 requires a cytoplasmic protein and a putative polytopic cytoplasmic membrane protein. Mol. Microbiol. 6, 95–105.

Pugsley, A.P. (1993). The complete general secretory pathway in Gram-negative bacteria. Microbiol. Rev. 57, 50–108.

Shirasu, K. & Kado, C.I. (1993). The *virB* operon of the *Agrobacterium tumefaciens* virulence regulon has sequence similarities to B, C, and D open reading frames downstream of the pertussis toxin-operon and to the DNA transfer-operons of broad-host-range conjugative plasmids. Nucleic Acids Res. 21, 353–354.

Spangrude, G.J., Braaten, B.A., & Daynes, R.A. (1984). Molecular mechanisms of lymphocyte extravasation. I. Studies of two selective inhibitors of lymphocyte recirculation. J. Immunol. 132, 354–362.

Tamura, M., Nogimori, K., Murai, S., Yajima, M., Ito, K., Katada, T., Ui, M., & Ishii, S. (1982). Subunit structure of islet-activating protein, pertussis toxin, in conformity with the A-B model. Biochemistry 21, 5516–5522.

Tamura, M., Nogimori, K., Yajima, M., Ase, K., & Ui, M. (1983). A role of the B oligomer moiety of islet-activating protein, pertussis toxin, in development of the biological effects on intact cells. J. Biol. Chem. 258, 6756–6761.

Thompson, D.V., Melchers, L.S., Idler, K.B., Schilperoort, R.A., & Hooykaas, P.J.J. (1988). Analysis of the complete nucleotide sequence of the *Agrobacterium tumefaciens* virB operon. Nucleic Acids Res. 16, 4621–4636.

Thorstenson, Y.R., Kuldau, G.A., & Zambryski, P.C. (1993). Subcellular localization of seven VirB proteins of *Agrobacterium tumefaciens*: implications for the formation of a T-DNA transport structure. J. Bacteriol. 175, 5233–5241.

Ward, J.E., Akiyoshi, D.E., Regier, D., Datta, A., Gordon, M.P., & Nester, E.W. (1988). Characterization of the *virB* operon from an *Agrobacterium tumefaciens* Ti plasmid. J. Biol. Chem. 263, 5804–5814.

Ward, J.E., Akiyoshi, D.E., Regier, D., Datta, A., Gordon, M.P., & Nester, E.W. (1990a). Correction: characterization of the *virB* operon from *Agrobacterium tumefaciens* Ti plasmid. J. Biol. Chem. 265, 4768.

Ward, J.E., Dale, E.M., Nester, E.W., & Binns, A.N. (1990b). Identification of a VirB10 protein aggregate in the inner membrane of *Agrobacterium tumefaciens*. J. Bacteriol. 172, 5200–5210.

Weiss, A.A., Johnson, F.D., & Burns, D.L. (1993). Molecular characterization of an operon required for pertussis toxin secretion. Proc. Natl. Acad. Sci. USA 90, 2970–2974.

Weiss, A.A. & Goodwin, M. (1989). Lethal infection by *Bordetella pertussis* mutants in the infant mouse model. Infect. Immun. 57, 3757–3764.

Welch, R.A. (1991). Pore-forming cytolysins of Gram-negative bacteria. Mol. Microbiol. 5, 521–528.

Wickner, W., Driessen, A.J.M., & Hartl, F.-U. (1991). The enzymology of protein translocation across the *Escherichia coli* plasma membrane. Annu. Rev. Biochem. 60, 101–124.

Witvliet, M.H., Burns, D.L., Brennan, M.J., Poolman, J.T., & Manclark, C.R. (1989). Binding of pertussis toxin to eucaryotic cells and glycoproteins. Infect. Immun. 57, 3324–3330.

Yajima, M., Hosoda, K., Kanbayashi, Y., Nakamura, T., Takahashi, I., & Ui, M. (1978). Biological properties of islets-activating protein (IAP) purified from the culture medium of *Bordetella pertussis*. J. Biochem. 83, 305–312.

Zambryski, P.C. (1992). Chronicles from the *Agrobacterium*-plant cell DNA transfer story. Annu. Rev. Plant Physiol. Plant. Mol. Biol. 43, 465–490.

Zambryski, P. (1988). Basic processes underlying *Agrobacterium*-mediated DNA transfer to plant cells. Annu. Rev. Genet. 22, 1–30.

TRANSPORT OF PROTEIN–NUCLEIC ACID COMPLEXES WITHIN AND BETWEEN PLANT CELLS

Vitaly Citovsky and Patricia Zambryski

Membrane Protein Transport
Volume 1, pages 39–57.
Copyright © 1995 by JAI Press Inc.
All rights of reproduction in any form reserved.
ISBN: 1-55938-907-9

I. INTRODUCTION

In eukaryotic organisms, nucleic acid molecules are constantly transported between
the nucleus and the cytoplasm. This transport includes nuclear export and import
of RNAs and nuclear import of retrotransposable elements and invading viruses.
Higher plants have an additional nucleic acid transport system that allows infectious
viral genomes to be transported between adjacent host cells. Only recently researchers have begun to unravel the molecular mechanisms by which intra- and
intercellular transport of nucleic acids occurs. In this review, we focus on two
processes of nucleic acid transport in plants: (1) transport of *Agrobacterium*
single-stranded (ss)DNA through nuclear pores, and (2) movement of tobacco
mosaic virus (TMV) genomic RNA through plant intercellular connections, the
plasmodesmata.

II. NUCLEAR IMPORT OF AGROBACTERIUM T-DNA

Agrobacterium tumefaciens is a soil phytopathogen that genetically transforms
plants by transferring a specific DNA segment (T-DNA) of its Ti-plasmid into the
plant cell genome. Most functions of the T-DNA transfer process are encoded by
the Ti-plasmid. Specifically, two genetic components are involved. The T-DNA
borders are two 25-bp direct repeats that delineate the segment of DNA to be
transferred. The T-DNA itself is sequence-nonspecific; any DNA between the
T-DNA borders is transported into the plant cell. The second component of the Ti
plasmid is the virulence (*vir*) region which encodes most of the protein machinery
of the T-DNA transport. Recent evidence suggests that the protein products of the
vir region are involved not only in the T-DNA transport from *Agrobacterium* to the
plant cell, but also in the nuclear import of the T-DNA following this transport
(reviewed by: Zambryski et al., 1989; Citovsky et al., 1992b; Zambryski, 1992).

A. The Structure of the T-DNA Transport Intermediate, the T-Complex

The first glimpse of the structure of the T-DNA transport intermediate came from
the study showing that the transported DNA is a linear single-stranded molecule,
designated the T-strand (Stachel et al., 1986). Subsequent studies demonstrated that
T-strands likely associate with two types of protein products of the *vir* region. One
molecule of the VirD2 protein is found covalently associated with the 5' end of the
T-strand (Herrera-Estrella et al., 1988; Ward and Barnes, 1988; Young and Nester,
1988; Howard et al., 1989). VirD2, in concert with VirD1, acts as an endonuclease
which specifically recognizes and cleaves the T-DNA borders. The borders are
cleaved on the bottom strand and serve as initiation and termination sites for the
generation of the T-strand (reviewed by: Zambryski et al., 1989; Citovsky et al.,
1992b; Zambryski, 1992). Following cleavage, VirD2 remains covalently attached

to the 5' end of the T-strand to serve as a pilot protein for the transported molecule (see below).

The second type of T-strand-associated protein is VirE2. VirE2 is a single-stranded (ss)DNA binding protein (SSB) which cooperatively coats the T-strand (Gietl et al., 1987; Christie et al., 1988; Citovsky et al., 1988; Das, 1988; Citovsky et al., 1989). The T-strand with its associated VirD2 and VirE2 proteins comprise the T-DNA transport complex, the T-complex (Howard et al., 1990; Howard and Citovsky, 1990). The T-complex is characterized by its thin and extended structure. Calculations based on electron microscopy observations and *in vitro* studies of VirE2 binding to ssDNA predict that nopaline-specific *Agrobacterium* produces a 20-kb T-strand packaged in a 3.6 µm-long and 2 nm-wide filament (Citovsky et al., 1989) containing one molecule of VirD2 and approximately 600 molecules of VirE2 (Citovsky et al., 1988). The combined molecular mass of this T-complex is estimated as 4.6×10^6 Daltons (Citovsky and Zambryski, 1993). How is such a large protein–nucleic acid complex, the T-complex, transported into the plant cell nucleus? As demonstrated in recent studies, the potential karyophilic properties of the protein components of the T-complex may provide an answer to this question.

B. Nuclear Import of VirD2, the 5' Pilot Protein of the T-Complex

The first component of the T-complex suggested to participate in its nuclear import is the 5' pilot VirD2 protein (Herrera-Estrella et al., 1990; Howard et al., 1992). This protein has been shown to localize to the cell nucleus in tobacco protoplasts (Howard et al., 1992). Nuclear import of VirD2 is mediated by a nuclear localization signal (NLS) sequence at the carboxyl terminus of the protein (Howard et al., 1992). This NLS belongs to the bipartite type exemplified by the NLS of *Xenopus* nucleoplasmin (Dingwall and Laskey, 1991; Howard et al., 1992) (Figure 1). VirD2 NLS can function independently of the rest of the protein; when fused to a reporter β-glucuronidase (GUS) enzyme, VirD2 NLS efficiently mediates nuclear import of the fusion protein in tobacco protoplasts (Howard et al., 1992).

Removal of the NLS sequence from the wild-type VirD2 results in reduction of *Agrobacterium* tumorigenicity (Shruvinton et al., 1992; Tinland et al., 1992), likely due to the reduced ability of the T-complex to enter the host cell nucleus. However, this decrease in *Agrobacterium* infection is only partial, ranging between 20 to 80% of the wild-type infectivity, depending on the host plant species (Shruvinton et al., 1992). It is likely, therefore, that another protein is involved in the nuclear import of the T-complex. Furthermore, the sheer size of the T-complex (4.6×10^6 Daltons in molecular mass and 3.6 µm in length) suggests that one molecule of the 5' end-bound VirD2 may not suffice for transport of the entire complex through the nuclear pore. Since VirE2 is the major protein component of the T-complex, it may promote nuclear import.

VirE2 (pTiC58) NSE 1	**K**l**R**pedRyiqte-**Ky**g**RR**
VirE2 (pTiA6) NSE 1	**K**l**R**pedRyvqte-**Ky**g**RR**
VirE2 (pTiC58) NSE 2	**Kt Ky**gsdtei---**K**l**Ks K**
VirE2 (pTiA6) NSE 2	**KRRy**ggetei---**K**l**Ks K**
VirD2 (pTiC58)	**KR**pRedddgepse**RKR**e**R**
nucleoplasmin	**KR**paat**KK**agqa-**KKKK**l

Figure 1. Amino acid sequence homology between bipartite NLSs of *Agrobacterium* VirE2 (nopaline C58 and octopine A6 strains), VirD2 and *Xenopus* nucleoplasmin. All basic residues are in uppercase; basic amino acids of the two domains of bipartite signals are bold and underlined. Adapted from Citovsky et al., 1992a.

C. Nuclear Import of VirE2, the T-Complex SSB

Similar to VirD2, nuclear localizing activity of VirE2 was demonstrated using its translational fusions with the reporter GUS enzyme (Citovsky et al., 1992a; Citovsky et al., 1994). Mutational analysis of VirE2 identified two independently active NLS sequences, designated NSE 1 and NSE 2. Each of these NLSs alone is able to promote nuclear import of the GUS reporter. However, unlike the VirD2 NLS, NSE 1 and NSE 2 function most efficiently when both are present in the full-length VirE2 (Citovsky et al., 1992a). Since both NLSs are required for the optimal nuclear import of VirE2 (Citovsky et al., 1992a), these signals are intrinsically weaker than the NLS of VirD2 which alone is sufficient for the maximal nuclear uptake of VirD2 (Howard et al., 1992).

Both NSE 1 and NSE 2 of VirE2 share homology with the NLS of VirD2 and belong to the bipartite type NLS (Citovsky et al., 1992a) (Figure 1). However, the NSE homology to the bipartite type NLS is not perfect. Generally, the first domain of a bipartite NLS has two adjacent basic residues, and the second domain contains at least three out of five basic amino acids (Dingwall and Laskey, 1991; Robbins et al., 1991). The NSE sequences of VirE2 have a modified first domain in which the two basic residues are separated by one amino acid; the consensus structure of the second domain, on the other hand, is preserved in the VirE2 NLSs (Citovsky et al., 1992a).

Mutational analysis of the SSB and NLS activities of VirE2 identified a carboxyl-terminal VirE2 mutant that was unable to bind ssDNA; because this mutant retained NSE 1 and NSE 2 signals, its nuclear localizing activity was not affected (Citovsky et al., 1992a). Removal of both NLS sequences or even uncharged amino acid substitutions of their basic residues inevitably caused the loss of both biological activities of the protein (Citovsky et al., 1992a, 1994). A more detailed analysis indicated that the NSE 2 signal may overlap the bona fide ssDNA binding domain; deletions or substitutions in NSE 2 completely blocked the ssDNA binding (Citovsky et al., 1994). NSE 1 appears to overlap the VirE2 sequence involved in binding cooperativity. Removal of NSE 1 or amino acid substitutions of its basic residues produced VirE2 mutants that bound ssDNA non-cooperatively (Citovsky et al., 1992a, 1994).

Nuclear localization activity of VirE2 suggests that this protein is involved directly in nuclear import of the T-complex. This in-planta function of VirE2 is further supported by experiments using transgenic tobacco plants expressing VirE2. While no crown gall tumors were produced by the VirE2-deficient strain of *Agrobacterium* on wild-type tobacco plants, inoculation of VirE2 transgenic plants with the same bacterial strain resulted in a high frequency of tumor formation (Citovsky et al., 1992a). Potentially, VirE2 expressed in transgenic plants assists in nuclear import of the invading T-strands produced by the VirE2-deficient *Agrobacterium*. In addition to its role in nuclear import, VirE2 as an SSB may act to protect the transported T-strands from cellular nucleases (Citovsky et al., 1989).

D. A Model for the T-complex Nuclear Import

In *Agrobacterium*, the cooperative binding of VirE2 is thought to unfold the T-strand during T-complex formation (Figure 2; Citovsky et al., 1989). Following its transfer into the plant cell cytoplasm, this unfolded protein–ssDNA complex is likely transported into the host cell nucleus by two consecutive steps (Figure 2): (1) targeting of the unfolded T-complex to the nuclear pore, possibly, following interaction between VirE2 and VirD2 NLSs with cytoplasmic NLS-binding proteins (NBPs); and (2) transport through the nuclear pore which is likely driven by a large number of VirE2 NLS signals positioned along the entire length of the T-complex.

As mentioned above, the T-complex is composed of three structural elements: one molecule of the T-strand, one molecule of VirD2, and more than 600 copies of VirE2. Since the T-strand is sequence-nonspecific, any DNA segment located between the T-DNA borders can function as T-DNA and be transported to plants. The T-strand, then, apparently does not possess specific nucleotide sequences for nuclear import; instead, it appears to be passively transported by the associated VirD2 and VirE2 proteins. Because VirD2 and VirE2 contain NLSs, both are expected to interact with the NBPs, the presumed receptor molecules that function to shuttle karyophillic proteins from the cytoplasm to the nuclear pore (reviewed by: Nigg et al., 1991; Dingwall and Laskey, 1992). VirD2 may act to initially target the T-complex to the nuclear pore in a polar fashion, mediating linear import of the T-strand in a 5′ to 3′ direction; such linear transfer may be necessary for the efficient integration of the transported DNA into the plant cell genome. Indeed, the requirement for polar transport of the T-DNA has been proposed in earlier genetic studies of *Agrobacterium*-mediated plant cell transformation (reviewed by Zambryski, 1992). Thus, VirD2 may be critical to determine the orientation of the T-complex nuclear import. The numerous VirE2 molecules with their NLSs may provide the

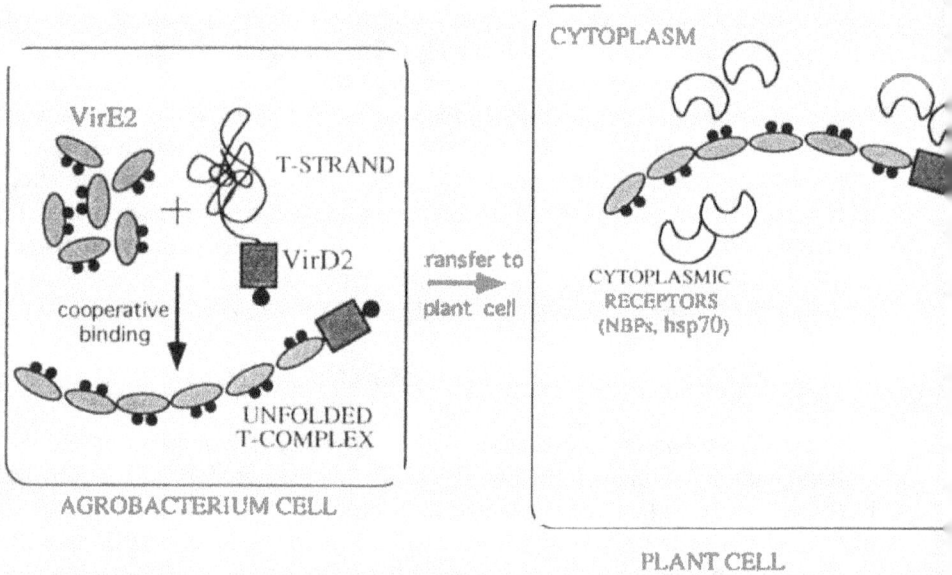

Figure 2. A model for nuclear transport of *Agrobacterium* T-complex. Following formatio[n] *Agrobacterium* cell, this complex is transported into the plant cell cytoplasm, VirE2, and receptors (e.g., NBPs, hsp70) and target the T-complex to the nuclear pore. Finally, the T-co[mplex] nucleus. Black circles on VirD2 and VirE2 molecules indicate NLS signals. Redrawn from Ci[...]

44

main driving force for translocation of the unfolded filament of the T-complex through the nuclear pore. For instance, the large number of NLSs supplied by the VirE2 molecules may greatly increase the probability of interaction between the T-complex and NBPs (Gerace and Burke, 1988) and, later, with the proteins of the nuclear pore. Furthermore, the presence of multiple NLSs along the length of the entire T-complex may maintain the active (open) conformation of the nuclear pore transporter which has been suggested to function as a double iris diaphragm (Akey, 1990; Dingwall, 1990; Akey, 1992; Dingwall and Laskey, 1992).

Since, in plant–pathogen interactions, the invading microorganism often adapts existing cellular machinery for its own needs, *Agrobacterium* likely employs an endogenous cellular pathway for the T-complex nuclear import. It is possible that plant cellular proteins analogous to VirE2 serve as molecular chaperones to coat and unfold nucleic acids and to target them to and through nuclear pores. Thus, nuclear transport of *Agrobacterium* T-complex may represent a generalized process by which ssDNA or RNA molecules move within the cell, i.e., as unfolded nucleic acid–protein complexes (see also below).

E. Molecular Pathways for the T-Complex Nuclear Import

Recent observations indicate that while both VirD2 and VirE2 NLSs share homology with many other NLS signals (Citovsky et al., 1992a; Howard et al., 1992), not all NLSs can function in nuclear transport of the T-complex. For example, VirD2 NLS can be successfully substituted with the NLS of the tobacco etch virus NIa protein; however, substitution with the SV40 large T-antigen NLS [which itself is active in plants (Lassner et al., 1991)] does not restore the virulence of *Agrobacterium* (Shruvinton et al., 1992). These observations suggest multiple pathways for nuclear transport. Potentially, these noncompeting pathways of nuclear transport may function independently to control traffic of diverse substrates through the same nuclear pore.

Other possible evidence for multiple and specific pathways for nuclear transport derives from the recent study of VirD2 and VirE2 nuclear accumulation in maize and tobacco leaf and mature (root base) and immature (root tip) root epidermal cells (Citovsky et al., 1994). Nuclear transport of VirD2 and VirE2 occurred efficiently only in leaf and immature root epidermis; both of these proteins remained cytoplasmic in mature epidermal cells of the root. That the GUS-VirD2 and GUS-VirE2 fusion proteins were produced from transfected DNA implies that the plasmid DNA entered the nucleus, was transcribed, and the pre-mRNA exported from the nucleus into the cytoplasm. However, that GUS-VirD2 and GUS-VirE2 themselves were not imported into the nucleus suggests that some pathway(s) for nuclear transport, (e.g., the T-complex import pathway) are repressed in the mature root tissue (Citovsky et al., 1994).

The mechanism by which some NLSs may be active in developing tissue and remain non-functional in mature tissue is unknown. Possibly, VirD2 and VirE2 nuclear import is mediated by specific cellular NBPs which may be absent in fully developed root epidermis. Potentially, this selective nuclear transport reflects a regulatory mechanism for developmentally specific gene expression; for example, only a certain subset of transcription factors may reach the cell nucleus at specific developmental stages of the tissue.

The pathway for VirE2 nuclear import was also studied *in vivo* using a variation of a "dominant negative mutation" strategy. Traditionally used in studies of gene function, this approach involves expression of a mutated gene that represses the activity of the wild-type gene (Herskowitz, 1987). In the case of VirE2, its deletion mutants were constitutively expressed in transgenic host plants; these plants were then tested for their ability to affect virulence of a wild-type *Agrobacterium* (Citovsky et al., 1994). When expressed in transgenic tobacco plants, VirE2 mutants which no longer bind ssDNA but retain some or all of their nuclear localizing activity, inhibited tumorigenicity of the wild-type *Agrobacterium* (Citovsky et al., 1994). Since these mutant proteins do not bind ssDNA, they are unable to form the T-complex; when present in the host cell, then, these VirE2 mutants presumably compete with the wild-type T-complex for the cellular nuclear import machinery.

Although the mutant VirE2 proteins expressed in the transgenic plants may reduce nuclear import of the T-complex, these plants are viable, implying that their normal cellular processes of nuclear transport are not significantly compromised. This result may indicate a number of different pathways for nuclear import in plant cells. Potentially, only one such pathway is utilized for the T-complex nuclear uptake.

F. Comparison between Nuclear Transport of Cellular RNA and the *Agrobacterium* T-Strands

Nuclear transport of *Agrobacterium* ssDNA and various types of cellular RNA probably does not occur by a single mechanism. However, it may be possible to identify major conserved structural and functional characteristics of the transport pathways and suggest a set of general features for the nucleic acid nuclear transport. To this end, we have compared the known requirements for nuclear transport of small nuclear (sn)RNA, heterogenous nuclear (hn)RNA, and *Agrobacterium* T-strands (reviewed by Citovsky and Zambryski, 1993). The comparison identified seven common features (Table 1), suggesting that nuclear import of *Agrobacterium* T-strands may reflect a generalized process of nucleic acid nuclear transport in plant as well as in animal cells.

1. *Association with specific proteins.* No nucleic acid is known to be transported as a free molecule. While each nucleic acid molecule is usually associated with a large number of protein molecules, the number of different protein species varies considerably. For example, *Agrobacterium* T-complex contains two classes of proteins (VirD2 and VirE2) (Citovsky et al., 1992a), and vertebrate hnRNPs may

Table 1. Common Features in Nuclear Transport of Cellular RNA and
Agrobacterium T-Strands[a]

	Transported Nucleic Acid		
Transport Feature	Agrobacterium T-Strand	snRNA	hnRNA
Association with Proteins	Yes	Yes	Yes
Unfolded Conformation	Yes	?	Yes
Localization Signals	Protein NLS	Protein NLS and tmG cap	Protein NLS and tmG cap (?)
Cytoplasmic Receptors	Yes	Yes	Yes
Opening of the Nuclear Pore	Yes	Yes	Yes
Polarity of Transport	Yes	?	Yes
Transport as a Single-Stranded Molecule	Yes	Yes	Yes

Note: [a]Adapted from Citovsky and Zambryski (1993).

contain more than 20 different proteins (Choi and Dreyfuss, 1984; Pinol-Roma et al., 1988).

2. *Unfolding of the transported nucleic acid–protein complex.* Generally, protein nuclear transport does not require unfolding of the transported molecule (Dingwall et al., 1982). Nucleic acid–protein complexes, however, are much larger structures than individual proteins. The sheer bulk of these complexes may require substantial conformational changes prior to transport. For example, *Agrobacterium* T-complex may unfold to a 2 nm-wide filament prior to transport (Citovsky et al., 1989; Citovsky et al., 1992a). Also, the Balbiani ring (BR) hnRNA–protein complex (RNP) unfolds during the translocation, changing its diameter from 50 nm to 25 nm (Mehlin et al., 1992); this reduced size of the BR hnRNP approximately corresponds to the maximal diameter of the fully open nuclear pore (23 nm) (Dworetzky and Feldherr, 1988). In addition to steric considerations, unfolding of transported complexes may be required to determine the polarity of the transport, i.e. to expose the leading end of the molecules during translocation (see below).

3. *Specific localization signals.* Protein binding to the transported nucleic acid provides specific targeting signals for transport. In some cases, protein NLSs are sufficient for nuclear transport [*Agrobacterium* T-complex (Citovsky et al., 1992a; Howard et al., 1992) and U6 snRNP (Fischer et al., 1991; Michaud and Goldfarb, 1991)]. In other instances, nuclear transport requires the presence of protein NLSs as well as specific signals contained in the nucleic acid itself [tmG cap U1-U5 snRNPs and some hnRNPs (reviewed by Izaurralde and Mattaj, 1992)].

4. *Cytoplasmic receptors.* Protein nuclear import initiates in the cytoplasm with binding of NLSs to the first cytoplasmic receptor, hsp70 (or its cognate hsc70), which may present the NLS in a locally unfolded form to the second type of

cytoplasmic receptors, the NLS binding proteins (NBPs) (reviewed by Dingwall and Laskey, 1992). The NBPs then direct the transported molecule to the nuclear pore where actual translocation across the nuclear envelope occurs (reviewed by: Nigg et al., 1991; Dingwall and Laskey, 1992). Since transported nucleic acids are always associated with karyophilic proteins (see above), the nucleic acid–protein complexes likely interact with cytoplasmic receptors as well.

　　5.　*"Opening" of the nuclear pore.* All known mechanisms for the active nuclear transport require opening of the nuclear pore to allow translocation. This active mechanism of transport is mediated by specific NLSs contained in the transported molecule(s) (reviewed by Garcia-Bustos et al., 1991). Undoubtedly, import of large protein–nucleic acid complexes, such as the T-complex or BR RNP, requires opening of the nuclear pore and the NLS signals of the specific proteins associated with the transported T-strand or RNA molecules likely mediate this process.

　　6.　*Polarity of transport.* Polar transport has been proposed for nuclear uptake of *Agrobacterium* T-complex (Citovsky et al., 1992a, b; Zambryski, 1992) and for nuclear export of BR hnRNPs (Mehlin et al., 1992). It is possible that other nucleic acid–protein complexes are also transported in a polar fashion. For example, polarity of TmG cap signals in pol II U snRNPs (Hamm et al., 1990) suggests these complexes are transported directionally. Vectorial transport of nucleic acid–protein complexes may be required for immediate processing of the emerging complex (e.g., integration, translation, etc.). Potentially, transport polarity is determined by a specific signal associated with one end of the transported nucleic acid molecule (e.g., VirD2 protein in *Agrobacterium* T-complex and tmG cap in snRNPs).

　　7.　*Transport of single-stranded nucleic acids.* Most of the transported nucleic acids (*Agrobacterium* T-strand, RNA, and genomic nucleic acids of many animal and plant viruses) are single-stranded. This could be a simple coincidence; however, transport of single-stranded nucleic acids has certain advantages over transport of the double helix: single strands are polar (as linear molecules), less rigid, and easily coated by signal sequence (such as NLS) containing SSBs.

III. TRANSPORT OF TMV RNA THROUGH PLASMODESMATA

Nuclear pores and plasmodesmata are the only known large biological pore structures involved in active bidirectional traffic of macromolecules. It will be interesting to see if nucleic acid transport through these, albeit structurally different, channels shares common requirements. The best studied example of plasmodesmal transport of nucleic acids is the cell-to-cell movement of TMV genomic RNA molecules.

　　TMV is a linear positive-sense ssRNA tobamovirus which replicates in the host cell cytoplasm. The TMV genome encodes at least three nonstructural proteins (P126, P183, and P30) and the coat protein (P17 or CP) (Goelet et al., 1982). P126 and P183 function in viral replication (Palikaitis and Zaitlin, 1986) and P30 potentiates movement of TMV through plasmodesmata. Several lines of evidence point to this role of P30: (1) temperature-sensitive TMV mutants defective in cell-to-cell spread have an altered P30 polypeptide (Jockusch, 1968; Peters and Murphy, 1975; Nishiguchi et al., 1978); (2) *in vitro* mutagenesis of P30 produces phenotypes restricted in cell-to-cell movement but not in replication (Meshi et al., 1987); (3) P30 expressed in transgenic plants can restore cell-to-cell spread of the movement-deficient mutant virus (Deom et al., 1987); and (4) TMV CP is dispensable for cell-to-cell spread (Siegal et al., 1962), suggesting that P30 is the only viral protein-mediating TMV movement between cells.

Until recently, the mechanism of P30 action had been a complete mystery. It began to unravel when P30 was reported to have at least two important biological activities: (1) P30, expressed in transgenic plants, was shown to increase the plasmodesmata molecular size exclusion limit (Wolf et al., 1989); and (2) purified P30 was shown to cooperatively bind single-stranded nucleic acids (Citovsky et al., 1990).

A. Interaction between P30 and Plasmodesmata

The P30 effect on plasmodesmal permeability was studied by microinjecting fluorescently labeled dextrans of increasing molecular weights into the leaf mesophyl tissue of transgenic tobacco plants expressing P30. Plasmodesmata size exclusion limit in these plants was calculated using Stokes radii of the injected dextran molecules (Wolf et al., 1989). This study showed that the size exclusion limit of plasmodesmata in transgenic plants that express P30 is higher (5–6 nm in diameter as measured using 10–20-kDa dextrans) than in normal plants (1.5 nm in diameter). A similar increase in plasmodesmal permeability was observed during tobacco rattle tobravirus (TRV) infection (Derrick et al., 1992). The interaction of P30 with plasmodesmata was further studied using direct microinjection of P30 [expressed in *E. coli* and purified to homogeneity (Citovsky et al., 1990)] into wild-type tobacco plants (Waigmann et al., 1994). When coinjected with fluorescently labeled dextrans, P30 increased the plasmodesmal size exclusion limit to 6–9 nm as measured using 20–40-kDa dextrans (Waigmann et al., 1994). In these experiments, P30-mediated movement of these large dextran molecules was detected 3 to 5 minutes after injection; 45 minutes after injection, the fluorescent dye was distributed between a large number of cells as far as 10 to 20 cells away from the initially microinjected cell. That fluorescent dextrans could spread so far from the site of injection may suggest actual movement of P30 itself between cells, because such P30 movement would be necessary to affect plasmodesmal permeability in noninjected cells (Waigmann et al., 1994). To promote this movement, microinjected P30 may have traveled across several cells to reach the cell microinjected with the labeled dextran. Alternatively, microinjected P30 may have triggered a putative intercellular signal transduction pathway that increased plasmodesmal permeability in numerous interconnecting cells.

Microinjection of various deletion mutants of P30 suggested that a protein domain, encompassed by amino acid residues 126 to 224, is required for P30-induced increase in plasmodesmal permeability (Waigmann et al., 1994). Presumably, increasing plasmodesmal size exclusion limit is a complex process. Consequently, several individual functions may reside in this large domain, such as interaction with a cytoplasmic shuttle protein and/or a plasmodesmal receptor that could be involved in localizing P30 to plasmodesmata. At the present time, the specific function of this P30 domain is undetermined.

The P30-induced increase in plasmodesmal size exclusion limit is the first experimental evidence that plasmodesmal permeability can be "upregulated". How P30 affects plasmodesmal size exclusion limit is unknown. Since this increase in plasmodesmal permeability is fast (Waigmann et al., 1994) and does not involve significant changes in plasmodesmata ultrastructure (Ding et al., 1992), it likely occurs by normal (albeit unknown) regulatory pathways rather than by physical disassembly of plasmodesmata. This finding has an intriguing implication on the general concept of plant intercellular communication. Traditionally, plasmodesmata have been considered to traffic only water, ions, and small metabolites. The ability of plasmodesmata to expand, however, indicates that in normal, healthy plants plasmodesmata may traffic macromolecules as well. Indeed, recent results suggest that cellular proteins are constantly transported through plasmodesmata between companion and sieve cells in the phloem (Fisher et al., 1992).

While P30 increases the size exclusion limit of plasmodesmata from 1.5 nm to 5–6 nm in transgenic plants (Wolf et al., 1989) and to 6–9 nm in microinjected wild-type plants (Waigmann et al., 1994), this increase in permeability is insufficient for cell-to-cell transport of the 10 nm-wide free-folded TMV RNA (Gibbs, 1976) that *in vivo* would be mediated by P30. One possibility is that, prior to translocation, viral RNA is unfolded by forming a protein–RNA transport complex with P30. Thus, P30 may not only act to increase plasmodesmal permeability, but it may also function as a molecular chaperone for the viral nucleic acid.

B. Interaction between P30 and Nucleic Acids

P30 has been identified as a single-stranded nucleic acid binding protein (Citovsky et al., 1990). Mutational analysis of P30 revealed two independently active single-stranded nucleic acid binding domains at the carboxyl-terminus of the protein (amino acid residues 111 to 268) (Citovsky et al., 1992). P30 binding is strong, cooperative, and sequence nonspecific (Citovsky et al., 1990). While preferential binding to single-stranded nucleic acids and binding cooperativity are characteristic of all known SSBs (Chase and Williams, 1986), unlike other SSBs P30 binds ssDNA and RNA with equal affinity (Citovsky et al., 1990). Binding of P30 to viral RNA was proposed to shape it into a transferable form (Citovsky and

Zambryski, 1991; Citovsky et al., 1992). Electron microscopy observations demonstrated that P30 can unfold the irregular collapsed structures of free single-stranded nucleic acids to form long and thin protein–nucleic acid complexes with a diameter of about 2 nm (Citovsky et al., 1992). The width of these complexes is compatible with the open plasmodesmal channel found in plants transgenic for P30 or microinjected with P30 (Wolf et al., 1989; Waigmann et al., 1994) and, by implication, modified during viral infection. Interestingly, cell-to-cell movement proteins of several other plant viruses [e.g., cauliflower mosaic virus (Citovsky et al., 1991), alfalfa mosaic virus (Schoumacher et al., 1992), red clover necrotic mosaic virus (Osman et al., 1992), and foxtail mosaic virus (Rouleau et al., 1993)] have been shown to function as ssDNA and RNA binding proteins, suggesting that formation of movement protein–nucleic acid complexes may be an important step in plasmodesmal transport of many plant viruses.

C. A Model for Plasmodesmal Transport of TMV RNA

Our current knowledge of P30 biochemistry and biology is integrated in the following model for TMV RNA transport through plasmodesmata (see Figure 2; Citovsky and Zambryski, 1993). In this model, P30 functions as a molecular chaperone to bind and unfold TMV RNA and target it to and through the plasmodesmal channel. Interestingly, unfolded P30-TMV RNA complexes may be functionally analogous to those for nuclear transport of *Agrobacterium* T-DNA (Citovsky et al., 1992a) and premessenger RNA (Mehlin et al., 1992) (see above).

The process begins with P30 binding to TMV RNA. Note that since replication and translation of many plant viruses are highly compartmentalized (Palikaitis and Zaitlin, 1986), movement proteins will likely associate with the viral nucleic acids rather than with host mRNAs. The binding of P30 to TMV RNA results in formation of long, thin, and unfolded ribonucleoprotein transport complexes (Figure 3; Citovsky et al., 1992). These complexes are then targeted to plasmodesmata. Potentially, as seen for nuclear import (reviewed by: Nigg et al., 1991; Dingwall and Laskey, 1992), cytoplasmic receptors may exist that transport and target P30–nucleic acid complexes to plasmodesmata. Again by analogy to nuclear transport, in addition to the putative P30 receptors, other cellular proteins (e.g., hsp70 and its cognate hsc70; reviewed by Dingwall and Laskey, 1992) may be involved in plasmodesmal targeting. After reaching the plasmodesmal channel, P30 acts to increase its permeability (Figure 3; Wolf et al., 1989; Waigmann et al., 1994).

The molecular mechanism by which P30 affects plasmodesmal permeability is unknown. It is well established that P30 can be specifically targeted to plasmodesmata in the plant cell walls. Electron microscopy studies of virus-infected plants reported specific association of P30 with plasmodesmata (Tomenius et al., 1987). Furthermore, in transgenic plants, P30 was found to accumulate in the central cavity of branched (secondary) but not in single (primary) plasmodesmata (Ding et al., 1992). However, that TMV is able to infect young apical leaves (Culver et al., 1991)

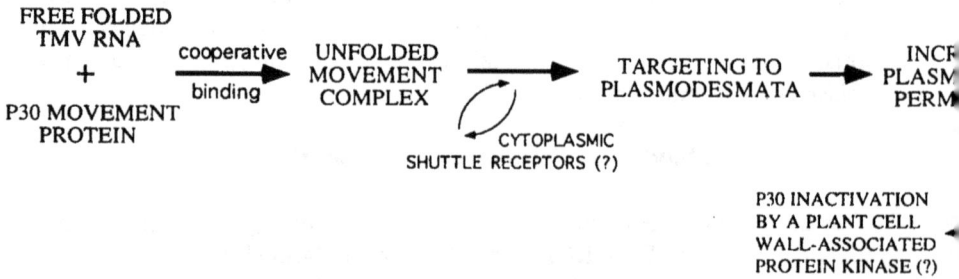

Figure 3. A model for plasmodesmal transport of TMV RNA. P30 cooperatively binds free, fol‹
P30-TMV RNA cell-to-cell transport complexes. These complexes are targeted to plasmodesm‹
cytoplasmic receptors. Following the P30-induced increase in plasmodesmal permeability,
through the plasmodesmal channel. During translocation, P30 may be partly inactivated fo‹
wall-associated protein kinase. Adapted from Citovsky and Zambryski, 1993.

which do not contain secondary plasmodesmata implies that P30 interacts with primary plasmodesmata as well. Thus, P30 interaction with primary plasmodesmata may be transient, while interaction with secondary plasmodesmata potentially results in irreversible deposition of P30 in the central cavity.

An insight into the possible mechanism of P30 accumulation in the secondary plasmodesmata was provided by recent study of P30 phosphorylation (Citovsky et al., 1993). The carboxyl-terminal part of P30 was shown to contain a major phosphorylation site, and recent results indicate that a plant cell wall-associated protein kinase phosphorylates P30 at amino acid residues Ser-257, Thr-261, and Ser-265. This protein kinase activity is developmentally expressed (Citovsky et al., 1993) correlating with the basipetal (tip-to-base) development of tobacco leaves and appears to parallel the development of secondary plasmodesmata (Ding et al., 1992). Since P30 specifically accumulates in secondary plasmodesmata (Ding et al., 1992), P30 phosphorylation may represent a mechanism for the host plant to deactivate this biologically potent protein by irreversible association with the cell walls (Figure 3).

Interaction of P30 with plasmodesmata ultimately results in transport of the TMV RNA into the neighboring cell. Two mechanisms for translocation are possible. The entire transport complex may cross into the neighboring cell. Alternatively, the transport complex may be partially unpacked during translocation. Partial uncoating of transported P30-TMV RNA complexes may resemble transport of premessenger RNPs through the nuclear pore since these complexes shed part of their proteins during nuclear export (Mehlin et al., 1992). Finally, following translocation, the viral movement protein may be completely displaced to allow translation and replication of the viral RNA.

IV. CONCLUSIONS

Nuclear import and plasmodesmal transport of nucleic acids in plants exhibit several structural and functional similarities. Both processes involve single-stranded nucleic acids which are transported as complexes with specialized transport proteins. These proteins cooperatively bind ssDNA and RNA, producing thin unfolded filaments of protein–nucleic acid transport complexes. In addition to shaping nucleic acid molecules in a transferable form, the transport proteins likely provide specific signals for targeting to and interaction with the nuclear pores or plasmodesmal channels.

ACKNOWLEDGMENTS

We thank Dr. Gail McLean for critical reading of the manuscript. The research relating to nuclear uptake of Agrobacterium T-complex was supported by NSF grant (89-15613), and that relating to plant virus movement by NIH grant (GM-45244-01).

REFERENCES

Akey, C.W. (1990). Visualization of transport-related configurations of the nuclear pore transporter. Biophys. J. 58, 341–355.

Akey, C.W. (1992). The nuclear pore complex. Curr. Op. Struct. Biol. 2, 258–263.

Chase, J.W. & Williams, K.R. (1986). Single-stranded DNA binding proteins required for DNA replication. Annu. Rev. Biochem. 55, 103–136.

Choi, Y.D. & Dreyfuss, G. (1984). Isolation of the heterogeneous nuclear RNA-ribonucleoprotein complex (hnRNP): a unique supramolecular assembly. Proc. Natl. Acad. Sci. USA 81, 1997–2004.

Christie, P.J., Ward, J.E., Winans, S.C., & Nester, E.W. (1988). The Agrobacterium tumefaciens virE2 gene product is a single-stranded-DNA-binding protein that associates with T-DNA. J. Bacteriol. 170, 2659–2667.

Citovsky, V., De Vos, G., & Zambryski, P. (1988). Single-stranded DNA binding protein encoded by the virE locus of Agrobacterium tumefaciens. Science 240, 501–504.

Citovsky, V., Wong, M.L., & Zambryski, P. (1989). Cooperative interaction of Agrobacterium VirE2 protein with single stranded DNA: implications for the T-DNA transfer process. Proc. Natl. Acad. Sci. USA 86, 1193–1197.

Citovsky, V., Knorr, D., Schuster, G., & Zambryski, P. (1990). The P30 movement protein of tobacco mosaic virus is a single strand nucleic acid binding protein. Cell 60, 637–647.

Citovsky, V., Knorr, D., & Zambryski, P. (1991). Gene I, a potential movement locus of CaMV, encodes an RNA binding protein. Proc. Natl. Acad. Sci. USA 88, 2476–2480.

Citovsky, V. & Zambryski, P. (1991). How do plant virus nucleic acids move through intercellular connections? BioEssays 13, 373–379.

Citovsky, V., Wong, M.L., Shaw, A., Prasad, B.V.V., & Zambryski, P. (1992). Visualization and characterization of tobacco mosaic virus movement protein binding to single-stranded nucleic acids. Plant Cell 4, 397–411.

Citovsky, V., Zupan, J., Warnick, D., & Zambryski, P. (1992a). Nuclear localization of Agrobacterium VirE2 protein in plant cells. Science 256, 1803–1805.

Citovsky, V.C., McLean, G., Greene, E., Howard, E., Kuldau, G., Thorstenson, Y., Zupan, J., & Zambryski, P. (1992b). Agrobacterium-plant cell interaction: induction of vir genes and T-DNA transfer. In: Molecular Signals in Plant-Microbe Communications (Verma, D. P. S., ed.), pp. 169–198, CRC Press, Inc.

Citovsky, V. & Zambryski, P. (1993). Transport of nucleic acids through membrane channels: snaking through small holes. Annu. Rev. Microbiol. 47, 167–197.

Citovsky, V., McLean, B.G., Zupan, J., & Zambryski, P. (1993a). Phosphorylation of tobacco mosaic virus cell-to-cell movement protein by a developmentally-regulated plant cell wall-associated protein kinase. Genes Dev. 7, 904–910.

Citovsky, V., Warnick, D., & Zambryski, P. (1994). Nuclear import of Agrobacterium VirD2 and VirE2 proteins in maize and tobacco. Proc. Natl. Acad. Sci. USA 91, 3210–3214.

Culver, J.N., Lindbeck, A.G.C., & Dawson, W.O. (1991). Virus-host interactions: induction of chlorotic and necrotic responses in plants by tobamoviruses. Annu. Rev. Phytopathol. 29, 193–217.

Das, A. (1988). Agrobacterium tumefaciens virE operon encodes a single-stranded DNA-binding protein. Proc. Natl. Acad. Sci. USA 85, 2909–13.

Deom, C.M., Shaw, M.J., & Beachy, R.N. (1987). The 30-kilodalton gene product of tobacco mosaic virus potentiates virus movement. Science 327, 389–394.

Derrick, P.M., Barker, H., & Oparka, K.J. (1992). Increase in plasmodesmatal permeability during cell-to-cell spread of tobacco rattle tobravirus from individually inoculated cells. Plant Cell 4, 1405–1412.

Ding, B., Haudenshield, J.S., Hull, R.J., Wolf, S., Beachy, R.N., & Lucas, W.J. (1992). Secondary plasmodesmata are specific sites of localization of the tobacco mosaic virus movement protein in transgenic tobacco plants. Plant Cell 4, 915–928.

Dingwall, C., Sharnick, S.V., & Laskey, R.A. (1982). A polypeptide domain that specifies migration of nucleoplasmin into the nucleus. Cell 30, 449–458.

Dingwall, C. (1990). Plugging the nuclear pore. Nature 346, 512–514.

Dingwall, C. & Laskey, R.A. (1991). Nuclear targeting sequences—a consensus? Trends Biochem. Sci. 16, 478–481.

Dingwall, C. & Laskey, R. (1992). The nuclear membrane. Science 258, 942–947.

Dworetzky, S.I. & Feldherr, C.M. (1988). Translocation of RNA-coated gold particles through the nuclear pores of oocytes. J. Cell Biol. 106, 575–584.

Fischer, U., Darzynkiewicz, E., Tahara, S.M., Dathan, N.A., Luhrmann, R., & Mattaj, I. (1991). Diversity in signals required for nuclear accumulation of U snRNPs and variety in the pathways of nuclear transport. J. Cell Biol. 113, 705–714.

Fisher, D.B., Wu, Y., & Ku, M.S.B. (1992). Turnover of soluble proteins in the wheat sieve tube. Plant Physiol. 100, 1433–1441.

Garcia-Bustos, J., Heitman, J., & Hall, M.N. (1991). Nuclear protein localization. Biochim. Biophys. Acta 1071, 83–101.

Gerace, L. & Burke, B. (1988). Functional organization of the nuclear envelope. Annu. Rev. Cell Biol. 4, 355–374.

Gibbs, A.J. (1976). Viruses and plasmodesmata. In: Intercellular Communication in Plants: Studies on Plasmodesmata (Gunning, B.E.S. & Robards, A.W., Eds.), pp. 149–164. Springer-Verlag, Berlin.

Gietl, C., Koukolikova-Nicola, Z., & Hohn, B. (1987). Mobilization of T-DNA from *Agrobacterium* to plant cells involves a protein that binds single-stranded DNA. Proc. Natl. Acad. Sci. USA 84, 9006–9010.

Goelet, P., Lomonossoff, G.P., Butler, P.J.G., Akam, M.E., Gait, M.J., & Karn, J. (1982). Nucleotide sequence of tobacco mosaic virus RNA. Proc. Natl. Acad. Sci USA 79, 5818–5822.

Hamm, J., Darzynkiewicz, E., Tahara, S.M., & Mattaj, I.W. (1990). The trimethylguanosine cap structure of U1 snRNA is a component of a bipartite nuclear targeting signal. Cell 62, 569–577.

Herrera-Estrella, A., Chen, Z., Van Montagu, M., & Wang, K. (1988). VirD proteins of *Agrobacterium tumefaciens* are required for the formation of a covalent DNA protein complex at the 5' terminus of T-strand molecules. EMBO J. 7, 4055–4062.

Herrera-Estrella, A., Van Montagu, M., & Wang, K. (1990). A bacterial peptide acting as a plant nuclear targeting signal: the amino-terminal portion of *Agrobacterium vir*D2 protein directs a β-galactosidase fusion protein into tobacco nuclei. Proc. Natl. Acad. Sci. USA 87, 9534–9537.

Herskowitz, I. (1987). Functional inactivation of genes by dominant negative mutations. Nature 329, 219–222.

Howard, E.A., Winsor, B.A., De Vos, G., & Zambryski, P. (1989). Activation of the T-DNA transfer process in *Agrobacterium* results in the generation of a T-strand protein complex: tight association with the 5' ends of T-strands. Proc. Natl. Acad. Sci. USA 86, 4017–4021.

Howard, E., Citovsky, V., & Zambryski, P. (1990). The T-complex of *Agrobacterium tumefaciens*. UCLA Symp. Mol. Cell Biol. New Ser. 129, 1–11.

Howard, E.A. & Citovsky, V. (1990). The emerging structure of the *Agrobacterium* T-DNA transfer complex. BioEssays 12, 103–108.

Howard, E., Zupan, J., Citovsky, V., & Zambryski, P. (1992). The VirD2 protein of *Agrobacterium tumefaciens* contains a C-terminal bipartite nuclear localization signal: implications for nuclear uptake of DNA in plant cells. Cell 68, 109–118.

Izaurralde, E. & Mattaj, I. (1992). Transport of RNA between nucleus and cytoplasm. Semin. Cell Biol. 3, 279–288.

Jockusch, H. (1968). Two mutants of tobacco mosaic virus temperature-sensitive in two different functions. Virology 35, 94–101.

Lassner, M.W., Jones, A., Daubert, S., & Comal, L. (1991). Targeting of T7 RNA polymerase to tobacco nuclei by an SV40 nuclear location signal. Plant Mol. Biol. 17, 229–234.

Mehlin, H., Daneholt, B., & Skoglund, U. (1992). Translocation of a specific premessenger ribonucleo-protein particle through the nuclear pore studied with electron microscope tomography. Cell 69, 605–613.

Meshi, T., Watanabe, Y., Saito, T., Sugimoto, A., Maeda, T., & Okada, Y. (1987). Function of the 30 kd protein of tobacco mosaic virus: involvement in cell-to-cell movement and dispensability for replication. EMBO J. 6, 2557–2563.

Michaud, N. & Goldfarb, D. (1991). Multiple pathways in nuclear transport: the import of U2 snRNP occurs by a novel kinetic pathway. J. Cell Biol. 112, 215–223.

Nigg, E.A., Baeuerle, P.A., & Luhrmann, R. (1991). Nuclear import-export: in search of signals and mechanisms. Cell 66, 15–22.

Nishiguchi, M., Motoyoshi, F., & Oshima, M. (1978). Behavior of a temperature sensitive strain of tobacco mosaic virus in tomato leaves and protoplasts. J. Gen. Virol. 39, 53–61.

Osman, T.A.M., Hayes, R.J., & Buck, K.W. (1992). Cooperative binding of the red clover necrotic mosaic virus movement protein to single-stranded nucleic acids. J. Gen. Virol. 73, 223–227.

Palikaitis, P. & Zaitlin, M. (1986). Tobacco mosaic virus: infectivity and replication. In: The Rod-Shaped Viruses (Van Regenmortel, M.H.V., & Fraenkel-Conrat, H., eds.), pp. 105–131, Plenum Press, New York.

Peters, D.L. & Murphy, T.M. (1975). Selection of temperature sensitive mutants of tobacco mosaic virus by lesion morphology. Virology 65, 595–600.

Pinol-Roma, S., Choi, Y.D., Matunis, M.J., & Dreyfuss, G. (1988). Immunopurification of heterogene-ous nuclear ribonucleoprotein particles reveals an assortment of RNA-binding proteins. Gen. Dev. 2, 215–227.

Robbins, J., Dilworth, S. M., Laskey, R.A., & Dingwall, C. (1991). Two interdependent basic domains in nucleoplasmin nuclear targeting sequence: identification of a class of bipartite nuclear targeting sequence. Cell 64, 615–623.

Rouleau, M., Bancroft, J.B., & Mackie, G.A. (1993). Purification and characterization of the potexviral movement protein ORF2. Abstracts of IXth International Congress of Virology, p. 99, Glasgow, Scotland.

Schoumacher, F., Erny, C., Berna, A., Godefroy-Colburn, T., & Stussi-Garaud, C. (1992). Nucleic acid binding properties of the alfalfa mosaic virus movement protein produced in yeast. Virology 188, 896–899.

Shruvinton, C.E., Hodges, L., & Ream, W. (1992). A nuclear localization signal and the C-terminal omega sequence in the *Agrobacterium tumefaciens* VirD2 endonuclease are important for tumor formation. Proc. Natl. Acad. Sci. USA 89, 11837–11841.

Siegal, A., Zaitlin, M., & Sehgal, O.P. (1962). The isolation of defective tobacco mosaic virus strains. Proc. Natl. Acad. Sci. USA 48, 1845–1851.

Stachel, S.E., Timmerman, B., & Zambryski, P. (1986). Generation of single-stranded T-DNA molecules during the initial stages of T-DNA transfer for *Agrobacterium tumefaciens* to plant cells. Nature 322, 706–712.

Tinland, B., Koukolikova-Nicola, Z., Hall, M.N., & Hohn, B. (1992). The T-DNA-linked VirD2 protein contains two distinct nuclear localization signals. Proc. Natl. Acad. Sci. USA 89, 7442–7446.

Tomenius, K., Clapham, D., & Meshi, T. (1987). Localization by immunogold cytochemistry of the virus coded 30 K protein in plasmodesmata of leaves infected with tobacco mosaic virus. Virology 160, 363–371.

Waigmann, E., Lucas, W., Citovsky, V., & Zambryski, P. (1994). Direct functional assay for tobacco mosaic virus cell-to-cell movement protein and identification of a domain involved in increasing plasmodesmal permeability. Proc. Natl. Acad. Sci. USA 91, 1433–1437.

Ward, E. & Barnes, W. (1988). VirD2 protein of *Agrobacterium tumefaciens* very tightly linked to the 5′ end of T-strand DNA. Science 242, 927–930.

Wolf, S., Deom, C.M., Beachy, R.N., & Lucas, W.J. (1989). Movement protein of tobacco mosaic virus modifies plasmodesmatal size exclusion limit. Science 246, 377–379.

Young, C. & Nester, E.W. (1988). Association of the VirD2 protein with the 5′ end of T-strands in *Agrobacterium tumefaciens*. J. Bacteriol. 170, 3367–3374.

Zambryski, P., Tempe, J., & Schell, J. (1989). Transfer and function of T-DNA genes from *Agrobacterium* Ti and Ri plasmids in plants. Cell 56, 193–201.

Zambryski, P. (1992). Chronicles from the *Agrobacterium*-plant cell DNA transfer story. Annu. Rev. Plant Physiol. Plant Molec. Biol. 43, 465–490.

SELECTIVE DEGRADATION OF CYTOSOLIC PROTEINS BY DIRECT TRANSPORT INTO LYSOSOMES

J. Fred Dice

Membrane Protein Transport
Volume 1, pages 59–75.
Copyright © 1995 by JAI Press Inc.
All rights of reproduction in any form reserved.
ISBN: 1-55938-907-9

ABSTRACT

Lysosomes are able to internalize cellular proteins in a variety of ways. One pathway is selective for cytosolic proteins containing peptide sequences biochemically related to Lys-Phe-Glu-Arg-Gln (KFERQ). This pathway is activated in confluent monolayers of cultured cells in response to deprivation of serum growth factors and applies to approximately 30% of cytosolic proteins. Intact animals also activate this proteolytic pathway in tissues such as liver, kidney, and heart in response to fasting.

We have reconstituted this lysosomal degradation pathway *in vitro* using highly purified lysosomes. Uptake and degradation of substrate proteins are stimulated by ATP and a member of the heat-shock 70-kDa protein family, the 73-kDa constitutive heat-shock protein (hsc73). This pathway is selective since ribonuclease A and ribonuclease S-peptide are good substrates, but ribonuclease S-protein is not. Furthermore, the uptake mechanism is saturable, and at 4 °C substrate proteins bind specifically to a protein component of lysosomal membranes, presumably a receptor or polypeptide transporter. A portion of cellular hsc73 is associated with lysosomes both on the cytoplasmic face of the lysosomal membrane and within the lysosomal lumen. Hsc73 within the lumen is required for polypeptide import into the organelle. These results indicate that the lysosomal polypeptide import process is strikingly similar to those for import of proteins for residence in other organelles.

I. INTRODUCTION

Intracellular protein degradation is important to organisms for a variety of reasons. Proteolysis allows cells to remove proteins that are no longer required due to changed position within the cell cycle or altered developmental or physiological status. The continual degradation and resynthesis of proteins ("protein turnover") also prevents proteins from accumulating nonenzymatic modifications such as glycosylation, deamidation, oxidation, and cross-linking (reviewed in Dice, 1989). Protein degradation during a fast provides critical amino acids to be used for the synthesis of important regulatory enzymes, for gluconeogenesis by the liver and kidneys, and as an energy substrate for skeletal muscle and other tissues (reviewed in Dice, 1987; Olson et al., 1992).

There are multiple pathways of intracellular proteolysis in eukaryotic cells (reviewed in Dice, 1987; Olson et al., 1992; Hershko and Ciechanover, 1992). The best studied cytosolic pathway is ATP- and ubiquitin-dependent and is responsible for the degradation of many abnormal proteins as well as many short-lived normal proteins and even certain long-lived proteins (Rechsteiner, 1991; Hershko and Ciechanover, 1992). Other cytosolic proteolytic pathways are independent of ubiquitin. Two calcium-activated proteases called calpains reversibly associate with membranes and the cytoskeleton and may be especially important in the limited proteolysis of calmodulin-binding proteins (Wang et al., 1989). Still other pathways of proteolysis are not cytosolic but are contained within organelles such as the endoplasmic reticulum and mitochondrion (reviewed in Olson et al., 1992).

II. LYSOSOMAL PATHWAYS OF PROTEOLYSIS

It is also well established that lysosomes play an important role in overall proteolysis (Mortimore, 1987; Pfeifer, 1987), and lysosomes appear to be able to internalize intracellular proteins in a variety of ways (Figure 1).

A. Endocytosis and Exocytosis

The lysosomal degradation of endocytosed extracellular proteins has been extensively studied (Schmidt, 1992). Many plasma membrane proteins and certain other proteins within the vacuolar apparatus are also degraded by lysosomes through endocytic pathways (Hare, 1990; Figure 1). Proteins traveling through exocytic

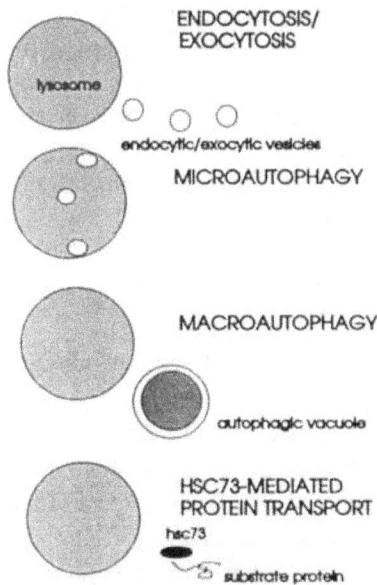

Figure 1. Schematic representations of pathways of lysosomal proteolysis. See the text for descriptions of the pathways.

pathways to reside on the cell surface or to be secreted from cells can also be diverted to lysosomes for degradation (Marzella and Glaumann, 1987; Olson et al., 1992).

B. Microautophagy

In well-nourished cells lysosomes appear to be able to internalize cytosolic proteins by a poorly understood process called microautophagy (Figure 1) in which the lysosomal membrane invaginates at multiple locations to form intralysosomal vesicles (Ahlberg et al., 1982; Marzella and Glaumann, 1987). These vesicles presumably disintegrate whereupon their contents can be accessed by lysosomal hydrolases. Microautophagy appears to be nonselective in that several proteins and inert particles are internalized at similar rates (Ahlberg et al., 1982).

C. Macroautophagy

When cultured cells reach confluence and in certain tissues of fasted animals an additional lysosomal pathway of proteolysis is stimulated (Cockle and Dean, 1982; Kominami, 1983; Knecht et al., 1986; Mortimore, 1987). Macroautophagy (Figure 1) involves formation of double-membraned autophagic vacuoles from ribosome-free regions of the rough endoplasmic reticulum (Dunn, 1990a), the smooth endoplasmic reticulum (Ueno et al., 1991), or a preexisiting organelle called the phagophore (Seglen et al., 1990). These autophagic vacuoles then acquire lysosomal membrane proteins, acidify, and finally acquire lysosomal hydrolases (Dunn, 1990b). Macroautophagy also appears to be nonselective in that many different organelles and proteins are sequestered at approximately the same rates (Kominami et al., 1983; Kopitz et al., 1990).

D. Direct Protein Transport

Specific cytosolic proteins also enter lysosomes for degradation when cultured confluent cells are deprived of serum growth factors and in certain animal tissues in response to fasting. This pathway is restricted to cytosolic proteins that contain peptide sequences biochemically related to Lys-Phe-Glu-Arg-Gln (KFERQ). The mechanism by which proteins with KFERQ-like peptide regions are targeted to lysosomes for degradation is similar in many respects to the import of newly synthesized proteins into the mitochondrion, endoplasmic reticulum, nucleus, or peroxisome. Therefore, proteins with KFERQ-like peptide regions may enter lysosomes by directly crossing a membrane bilayer rather than by vesicular pathways (Figure 1).

E. Activation of Lysosomal Proteolytic Pathways

Endocytosis and exocytosis appear to be active under many different conditions with the notable exception of the mitotic phase of the cell cycle (Warren, 1993). However, the other pathways of lysosomal proteolysis are more acutely regulated and are active during different growth conditions of cultured cells. Microautophagy may be the primary pathway of lysosomal proteolysis in cells during rapid cell division. Macroautophagy is maximally stimulated when cultured cells become confluent and is not further stimulated by serum withdrawal (Knecht et al., 1986). Instead, serum withdrawal activates the direct protein transport pathway of lysosomal proteolysis, the subject of this review.

Different lysosomal proteolytic pathways are also activated in liver in response to feeding and fasting. Microautophagy appears to be the major lysosomal proteolytic pathway immediately after a meal, but macroautophagy is activated within a few hours following feeding (Mortimore, 1987). In response to longer periods of starvation, the direct protein transfer pathway is activated (Simon et al., 1991; A.M. Cuervo, S.R. Terlecky, J.F. Dice, and E. Knecht, unpublished results).

III. RIBONUCLEASE A AS A PROBE FOR SELECTIVE LYSOSOMAL PROTEOLYSIS

We used red cell-mediated microinjection to introduce specific radiolabeled proteins into the cytosol of confluent cultures of human fibroblasts (Neff et al., 1981; McElligott and Dice, 1984). Degradation of the injected protein could be followed by monitoring either the loss of radioactivity from the cell monolayer or the appearance of acid-soluble radioactivity in the culture medium (McElligott and Dice, 1984). These two measurements do not give identical half-lives because some acid-insoluble peptide degradation fragments as well as intact protein are also released from cells (Isenman and Dice, 1989, 1993). However, conclusions regarding serum-regulated degradation are evident using either measurement.

A. Selectivity of the Proteolytic Pathway

Radiolabeled ribonuclease A (RNase A[1]) is degraded with a half-life of approximately 100 hours in serum-supplemented cells and 50 hours in serum-deprived cells (Figure 2; Backer et al., 1983). The half-lives of certain other microinjected proteins are also reduced in response to serum withdrawal, but half-lives of other microinjected proteins are unaffected by serum (Table 1).

This selectivity in proteolysis is particularly striking in experiments where [125I]RNase A and [131I]RNase S-protein (residues 21–124 of RNase A) were coinjected into the same cells (Backer et al., 1983). [125I]RNase A was degraded

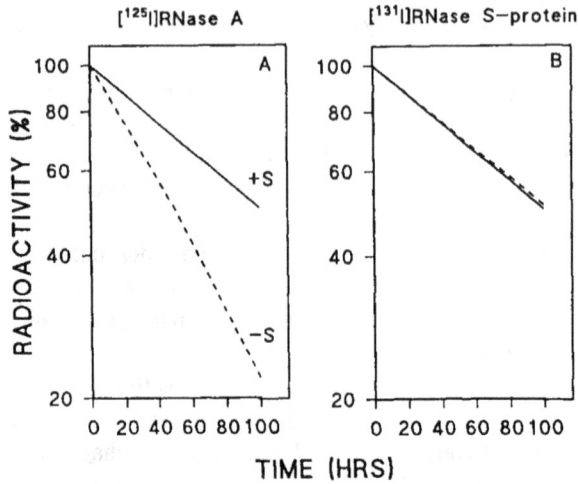

Figure 2. Degradation of [^{125}I]RNase A and [^{131}I]RNase S-protein microinjected into the same cells. S = 10% calf serum.

more rapidly in response to serum withdrawal while [^{131}I]RNase S-protein was degraded at the same rate in the presence or absence of serum (Figure 2).

B. Targeting Signal for Selective Proteolysis

The amino terminal 20 amino acids of RNase A (RNase S-peptide) microinjected by itself is degraded in a serum-dependent manner (Backer et al., 1983; Figure 3)

Table 1. Degradation of Long-Lived Proteins Microinjected into Human Cells[a]

	T1/2 (h)	
Protein	+ Serum	− Serum
Group 1: Regulated Degradation		
RNase A	100	50
Aspartate amino-transferase	80	40
Pyruvate kinase	296	121
Hemoglobin	213	106
Group 2: Nonregulated Degradation		
RNase S-Protein	100	100
Ovalbumin	55	49
Lysozyme	59	59
Insulin A-chain	104	104

Notes: [a]IMR-90 human diploid fibroblasts were recipient cells in all cases except for pyruvate kinase which was injected into HeLa cells. References to half-life values have been previously cited in Dice, 1990.

PEPTIDE	REGULATED DEGRADATION
KETAAAKFERQHMDSSTSSA	YES
KETAAAKFERQHMD	YES
AAKFERQHM	YES
ETAAAKF	NO
KETAAAKFER	NO
KFERQ	YES

Figure 3. Degradation characteristics of RNase S-peptide and smaller peptides. The first peptide listed is RNase S-peptide. Half-lives were measured in the presence and absence of serum. An approximate twofold increase in degradation rate in the absence of serum was considered regulated degradation.

confirming that it contains the essential information for serum-regulated degradation. Furthermore, covalent attachment of RNase S-peptide to heterologous proteins using water-soluble carbodiimides also causes their degradation rates to increase in response to serum withdrawal (Backer and Dice, 1986; Table 2).

To further define the crucial amino acid residues within RNase S-peptide, we microinjected various synthesized fragments of RNase S-peptide and determined whether or not their degradation was serum-regulated (Figure 3). Degradation of certain fragments (amino acids 1–14 and 3–13) were serum-regulated while degradation of other fragments (amino acids 2–8 and 1–10) were not (Dice et al., 1986). We then took advantage of the observation that red blood cell proteases cleave [³H]RNase S-peptide during loading. A radiolabeled pentapeptide derived from RNase S-peptide was degraded faster in response to serum withdrawal, and we were

Table 2. Effect of RNase S-Peptide Sequences on Serum-Regulated Degradation of Test Proteins[a]

Protein	T1/2 (h)	
	+ Serum	– Serum
Lysozyme	59	59
RNase S-peptide-lysozyme	47	21
Insulin A-chain	104	104
RNase S-peptide-insulin A-chain	100	44
β-Galactosidase[b]	134	154
RNase S-peptide-β-galactosidase[b]	172	130
RNase S-peptide-linker-β-galactosidase[b]	178	78

Notes: [a]Half-life values can be found in Backer and Dice, 1986.
[b]L.J. Terlecky, M. Kirven-Brooks, and J.F. Dice, unpublished results.

able to identify this pentapeptide as residues 7–11, KFERQ. [^{3}H]KFERQ microinjected into fibroblasts was degraded in a serum regulated fashion (Dice et al., 1986; Figure 3). Additionally, comicroinjection of [^{125}I]RNase A and excess nonradioactive KFERQ blocked the enhanced degradation of [^{125}I]RNase A in response to serum withdrawal (Chiang and Dice, 1988).

In order to experimentally determine the importance of individual amino acids within the KFERQ region, we turned to a recombinant DNA approach. We synthesized an oligonucleotide that encoded a slightly modified RNase S-peptide (Goff et al., 1987) and fused it in frame to the *E. coli* β-galactosidase gene. β-Galactosidase introduced into human fibroblasts was degraded at the same rate in the presence and absence of serum. In addition, the direct RNase S-peptide-β-galactosidase fusion protein was degraded at the same rate in the presence and absence of serum. The KFERQ region in this construction proved to be hidden within the active tetramer. When a linker region coding for 24 amino acids was added between the RNase S-peptide and the β-galactosidase sequences, the KFERQ region was exposed, and this fusion protein was degraded more rapidly in response to serum withdrawal (L. Jeffreys-Terlecky, M. Kirven-Brooks, and J.F. Dice, unpublished results; Table 2). We are currently examining the effects of mutations in each of the codons of the KFERQ region to experimentally determine the types of peptides that will lead to enhanced degradation in response to serum withdrawal.

C. Lysosomal Degradation of Microinjected RNase A

Several lines of evidence indicated that the pathway of degradation of microinjected RNase A was lysosomal. For example, degradation of RNase A in the absence of serum was partially inhibited by the lysosomotropic agent, ammonium chloride (McElligott et al., 1985). In addition, as mentioned earlier, degradation of microinjected RNase A results in secretion not only of free amino acids but also peptide degradation products from cells. The same peptides derived from RNase A are released from cells after microinjection and after lysosomal hydrolysis following endocytosis (Isenman and Dice, 1989). Since the lysosomal degradation of proteins following endocytosis is well established (Schmidt, 1992), lysosomes are also likely to be responsible for the degradation of microinjected RNase A.

Additional evidence in support of lysosomal degradation of microinjected RNase A came from studies in which RNase A was labeled with [^{3}H]raffinose, an inert trisaccharide. The degradation products of [^{3}H]raffinose-tagged proteins are trapped in the compartment where they are generated, so their location serves as a cumulative marker of the site of degradation of a protein. All of the degradation products from microinjected [^{3}H]raffinose-RNase A were localized to lysosomes for cells grown both in the presence (McElligott et al., 1985) and absence of serum (M.A. McElligott and J.F. Dice, unpublished results). However, in the presence of serum, RNase A and RNase S-protein are both taken up by lysosomes probably by micro- or macroautophagy. In the absence of serum, RNase A is selectively taken up by lysosomes by the direct protein transport pathway.

IV. GENERALITY OF THIS PATHWAY OF PROTEIN DEGRADATION

To examine whether peptide regions similar to KFERQ exist in intracellular proteins, we raised polyclonal antibodies to KFERQ and affinity-purified IgGs specifically directed toward the pentapeptide (Chiang and Dice, 1988). The anti-KFERQ IgGs immunoprecipitated approximately 30% of [^{3}H]leucine-labeled cytosolic proteins from human fibroblasts.

We followed loss of radioactivity from total, immunoreactive, and nonimmunoreactive cytosolic proteins by incubating radiolabeled cells in medium containing excess unlabeled leucine to suppress reutilization of the isotope. The immunoprecipitable proteins were preferentially degraded in response to serum withdrawal. In contrast, nonimmunoprecipitable proteins were degraded at the same rate in the presence or absence of serum. These results showed that the approximately twofold enhanced degradation in total cytosolic protein in response to serum withdrawal was actually due to more strikingly enhanced degradation of 30% of the cytosolic proteins containing peptide regions biochemically related to KFERQ.

V. ROLES OF HEAT-SHOCK 70-kDa PROTEINS (HSP70S) IN SELECTIVE LYSOSOMAL PROTEOLYSIS

A possible mechanism for selectivity in lysosomal proteolysis would be the recognition by an intracellular protein of the KFERQ-like peptide regions in proteins that are degraded more rapidly in response to serum withdrawal. A protein of 73 kDa, purified on an RNase S-peptide affinity column, was designated prp73 for peptide recognition protein of 73 kDa (Chiang et al., 1989). This protein bound to RNase S-peptide with a K_d of 8 μM (Terlecky et al., 1992).

A monoclonal antibody that recognizes many members of the hsp70 family (Kurtz et al., 1986) reacted with prp73 purified from human fibroblast cytosol. In addition, sequence data obtained from purified prp73 tentatively identified it as the constitutively expressed heat-shock cognate protein of 73 kDa (hsc73) (Chiang et al., 1989). This was an exciting discovery since hsp70 family members had been implicated in the transport of precursor proteins into mitochondria and endoplasmic reticulum (Chirico et al., 1988; Deshaies et al., 1988). More recent experiments also demonstrate roles of hsc73 in the import of proteins into the nucleus (Dingwall and Laskey, 1992) and peroxisome (Walton et al., 1993).

Hsc73 and prp73 were functionally equivalent in several assays. Both proteins bound to RNase A, RNase S-peptide, KFERQ, aspartate aminotransferase, and pyruvate kinase—proteins and peptides that contain KFERQ-like peptide motifs. Neither hsc73 nor prp73 bind to ovalbumin, lysozyme, ubiquitin, or β-galactosidase—proteins that lack KFERQ-like peptide motifs (Terlecky et al., 1992).

Such selectivity in substrate binding of hsc73 is somewhat controversial. Some evidence suggests that the binding is rather nonspecific in that hsc73 transiently associates with many different nascent chains as they emerge from polysomes (Beckman et al., 1990), and hsc73 binds to several denatured proteins but not to their native counterparts (Palleros, et al., 1991). On the other hand, Flynn et al., (1989) demonstrated that the ability of two different peptides to interact with hsc73 differed by 14-fold, and preferential binding of hsc73 to certain peptides has been shown by others (DeLuca-Flaherty et al., 1990; Hightower et al, 1994). Perhaps hsc73 is able to weakly interact with many different polypeptides, and even nonpeptide molecules (Alvares et al., 1990), but higher affinity binding may require specific peptide characteristics.

VI. SELECTIVE DEGRADATION OF PROTEINS BY ISOLATED LYSOSOMES

To try to elucidate the mechanism of degradation of KFERQ motif-containing proteins, an *in vitro* assay using lysosomes isolated from human fibroblasts was developed (Chiang et al., 1989; Terlecky et al., 1992; Terlecky and Dice, 1993). Maximal degradation of [^{3}H]RNase S-peptide and [^{3}H]RNase A by isolated

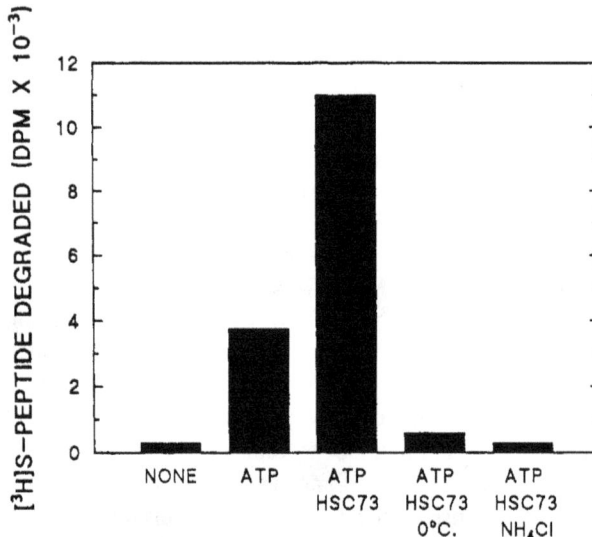

Figure 4. Import and degradation of [^{3}H]RNase S-peptide by isolated lysosomes. Degradation to acid-soluble radioactivity is shown after 2 hr at 24 °C. Other experimental details are as described in Chiang et al. (1989) and Terlecky et al. (1992).

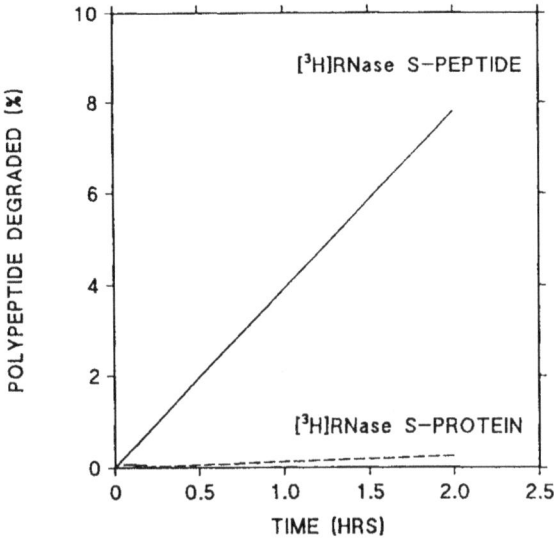

Figure 5. Import and degradation of [³H]RNase S-peptide and [³H]RNase S-protein by isolated lysosomes. Degradation was measured in the presence of hsc73 and ATP.

lysosomes required both ATP and hsc73 (Figure 4; Chiang et al., 1989; Terlecky et al., 1992). Hsc73 and prp73 functioned identically in stimulating degradation of [³H]RNase S-peptide by isolated lysosomes under conditions where three other members of the hsp70 family had no activity (Terlecky et al., 1992). This degradation was selective because [³H]RNase S-protein was degraded little, if at all, under the same conditions (Figure 5; Terlecky and Dice, 1993).

Degradation of [³H]RNase S-peptide could be inhibited by reducing the temperature (Figure 4; Chiang et al., 1989), and degradation appeared to occur within intact lysosomes because it could be inhibited by ammonium chloride (Figure 4; Chiang et al., 1989) and leupeptin (Terlecky and Dice, 1993). Hydrolases released from damaged lysosomes could not account for the proteolysis because the incubation buffer maintained the pH at 7.2, and there was no detectable RNase S-peptide hydrolyzing activity at this pH (Terlecky and Dice, 1993).

Lysosomal uptake of [³H]RNase S-peptide was saturable (K_m = 5 μM). At 4 °C and in the presence of hsc73 [³H]RNase S-peptide specifically bound to lysosomal membranes, and this binding was reduced by prior mild trypsinization (Terlecky and Dice, 1993). No such binding was observed for RNase S-protein. Presumably, this RNase S-peptide binding component is a receptor or a polypeptide transporter.

Lysosomes isolated from serum-deprived cells were twice as active in protein uptake *in vitro* as were lysosomes derived from serum-supplemented cells. Correlated with this increased activity was an increased amount of hsc73 in the lysosomal fraction (Terlecky and Dice, 1993). Some of this hsc73 appeared to be associated with the lysosome surface because it could be removed with trypsin. However, most of the lysosomal hsc73 appeared to be in the lumen of the lysosome since it was not digested by trypsin unless the lysosomal membrane was disrupted. Indirect immunofluorescence using laser scanning confocal microscopy confirmed the colocalization of a fraction of hsc73 with lysosomal marker enzymes (S.R. Ter-lecky, F. Agarraberes, and J.F. Dice, unpublished results). These studies also demonstrated that certain lysosomes did not contain hsc73 and that serum with-drawal caused a marked morphological change in hsc73-containing lysosomes; they fused to become a tubular network. The functional significance of these morphological changes is not yet clear.

Lysosomes isolated from rat liver have recently been reported to selectively take up and degrade glyceraldehyde-3-phosphate dehydrogenase (GAPDH) (Aniento et al., 1993). This process appeared to be very similar to the uptake and degradation of RNase A by fibroblast lysosomes in that it was selective and stimulated by hsc73 and ATP (A.M. Ceurvo, S.R. Terlecky, J.F. Dice, and E. Knecht, unpublished results). Indeed, uptake of GAPDH by rat liver lysosomes could compete with RNase A or RNase S-peptide, and uptake of RNase A could compete with GAPDH. RNase S-protein and ovalbumin showed no competition in these assays. Interest-ingly, in this rat liver lysosome system an import intermediate of RNase A was detected in which most of the molecule had entered the lysosome while a small portion (2 kDa of the carboxy terminus) remained outside. (A.M. Ceurvo, S.R. Terlecky, J.F. Dice, and E. Knecht, unpublished results).

The hsp70 in the lumen of fibroblast and rat liver lysosomes was identified as hsc73 using a monoclonal antibody (mAb), 13D3 (Maekawa et al., 1989) that was shown to be specific for hsc73 (Terlecky et al., 1992; S.R. Terlecky, F. Agarraberes, and J.F. Dice, unpublished results). Other experiments showed that mAb 13D3 had the unusual property for an IgM in that it also immunoprecipitated native hsc73. Other researchers have also reported hsp70s within lysosomes (Mayer et al., 1991; Domanico et al., 1993).

Whether or not the intralysosomal hsc73 played a role in this pathway of proteolysis was assessed by allowing cells to endocytose large amounts of mAb 13D3 in order to neutralize the intralysosomal hsc73. This treatment had no effect on the intralysosomal degradation of endocytosed [3H]RNase A, thus the activity of lysosomal proteases was not affected (S.R. Terlecky, F. Agarraberes, and J.F. Dice, unpublished results). However, when cells were metabolically labeled with [3H]leucine and then protein degradation followed, the endocytosed 13D3 was found to completely block the enhanced protein degradation in response to serum withdrawal without altering proteolysis in the presence of serum (Figure 6). Endocytosis of mAb P32, a control IgM, had no effect on degradation, and endocytosis of 13D3 in the presence of an equimolar amount of hsc73 also had no effect on proteolysis (Figure 6; S.R. Terlecky, F. Agarraberes, and J.F. Dice, unpublished results). The intralysosomal hsc73 is likely to be required to pull the substrate proteins across the lipid bilayer, a role that has been shown for other hsp70

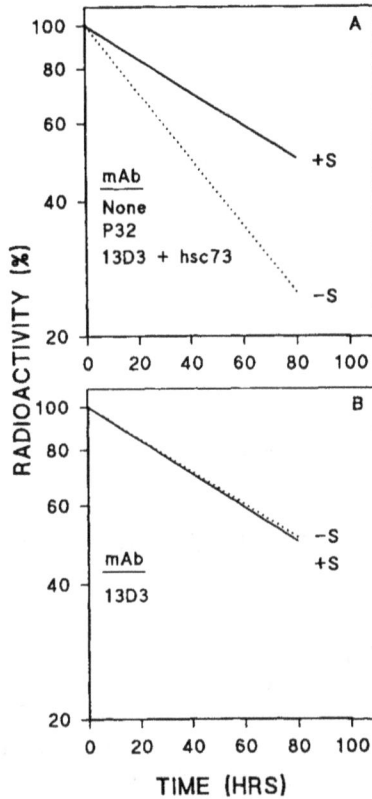

Figure 6. Effect of endocytosed IgMs on intracellular proteolysis. Confluent cultures of human fibroblasts were metabolically labeled with [^{3}H]leucine for 2 days. Prior to beginning degradation measurements, the cells were allowed to endocytose IgMs for 16 hr. (A) Enhanced degradation in the absence (–S) compared to the presence (+S) of serum was evident for cells that endocytosed no IgMs, a control IgM (P32), or 13D3 with an equimolar amount of hsc73. (B) Enhanced degradation in the absence of serum was blocked by endocytosis of 13D3.

family members within mitochondria and the endoplasmic reticulum (Kang et al., 1990; Vogel et al., 1990; Nicchita and Blobel, 1993).

Our results suggest the following model (Figure 7) for the steps in the selective entry of proteins into lysosomes: (1) binding of hsc73 to cytosolic proteins containing KFERQ motifs; (2) binding of the substrate protein to a lysosomal membrane protein; (3) import of the substrate protein by the action of hsc73 within the lysosomal lumen; and 4) intralysosomal degradation of the protein to peptides and free amino acids.

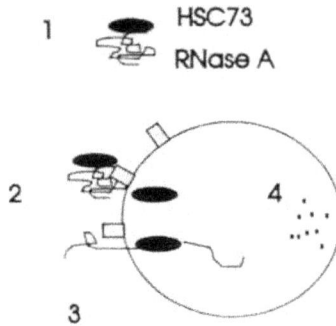

Figure 7. A possible pathway for the selective degradation of polypeptides by lysosomes. See the text for descriptions of steps 1–4.

VII. FUTURE DIRECTIONS

The mechanisms by which hsc73 stimulates lysosomal degradation of proteins containing KFERQ-like peptide regions are not yet known. Hsc73 stimulates import of precursor proteins into mitochondria at least in part by preventing misfolding into a transport-incompetent conformation (Hendrick and Hartl, 1993). Such a role seems unlikely for the stimulation of lysosomal uptake of mature, folded proteins such as RNase A. Perhaps hsc73 unfolds proteins prior to lysosomal import. Alternatively, hsc73 may participate in binding of the substrate proteins to the receptor or protein transporters in the lysosomal membrane. How hsc73 enters the lysosome remains to be established. It may enter along with substrate proteins, but hsc73 could itself be a substrate for this selective lysosomal uptake pathway since two KFERQ motifs are contained within the hsc73 sequence (Terlecky et al., 1992). Hsc73 could also enter lysosomes by nonselective microautophagy or macroautophagy. Hsc73 would then have to be relatively resistant to intralysosomal hydrolysis since it is more abundant in lysosomes than are other more highly expressed cytosolic proteins such as hsp90 (Terlecky and Dice, 1993).

Another important issue to resolve is how hsc73 actions are regulated. This protein appears to facilitate protein transport into mitochondria, endoplasmic reticulum, nuclei, and peroxisomes (Chirico et al., 1988; Deshaies et al., 1988; Dingwall and Laskey, 1992; Hendrick and Hartl, 1993; Walton et al., 1993) as well as facilitating the lysosomal degradation pathway described here. Possible ways to regulate hsc73 functions include posttranslational modifications and/or binding of cofactors that alter peptide binding properties or subcellular localization. Perhaps only the lysosome-associated hsc73 is active in the protein degradation pathway. However, it is also possible that components other than hsc73 in the various polypeptide import pathways are regulated. For example, activation of the lysosomal degradation pathway may be due to changes in the substrate proteins that expose the KFERQ-like regions and/or due to activation of the putative receptor or peptide transporter on the lysosome surface. Answering these questions is likely to uncover new aspects of the roles of hsp70s as molecular chaperones.

ACKNOWLEDGMENTS

Research in the author's laboratory was supported by NIH grant AG06116. I thank former (Nicola Neff, Lizabeth Bourret, Paul Miao, Mary Ann McElligott, Jon Backer, Elizabeth Spenser, Hui-Ling Chiang, Stephen Goff, Sharla Short, Tracy Olson, Sabine Freundlieb, Stan Terlecky, and Laura Terlecky) and current (Fernando Agarraberes, Ron Skurat, Melissa Kirven-Brooks, Lois Isenman, and Marilyn Negron) members of the lab for providing the information presented in this chapter.

ABBREVIATIONS

RNase A:	bovine pancreatic ribonuclease A
RNase S-peptide:	residues 1–20 of RNase A
RNase S-protein:	residues 21–124 of RNase A
hsp70:	heat-shock protein of 70 kDa
prp73:	peptide recognition protein of 73 kDa
hsc73:	heat-shock cognate protein of 73 kDa
GAPDH:	glyceraldehyde-3-phosphate dehydrogenase

REFERENCES

Ahlberg, J., Marzella, L., & Glaumann, H. (1982). Uptake and degradation of proteins by isolated rat liver lysosomes. Suggestion of a microautophagic pathway of proteolysis. Lab. Invest. 47, 523–532.

Alvares, K., Carrillo, A., Yuan, P.M., Kawang, H., Morimoto, R.I., & Reddy, J.K. (1990). Identification of a cytosolic peroxisome proliferator binding protein as a member of the heat shock protein HSP70 family. Proc. Nat. Acad. Sci. USA 87, 5293–5297.

Aniento, F., Roche, E., Cuervo, A.M., & Knecht, E. (1993). Uptake and degradation of glyceraldehyde-3-phosphate dehydrogenase by rat liver lysosomes. J. Biol. Chem. 268, 19463–10470.

Backer, J.M., Bourret, L., & Dice, J.F. (1983). Regulation of catabolism of microinjected ribonuclease A requires the amino terminal twenty amino acids. Proc. Nat. Acad. Sci. USA 80, 2166–2170.

Backer, J.M. & Dice, J.F. (1986). Covalent linkage of ribonuclease S-peptide to microinjected proteins causes their intracellular degradation to be enhanced during serum withdrawal. Proc. Nat. Acad. Sci. USA 83, 5830–5834.

Beckmann, R.P., Mizzen, L.A., & Welsh W.J. (1990). Interaction of hsp70 with newly synthesized proteins: Implications for protein folding and assembly. Science 248, 850–854.

Chiang, H.-L. & Dice, J.F. (1988). Peptide sequences that target proteins for enhanced degradation during serum withdrawal. J. Biol. Chem. 263, 6797–6805.

Chiang, H.-L., Terlecky, S.R., Plant, C.P., & Dice, J.F. (1989). A role for a 70-kilodalton heat shock protein in lysosomal proteolysis of intracellular proteins. Science 246, 282–285.

Chirico, W.J., Waters, M.G., & Blobel, G. (1988). 70-kDa heat shock related proteins stimulate protein translocation into microsomes. Nature 332, 805–810.

Cockle, S.M. & Dean, R.T. (1982). The regulation of proteolysis in normal fibroblasts as they approach confluence. Evidence for participation of the lysosomal system. Biochem. J. 208, 243–249.

DeLuca-Flaherty, C., McKay, D.B., Parnumr P., & Hill, B.L. (1990). Uncoating protein (hsc70) binds a conformationally labile domain of clathrin light chain LC$_a$ to stimulate ATP hydrolysis. Cell 62, 875–887.

Deshaies, R.J., Koch, B.D., Werner-Washburne, M., Craig, E.A., & Schekman, R. (1988). 70 kDa stress protein homologues facilitate translocation of secretory and mitochondrial precursor polypeptides. Nature 332, 800–805.

Dice, J.F., Chiang, H.-L., Spenser, E.P., & Backer, J.M. (1986). Regulation of catabolism of microinjected ribonuclease A, Identification of residues 7–11 as the essential pentapeptide. J. Biol. Chem. 262, 6853–6859.

Dice, J.F. (1987). Molecular determinants of protein half-lives in eukaryotic cells. FASEB J. 1, 349–357.

Dice, J.F. (1989). Altered intracellular protein degradation in aging: A possible cause of proliferative arrest. Expt. Gerontology 24, 451–459.

Dice, J.F. (1990). Peptide sequences that target cytosolic proteins for lysosomal proteolysis. Trends Biochem. Sci. 15, 305–309.

Dingwall, C. & Laskey, R. (1992). The nuclear membrane. Science 258, 942–947.

Domanico, S.Z., DeNagel, D.C., Dahlseid, J.N., Green, J.M., & Pierce, S.K. (1993). Cloning of the gene encoding peptide-binding protein 74 shows that it is a new member of the heat shock protein 70 family. Mol. Cell. Biol. 13, 3598–3610.

Dunn, W.A. (1990a). Studies on the mechanisms of autophagy: Formation of the autophagic vacuole. J. Cell. Biol. 110, 1923–1933.

Dunn, W.A. (1990b). Studies on the mechanisms of autophagy: Maturation of the autophagic vacuole. J. Cell. Biol. 110, 1935–1945.

Flynn, G.C., Chappell, T.G., & Rothman, J.E. (1989). Peptide binding and release by proteins implicated as catalysts in protein assembly. Science 245, 385–390.

Hare, J.F. (1990). Mechanisms of membrane protein turnover. Biochim. Biophys. Acta 1031, 71–90.

Hendrick, J.P. & Hartl, F.-U. (1993). Molecular chaperone functions of heat shock proteins. Ann. Rev. Biochem. 62, 349–384.

Hightower, L. (1994). Interactions of vertebrate hsc70 and hsp70 with unfolded proteins and peptides. In: The Biology of Heat Shock Proteins and Molecular Chaperones (Morimoto, R.I., Tissieres, A., & Georgopoulos, C., Eds.). Cold Spring Harbor Laboratory Press, Cold Spring Harbor, New York, in press.

Isenman, L.D. & Dice, J.F. (1989). Secretion of intact proteins and peptide fragments by lysosomal pathways of protein degradation. J. Biol. Chem. 264, 21591–21596.

Isenman, L.D. & Dice, J.F. (1993). Selective release of peptides from lysosomes. J. Biol. Chem. 268, in press.

Kang, P.-J., Ostermann, J., Shilling, J., Neupert, W., Craig, E.A., & Pfanner, N. (1990). Requirement for hsp70 in the mitochondrial matrix for translocation and folding of precursor proteins. Nature 348, 137–143.

Knecht, E., Hernandez-Yago, J., & Grisolia, S. (1984). Regulation of lysosomal autophagy in transformed and non-transformed mouse fibroblasts under several growth conditions. Exp. Cell. Res. 154, 224–232.

Kominami, E., Hashida, E., Khairallah, E. A., & Katunuma, N. (1983). Sequestration of cytoplasmic enzymes in an autophagic vacuole-lysosomal system induced by injection of leupeptin. J. Biol. Chem. 258, 6093–6100.

Kopitz, J., Kisen, G.O., Gordon, P.B., Bohley, P., & Seglen, P.O. (1990). Non-selective autophagy of cytosolic enzymes in isolated rat hepatocytes. J. Cell Biol. 111, 941–954.

Kurtz, S., Rossi, J., Petko, L., & Lindquist, S. (1986). An ancient developmental induction: Heat shock proteins induced in sporulation and oogenesis. Science 231, 1154–1157.

Maekawa, M., O'Brien, D.A., Allen, R.L., & Eddy, E.M. (1989). Heat-shock cognate protein (hsc71) and related proteins of mouse spermatogenic cells. Biol. Repro. 40, 843–852.

Marzella, L. & Glaumann, H. (1987). Autophagy, microautophagy, and crinophagy as mechanisms for protein degradation. In: Lysosomes: Their Role in Protein Breakdown (Glaumann, H. & Ballard, F.J., Eds.), pp. 319–367. Academic Press, New York.

Mayer, R.J., Lowe, J., Landon, M., McDermott, H., Tuckwell, J., Doherty, F., & Lazlo, L. (1991). Ubiquitin and the lysosomal system: Molecular pathological and experimental findings. In: Heat Shock. (Maresca, B. & Lindquist, S., Eds.), pp. 299–314. Springer-Verlag, Heidelberg.

McElligott, M.A. & Dice, J.F. (1984). Microinjection of cultured cells using red cell-mediated fusion and osmotic lysis of pinosomes: A review of methods and applications. Biosci. Rep. 4, 451–466.

McElligott, M.A., Miao, P., & Dice, J.F. (1985). Lysosomal degradation of ribonuclease A and ribonuclease S-protein microinjected into human fibroblasts. J. Biol. Chem. 260, 11986–11993.

Mortimore, G.E. (1987). Mechanism and regulation of induced and basal protein degradation in liver. In: Lysosomes: Their Role in Protein Breakdown. (Glaumann, H. & Ballard, F.J., Eds.), pp. 415–444. Academic Press, New York.

Nicchitta, C.V. & Blobel, G. (1993). Lumenal proteins of the mammalian endoplasmic reticulum are required to complete protein translocation. Cell 73, 989–998.

Olson, T.S., Terlecky, S.R., & Dice, J.F. (1992). Pathways of intracellular protein degradation in eukaryotic cells. In: Stability of Protein Pharmaceuticals: In Vivo Pathways of Degradation and Strategies for Protein Stabilization (Ahern, T.J. & Manning, M.C., Eds.), pp. 89–118. Plenum Publishing, New York.

Palleros, D.R., Welsh, W.J., & Fink, A.L. (1991). Interaction of hsp70 with unfolded proteins: Effects of temperature and nucleotides on the kinetics of binding. Proc. Nat. Acad. Sci. USA 88, 5719–5723.

Pfeifer, U. (1987). Functional morphology of the lysosomal apparatus. In: Lysosomes: Their Role in Protein Breakdown (Glaumann, H. & Ballard, F.J., Eds.), pp. 3–59. Academic Press, New York.

Rechsteiner, M. (1991). Natural substrates of the ubiquitin proteolytic pathway. Cell 66, 615–618.

Schmidt, S.L. (1992). The mechanism of receptor-mediated endocytosis: More questions then answers. Bioessays 14, 589–596.

Terlecky, S.R., Chiang, H.-L., Olson, T.S., & Dice, J.F. (1992). Protein and peptide binding and stimulation of *in vitro* lysosomal proteolysis by the 73-kDa heat shock cognate protein. J. Biol. Chem. 267, 9202–9209.

Terlecky, S.R. & Dice, J.F. (1993). Polypeptide import and degradation by isolated lysosomes. J. Biol. Chem., in press.

Ueno, T., Muno, D., & Kominami, E. (1991). Membrane markers of endoplasmic reticulum preserved in autophagic vacuolar membranes from leuptpin-administered rat liver. J. Biol. Chem. 266, 18995–18999.

Vogel, J.P., Misra, L.M., & Rose, M.D. (1990). Loss of BiP/GRP78 function blocks translocation of secretory proteins in yeast. J. Cell Biol. 110, 1855–1895.

Walton, P.A., Morello, J.P., & Welch, W.J. (1993). Inhibition of the peroxisomal import of a microinjected protein by coinjection of antibodies to members of the 70-kDa heat shock protein family. J. Cell. Biochem. Suppl 17C:22.

Wang, K.K.W., Villalobo, A., & Roufogalis, B.D. (1989). Calmodulin-binding proteins as calpain substrates. Biochem. J. 158, 401–407.

THE CONFORMATION AND PATH OF NASCENT PROTEINS IN RIBOSOMES

Boyd Hardesty, Ada Yonath, Gisela Kramer,

O.W. Odom, Miriam Eisenstein,

Francois Franceschi, and Wieslaw Kudlicki

Membrane Protein Transport
Volume 1, pages 77–107.
Copyright © 1995 by JAI Press Inc.
All rights of reproduction in any form reserved.
ISBN: 1-55938-907-9

I. INTRODUCTION

Recent advances have suggested that nascent proteins are extended into a tunnel or cavity within the large ribosomal subunit as they are formed by the successive addition of specific amino acids to their N-terminus. This process appears to be associated with the acquisition of secondary and tertiary structure that is important for folding into the native conformation or transport of the newly formed protein into membranes or other subcellular structures. Our purpose here is to briefly review those aspects of the structure and function of ribosomes that contribute to these processes.

Davis and his co-workers (Smith et al., 1978) postulated that the nascent peptide provides the basis for the intimate relationship between active ribosomes and membranes in bacteria. Their conclusion, based on the observation that treatment with the aminoacyl-tRNA analog, puromycin, caused the release of the ribosomes from the membrane, was that the nascent peptide itself provides the primary point of attachment of the ribosome to the membrane. Such observations prompted the question, treated in other chapters, of how ribosomes bearing the nascent peptides of certain proteins are recognized, bound to membranes, and folded into their native conformation in prokaryotic and eukaryotic cells. Parallel questions involve the conformation of nascent peptides as they are formed on ribosomes and the role of the ribosome in folding nascent peptides into their native conformations. Chantrenne (1961) appears to have been the first to clearly enunciate the proposition that specific folding of a peptide into its native conformation may occur vectorially as it is formed on a ribosome. His statement, remarkable for 1961, is appropriate here as background.

Complete uncoiling is usually irreversible, for there are too many possibilities of formation of bonds between different parts of the chain or between chains. If the polypeptides were released from the assembly line as randomly coiled molecules, their remolding into the right configuration would be difficult. On the contrary, if the end of the polypeptide is allowed to fold spontaneously as it comes off the template, then folding might go on in a regular and unique manner as the chain grows. The tertiary structure would then be strictly determined at each step of its formation, by the nature of the amino acid residues and by the shape and position of the already formed parts of the molecule. Folding might conceivably be influenced by other substances present in the vicinity of the template.

This hypothesis has received some limited experimental attention over the years; however, the major effort, prompted by Anfinsen's seminal observations that denatured ribonuclease A would refold spontaneously *in vitro* into the conformation of the native enzyme has focused on renaturation and folding of protein from their denatured state (Matthews, 1993). Our objective here is to consider reports dealing with the conformation of nascent peptides on ribosomes and the role of the ribosome in translocation and in folding of proteins into their native conformation. Special emphasis will be given to the mechanism and structural features of ribosomes that generate the nascent peptide and then facilitate its folding or translocation into and through membranes.

II. RIBOSOME STRUCTURE AND FUNCTIONS

A. General Features

The synthesis of proteins in all living cells is carried out by ribosomes that are composed of two structurally different subunits which associate upon initiation of protein biosynthesis. Ribosomes are massive entities comparable in size to large multienzyme complexes. They have unique structures composed of RNA and proteins of specific primary sequence. These remarkable structures function as molecular machines to carry out the chemical–mechanical steps of movement of the mRNA that is strictly coupled with decoding and binding of tRNA, synthesis of peptide bonds, and protection of the nascent peptide during the early stages of its folding. Although the detailed structure of ribosomes from different organisms differ considerably, the basic features of structure and function are highly conserved throughout all living systems. A typical prokaryotic ribosome contains about one-quarter million atoms and is of a molecular weight of approximately 2.3 million, about one-third of which is made up of 56 different proteins. The remainder is composed of three RNA chains, two of which occur in the large subunit.

The results of intensive biochemical, biophysical, and genetic analyses led to a low resolution, consensus model for the overall shape and quaternary structure of the ribosome, for the spatial proximities of various ribosomal components, and for the approximate positions of some of the reaction sites (Stöffler and Stöffler-Meilicke, 1986). Binding and decoding of the mRNA through codon–anticodon interaction with tRNA takes place on the small subunit, whereas peptide bond formation takes place in a specific domain of the large ribosomal subunit called the peptidyl transferase center. This center, which provides the attachment site for the 3'-end of the tRNA and the nascent peptide, is located on the interfacing surface of the large subunit below a physically distinct structure known as the central protuberance. The attachment site for an amino acid on ribose at the 3'-end of a tRNA is in the order of 85 Å from the anticodon at the opposite end of the L-shaped tRNA molecule (Kim et al., 1974; Robertus et al., 1974). This is reflected in the distance between the decoding site on the small subunit and the peptidyl transferase center at the base of the central protuberance on the large subunit.

An understanding of the molecular mechanism of protein biosynthesis is still hampered by the lack of high resolution structural information and corresponding molecular models. Development of these models and the mechanism they represent are strictly limited by the techniques and procedures that are available. X-ray diffraction holds great promise for the future. To this end, three-dimensional crystals, diffracting in the best cases to 2.9 Å resolution, have been grown and are being investigated (von Boehlen et al., 1991; Franceschi et al., 1993). Much of the information considered below is derived from two types of procedures that have been extensively utilized by the authors, namely, image reconstruction from electron photomicrographs of crystalline arrays and fluorescence techniques. These are briefly described and considered below.

B. Three-Dimensional Image Reconstruction from Crystalline Arrays

Traditional electron microscopy, which used to be the choice method for viewing isolated ribosomal particles, suffers from several inherent shortcomings originating from its subjective nature, from the limited information emerged from observing surfaces in projection, and from shape distortions introduced by the probable flattening on the microscopical grids or/and from the microscope vacuum. To avoid a significant part of these shortcomings and to enhance the level of objectivity, we performed three-dimensional reconstruction from cystalline arrays. In this procedure, diffraction patterns rather than subjective imaging provide the raw data for the reconstruction, and the reliability of the resulting models is determined by well-established crystallographic criteria, rather than by decisions based on visual inspections. Furthermore, the distortions introduced by the flat electron microscope grids are reduced or eliminated, as the two-dimensional arrays are held by crystalline forces.

A three-dimensional model for crystalline 50S subunits was elucidated at 26–28 Å resolution from tilt series (±60° with intervals of 2–5°) of negatively stained crystalline arrays observed at ambient temperature (Figure 1; Yonath et al., 1987a,b) and cryo-temperature (Y. Fujiyoshi, private communication, and Yonath and Berkovitch-Yellin, 1993). These studies have been extended recently using unstained monolayers embedded in vitreous ice that diffract to 15–18 Å (Avila-Sakar et al., 1993).

The high reliability of the so obtained models can be assessed by the remarkable similarities in shape and size of these images and those obtained by the optimized series expansion (Vogel, 1983) or random-conical reconstruction (Radermacher et

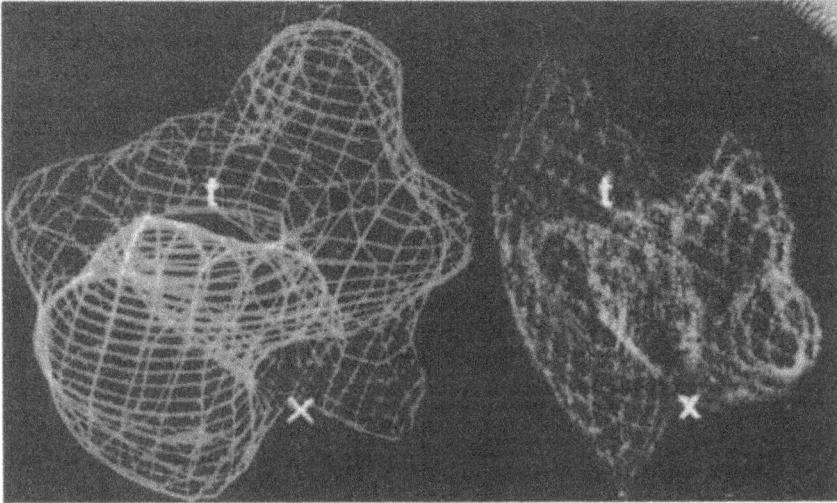

Figure 1. Two views of the model of the 50S subunit, as reconstructed from negatively stained crystalline arrays of these subunits from *B. stearothermophilus* at 28 Å resolution (Yonath et al., 1987). **t** points at the entrance to the internal tunnels and **X** shows the funnel-shape exit from the longest tunnel. This region may provide the site for the initial folding of nascent chains. The *left* views were taken from the peptidyl-transferase center, and the *right* is seen from the side.

al., 1992), both performed on samples embedded in amorphous ice. One of the deviations between our reconstructed models and those observed by traditional electron microscopy, namely a nonstretched rather than extended L7/L12 stalk, indicates that the crystalline arrays or the vitreous ice-embedded particles suffer less distortions than the isolated particles.

These reconstructed 50S particles as well as 70S ribosomes, at 47 Å resolution, revealed prominent features which were not observed in previous studies of prokaryotic ribosomes, such as tunnels and hollows (Arad et al., 1987; Yonath·et al., 1987b). Some of these features were not resolved in traditional or sophisticated electron microscopy, but were seen in reconstructions from *in situ* arrays of eukaryotic ribosomes (Milligan and Unwin, 1986) and in the random-conical reconstruction obtained recently from noncrystalline 70S ribosomes embedded in amorphous ice (Frank et al., 1991).

Modeling experiments were performed on an Evans and Sutherland PS390 computer-graphics terminal using the program FRODO (Jones, 1978), according to the algorithm described in Berkovitch-Yellin et al. (1990). The ribosomes and their subunits were represented by their reconstructed images, obtained from the diffraction patterns collected from tilt series of negatively stained crystalline arrays.

To account for possible shrinkage during the preparation for electron microscopy and due to the microscope vacuum, these were slightly expanded, as described in Eisenstein et al. (1991).

The coordinates used for representing the tRNA molecules were taken from PDB (Protein Data Bank) (Abola et al., 1987) entry 6TNA (Sussman et al., 1978). Significant flexibility was introduced into the hinge regions of this structure to imitate its bound state as it is being aminoacylated (Moras, 1989). We thank Dr. L. Liljas (Uppsala) and his collaborators for the gift of the coordinates of the MS2 coat protein. As no structure of an entire antibody molecule was found in the PDB, it was approximated by juxtaposing an Fc and two Fab fragments (entries 1FC1 and 3FAB in the PDB) in a manner suggested by the low resolution structure of the T-shape IgG$_1$ molecule (Silverston et al., 1977). A somewhat different low resolution structure was determined for IgG, namely the Y-shape conformation (Huber et al., 1976). The main difference between these two structures, the angle between the arms of the Fab fragments, was found to be negligible at the low resolution of our current studies.

C. Functional Assignment in the Reconstructed Models

Despite the rather low resolution of the reconstructed images (47 and 28 Å, for 70S and 50S particles, respectively) and the shortcomings of electron microscopy studies, they provide a valuable tool for further understanding of the function of the ribosome. For their interpretation, the significant similarities in specific features of corresponding regions in the reconstructed models of the 50S and 70S particles were exploited. These efforts led to: (1) suggestions for the positioning of the 50S subunit within the associated 70S ribosome and possible conformational changes occurring by the association and dissociation of the ribosomal subunits; (2) the approximations of a model for associated 30S subunit; and (3) tentative assignments of biological functions to some structural features (Yonath and Wittmann, 1987b, 1989; Yonath et al., 1990; Berkovitch-Yellin et al., 1990, 1992; Franceschi et al., 1993; Yonath and Berkovitch-Yellin, 1993).

A free space, estimated as 15–20% of the total volume was detected in all models reconstructed from crystalline 70S ribosomes from *B. stearothermophilus* (Figure 2; in: Arad et al., 1987; Yonath et al., 1987b; Berkovitch-Yellin et al., 1990) and from single 70S ribosomes from *E. coli* by the random-conical procedure (Frank et al., 1991). This free space was interpreted as the separation between the small and the large subunit.

Holes, channels, and elongated internal tunnels were detected within the models of the large subunits from prokaryotic (Vogel, 1983; Yonath et al., 1987a,b; Radermacher et al., 1989, Frank et al., 1991) and eukaryotic (Milligan and Unwin, 1986) sources, obtained by a variety of reconstruction methods (optimized series expansion of single particles, diffraction of crystalline arrays, and the random-conical approach, respectively).

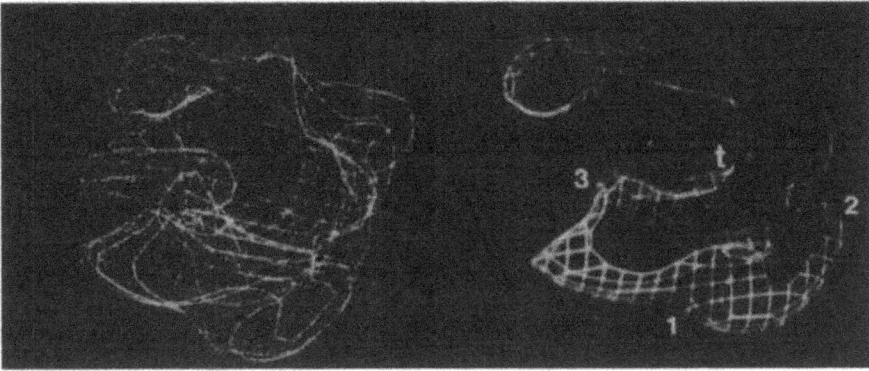

Figure 2. (*Left*) A computer graphics display of the outer contours of the most deviating models of 70S ribosomes reconstructed from negatively stained crystalline arrays of 70S ribosomes from *B. stearothermophilus* at 47 Å resolution (Arad et al., 1987). (*Right*) The net represents a slice of about 50 Å in thickness, through the model shown as solid lines on the left. t points at the common entrance to the three internal tunnels, and the numbers 1, 2 and 3 show the exits sites from tunnels t1, t2 and t3, respectively. Superimposed on it is a slice of about the same thickness of the 50S subunit, shown in solid lines.

One of these tunnels (t1 in Figure 2) appears in all the reported reconstructions from crystalline 50S and 70S particles from *Bacillus stearothermophilus* (Yonath et al., 1987a,b, 1990; Yonath and Wittmann, 1989; Berkovitch-Yellin et al., 1990) and *Thermus thermophilus* (Y. Fujiyoshi, personal communication). A feature similar to this tunnel was also detected within the density map obtained at 30 Å resolution from neutron diffraction data, collected from three-dimensional crystals of 50S subunits from *Haloarcula marismortui* (Eisenstein et al., 1991). This tunnel is about 100–120 Å in length and up to 25 Å in diameter, dimensions suitable for accommodating peptide chains of more than 30 amino acids at various conformations. Therefore, based on a large body of evidence derived from biochemical, fluorescence, functional, and immunoelectron microscopy experiments this tunnel was assumed to be the protected path for nascent proteins (reviewed in Eisenstein et al., 1993).

The original and most feasible interpretation of the reconstructed 70S ribosome was based solely on objective structural arguments. The 50S subunit was placed in a position which allows matching of the direction of its main tunnel t1 in the 70S ribosome (Figure 2), and in an orientation allowing the best fit of the external shapes of the 50S subunit and the part of the 70S ribosome assigned to it. This tunnel originates at the intersubunit free space and terminates at the distal, outer surface of the 50S subunit. On the opposite side of the free space, at the interface with the small subunit, a distinct region of crowded rRNA was revealed by preferential

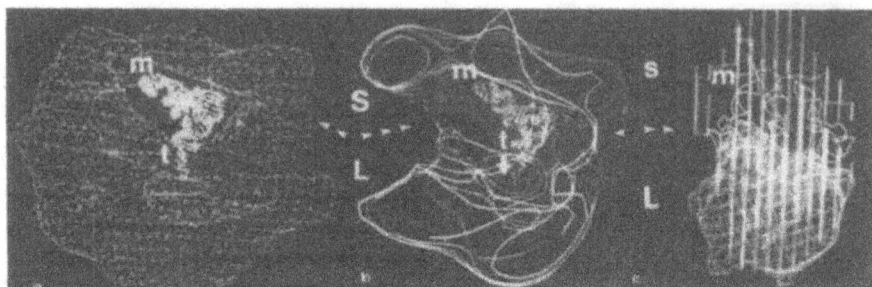

Figure 3. The positioning of the tRNA molecules based on the original interpretation. The arrow heads indicate the approximate directions of the interface between the S (small = 30S) and L (large = 50S) subunits. **t** points at the entrance to the tunnel. The presumed positions of the bound part of the mRNA chain are marked by m. A model built tRNA is shown in the thicker (**a**) and thinner (**b**) models (Figure A2). (**c**) A perpendicular view of (**a**), in which the contours of the 70S ribosome are shown in lines, and the 50S subunit as a net. In the free space at the intersubunit interface we placed three binding sites for tRNA molecules, two with the CCA end very close to the entrance of the tunnel, and the third, somewhat further away, presumably in a position allowing less tight contacts (the three tRNA molecules are bound to the ribosomes simultaneously).

staining. A groove within this region was suggested to be the path occupied by mRNA, in accord with evidence showing that during translation a segment of about 30–40 nucleotides of mRNA is masked by the ribosome (Kang and Cantor, 1985). Although the resolution of the current reconstruction images is too low for the determination of the dimensions of this groove, a rough estimate indicated that it may accommodate a stretch of mRNA comparable in length to that of the masked mRNA segment at a random, U-shaped, or helical conformation.

The dimensions of the tRNA allow its placement in the intersubunit space, so that its anticodon loop associates with the mRNA, and its CCA-terminus is positioned so that the newly formed peptidyl group extends into the entrance of the tunnel. In this orientation the tRNA molecule may also form several noncognate interactions (Figure 3). At the current resolution of the reconstructions, the two crystallographically determined orientations of tRNA—the native-closed and bound-open one (for review see Moras, 1989)—are indistinguishable.

The assignment of the intersubunit free space as the location of the various enzymatic activities of protein biosynthesis and the positioning of the tRNA molecules in it, between the two subunits, is in agreement with a large volume of circumstantial evidence accumulated during the last two decades (reviewed in Yonath and Berkovitch-Yellin, 1993, and discussed in Spirin et al., 1993). It is further supported by recent findings which show that, upon binding to 70S ribosomes, the entire P-site tRNA molecule is inaccessible even to hydroxyl radicals (Huettenhofer and Noller, 1992).

Steric considerations showed that the intersubunit void is spacious enough to provide up to three tRNA binding sites, along with other nonribosomal components that participate in peptide synthesis. It is noteworthy that these tRNA molecules may assume various relative orientations, ranging from parallel, the lowest space-requiring arrangement, to perpendicular (Figure 3), the most spacious one.

The intersubunit free space revealed in the random-conical model was also suggested to host the peptidyl transferase activity (Mitchell et al., 1992), although the assignments of the two subunits (Frank et al., 1991) differ significantly from that used in the original interpretation of the reconstructed model, and no effort was made for the localizations of the paths of the nascent chain. Subsequent effort for alternative interpretations (Franceschi et al., 1992; Yonath and Berkovitch-Yellin, 1993) indicated the considerable uncertainties in the current models, which due to their limited resolution do not possess the level of details required for unequivocal positioning of the two subunits. However, as the intersubunit free space is of dimensions which permit hosting the components participating in protein biosynthesis in several sterically reasonable arrangements, the assignment of the intersubunit void as the site of peptide synthesis does not critically depend on the accurate location of the small and the large subunits.

III. TECHNIQUES TO STUDY SYNTHESIS OF NASCENT POLYPEPTIDE CHAINS AND THEIR FOLDING

A. Fluorescence Techniques

Many of the investigations in the Hardesty laboratory have involved fluorescence measurements. Fluorescence techniques have several advantages for such studies. (1) They can be carried out at very low concentrations. With highly sensitive probes such as coumarin or fluorescein, measurements can be made easily at a probe concentration of 10 nM or below. (2) They can be performed in solution on active systems. For ribosomes, peptide elongation can be blocked at different points and analyzed at different steps in the reaction cycle.

Five different phenomena involving fluorescence can be used to monitor the interaction of labeled components (Hardesty et al., 1992). These are: quantum yield or fluorescence intensity, emission maximum, fluorescence anisotropy, nonradiative energy transfer, and fluorescence quenching by solute molecules.

Quantum yield (Q_f) is equal to the ratio of photons emitted to photons absorbed by the system. The interaction of a fluorophore with its environment (solvent or solute molecules such as proteins and nucleic acids) may have a large effect on Q_f. A change in Q_f is a direct indication of a change in the environment of the probe and has been used, for instance, to monitor binding of a component to ribosomes or movement of the component during a reaction step of protein synthesis.

Like the quantum yield, the emission maximum (λ_{max}) of most probes is sensitive to the local environment and a shift in the emission maximum can be used to quantitate interaction of one component with another. Generally, emitted light is more sensitive to perturbation by environmental changes than absorption. Coumarins are particularly sensitive and are the probes that we have used most frequently for this purpose. Generally λ_{max} for coumarin and pyrene probes shift towards the blue as the polarity of the solvent decreases, whereas the opposite effect is observed with fluorescein derivatives. Antibodies that we have prepared against coumarin maleimide (CPM) derivatives of proteins and nucleic acids cause a large decrease in λ_{max} and an increase in Q_f to near 1.0, whereas antifluorescein IgG decreases the Q_f of fluorescein probes to which it binds to near zero.

Fluorescence anisotropy (A) or fluorescence polarization (P) can be used as a measure of movement of a fluorophore. If plane-polarized light is used to excite a randomly oriented fluorescent probe held rigidly as in a glass and if the same electronic transitions are involved in absorption and emission (the absorption and emission transition moments are parallel), then the emitted light will be polarized. If, however, the fluorophore molecules can reorient randomly in the interval between the time of excitation and emission, then the emitted light will be completely depolarized. The fluorescence lifetime for many of the fluorophores commonly used in biochemistry is in the time range in which smaller proteins and nucleic acids tumble in solution. Thus binding of specifically labeled protein or RNA such as a tRNA to a relatively massive ribosome generally results in an increase in fluorescence anisotropy (A) or fluorescence polarization (P). The limiting value for A is 0.4, while the corresponding value for P is 0.5.

Heavy ions such as iodide or cesium, oxygen in the triplet ground state, and some organic molecules like acrylamide and methyl viologen cause quenching of fluorescence when they come in very close proximity or collide by Brownian movement with excited fluorophores. Changes in the system that alter the accessibility of the fluorophore to a quenching atom or molecule may be detected by a change in fluorescence intensity. Thus the quenching constant, K_q, for a fluorophore buried within a protein or ribosome may be very different than for a fluorophore free in solution. Changes in charge that repel or attract the quenching molecule to the fluorophore may also cause changes in K_q. We have used this technique to estimate the accessibility of probes attached to the N-termini of nascent peptide chains within ribosomes.

Distance between probes can be measured by non-radiative energy transfer. If a fluorophore in an excited state is adequately close and properly oriented to the electronic transition axis of a molecule at a lower energy state with which it can interact, there may be a direct transfer of energy without emission of a photon. The emission spectrum of the donor fluorophore must overlap the absorption spectrum of the acceptor. If the latter is fluorescent, energy transfer will result in decrease and increase, respectively, of fluorescence from the donor and acceptor probes. Stryer and his co-workers tested the validity of Förster's basic theory and demonstrated the use of the system as a "spectroscopic ruler" (Stryer and Hoagland, 1967). Distances in the range of 20 to 80 Å can be measured with appropriate choice of commonly used fluorophores.

B. Preparation of Fluorescent Materials

In most of the studies considered here, nascent peptides were fluorescently labeled at their N-terminal amino acid by initiating protein synthesis with a fluorescent derivative of *N*-mercaptoacetylaminoacyl-tRNA. The procedure for synthesizing the labeled tRNA derivative was derived in the Hardesty laboratory (Odom et al., 1990) and involves enzymatic aminoacylation of the desired tRNA, mercaptoacetylation with the succinimidyl monoester of dithiodiglycolic acid (the disulfide of mercaptoacetic acid), reduction of the disulfide with dithioerythritol, and reaction of the resulting sulfhydryl group with an activated fluorophore, most often CPM (3-[4-malemidophenyl]-7-diethylamino-4-methylcoumarin). The CPM-SAc-aminoacyl-tRNA is purified by reversed-phase high pressure liquid chromatography on a C_3 column. In a few cases Lys-tRNA labeled at its ε-amino group (Johnson et al., 1976) or Cys-tRNA (Hardesty et al., 1993) labeled at its sulfhydryl group have been prepared and incorporated internally into nascent peptides.

Ribosomes containing several species of amino acid homopolymers as well as some natural peptides have been prepared. Homopolymers which have been synthesized include polylysine, polyphenylalanine, polyserine, and polyalanine. Natural *E. coli* tRNA was used with poly(adenylic acid) for the synthesis of polylysine, but the other homopolymers all utilize poly(U), polyuridylic acid. This was made possible in the case of polyalanine and polyserine by construction of $tRNA^{Ala}$ and $tRNA^{Ser}$ containing AAA anticodons, which could be enzymatically acylated with alanine or serine, respectively, but which recognized polyuridylic acid as a template (Picking et al., 1991a; 1991b). Synthesis was initiated with CPM-SAc-AcLys-tRNA for polylysine, CPM-SAcPhe-tRNA for polyphenylalanine, and modified CPM-SAcAla-tRNA and CPM-SAcSer-tRNA, respectively, for polyalanine and polyserine. In the case of natural polypeptides, such as phage MS2 coat protein, initiation was obtained with N-(CPM-SAc)Met-tRNA_f.

To obtain ribosomes with different lengths of nascent peptides, synthesis was arrested in various ways. For the homopolymers the synthesis was allowed to proceed for increasing amount of time at 20 °C and was stopped at the desired time by lowering the temperature to 0 °C. The average chain length was determined by dividing the pmol of the radioactive amino acid incorporated into the body of the peptide by the pmol of fluorescently labeled, radioactive amino acid incorporated at the N-terminus during initiation of the peptide. The lengths of natural peptides were controlled by deleting certain amino acids occurring at known positions or by hybridization of a complementary oligodeoxynucleotide to a specific sequence of the mRNA. In all cases after translational arrest, material not bound to ribosomes was removed by gel filtration on Sephacryl S300.

C. *In Vitro* System for Coupled Transcription/Translation

A number of proteins with enzymatic activity have been synthesized *in vitro* by coupled transcription/translation. Generally, an *E. coli* S30 extract was prepared according to Zubay (1974), then centrifuged to isolate the ribosomes. The ribosomal pellets contain all protein components necessary for efficient protein synthesis (Kudlicki et al., 1992). For *in vitro* expression of DNA sequences under the T7 or SP6 promoter, the respective plasmid in nonlinearized form was added to the ribosome fraction together with the required RNA polymerase, rifampicin to inhibit *E. coli* RNA polymerase, total *E. coli* tRNA, and all necessary low molecular weight components.

After completion of protein synthesis, the isotopically labeled product was analyzed by SDS-PAGE (polyacrylamide gel electrophoresis in the presence of sodium dodecylsulfate), the amount formed was quantitated by TCA (trichloroacetic acid) precipitation, and the product was assayed for enzymatic activity. By use of a labeled amino acid of known specific radioactivity, the molar amount of the amino acid that is incorporated can be calculated. In turn, the molar amount of product peptide that is formed can be calculated if a single polypeptide species of known amino acid composition is formed. With rifampicin and a gene that can be transcribed by SP6 or T7 RNA polymerase, the transcription system can be constituted so that only one mRNA species is synthesized; however, the total peptide product that is formed includes some proportion of incomplete chains either as peptidyl-tRNA attached to the ribosomes or those that have been prematurely aborted and released into the soluble phase. Typically, the incomplete chains in the soluble fraction constitute less than 10% of the total. Thus, with this relatively small correction, the amount of full-length enzyme protein in the soluble phase can be calculated. In turn, its specific enzymatic activity can be calculated and compared with the specific enzymatic activity of the native enzyme as a measure of correct folding of the synthesized polypeptide.

IV. THE PATH OF NASCENT PEPTIDES

A. Peptidyl Transferase Reaction

A nascent peptide is bound as peptidyl tRNA in the peptidyl transferase center in such a way that it can be chemically transferred to the free amino group of an L-amino acid attached to the next incoming tRNA. By this peptidyl transferase reaction, the peptide is extended by one amino acid. The chemical reaction involves

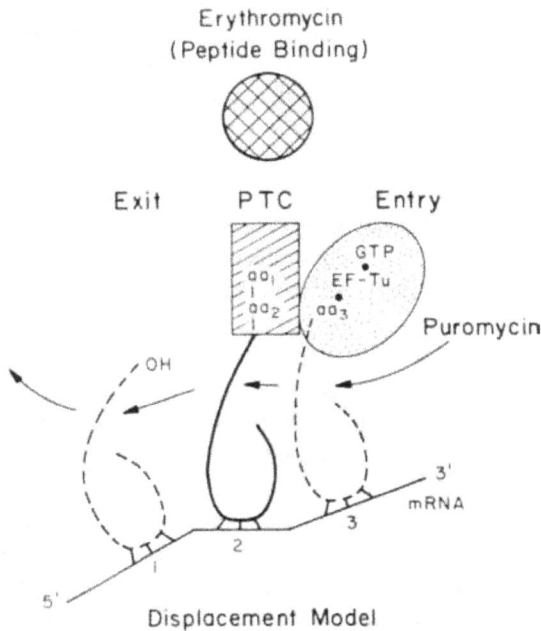

Figure 4. The displacement model for peptide elongation.

the transfer of the carboxyl group of the nascent peptide form an ester linkage with the 3′ hydroxyl of the terminal ribose of the tRNA to the α-carbon amino group of an incoming aminoacyl-tRNA. An amide or peptide bond is formed in what appears to be an SN2-type displacement reaction with nucleophilic substitution through a tetrahedral intermediate. A molecule of deacylated tRNA is generated as the peptide is elongated by one amino acid and the incoming tRNA becomes the new peptidyl-tRNA (Figure 4). This chemical reaction is associated with physical movement of the 3′-end of the newly formed deacylated tRNA of about 20 Å, relative to the ribosome, whereas the nascent peptide itself remains in essentially the same position in the peptidyl transferase center (Hardesty et al., 1986, 1990). These observations led to the postulation of the "displacement model" of ribosome function and peptide synthesis. Thus in this sense it might be more appropriate to refer to the tRNA transferase reaction rather than the peptidyl transferase reaction.

Originally, Watson (1964) postulated that the peptidyl transfer reaction must involve two sites or positions to which tRNA can be bound in the large subunit. These were the peptidyl or P site to which peptidyl-tRNA was bound before transfer of the peptide and an acceptor or A site to which incoming aminoacyl tRNA was bound. However, the mechanism described above requires that there should be three tRNA binding sites (Figure 4): (1) the peptidyl transferase center to which peptidyl tRNA is always bound; (2) a site into which deacylated tRNA is moved after the peptidyl transferase reaction is completed; and (3) a site to which an incoming aminoacyl-tRNA is bound before a new peptide bond is formed. The number of tRNA binding sites has been a matter of considerable controversy. Noller and his co-workers reported results that indicated only two tRNA molecules can bind through codon–anticodon interaction with the mRNA on the small subunit (Moazed & Noller, 1989a). However, fluorescence techniques using an N-acylcoumarin derivative of Phe-tRNA, showed unequivocally that this peptidyl-tRNA analog, a molecule of deacylated tRNA, and a molecule of puromycin could be simultaneously bound to the same ribosome (Odom et al., 1990). Puromycin is an analog of the 3′ end of aminoacyl-tRNA and presumably binds to the site to which aminoacyl-tRNA is bound on the large subunit. Together, these considerations indicated that there are three sites with which tRNA can interact on the large ribosomal subunit but only two sites on the small subunit. Noller and his co-workers proposed a hybrid model of ribosomal function (Moazed and Noller, 1989b) that accommodates these observations.

An essential feature of the peptidyl transferase reaction is that it does not involve a catalytic center or functional group that participates directly in the chemical reaction as an intermediate, as occurs during catalysis by most enzymes. Rather, it is the ribosome itself as a functional unit that mediates the alignment of the two tRNAs in such a way that the displacement reaction takes place. Indeed, a remarkable range of covalent derivatives other than peptides or amides can be formed in the peptidyl transferase center. These include esters (Fahnestock et al., 1970), thioesters (Gooch and Hawtrey, 1975), thioamides (Victorova et al., 1976), and phosphinoamides (Taressova et al., 1981). It is noteworthy that D-amino acid esters of tRNA cannot function in the peptidyl transferase reaction.

As described in greater detail below, analysis of quenching of fluorescence from a fluorophore covalently attached to the N-terminus of a nascent peptide by methyl viologen (MV^{2+}) led to the conclusion that the peptidyl transferase center itself and much of the ribosomal environment encountered by the nascent peptide as it is extended is due to RNA. The results indicated that positively charged MV^{2+} molecules are bound to ionized phosphate in the backbone of the ribosomal RNA within a few angstroms of the N-terminal probe (Picking et al., 1992). They appear to indicate that the peptidyl transferase center and the path encountered by the nascent peptides as it transverses the ribosome is largely or totally comprised of RNA.

The mechanism of the peptidyl transferase reaction described above and the RNA environment found at the peptidyl transferase center as well as the path for the nascent peptide have implications for the early steps of evolution. RNA molecules are capable of catalyzing the formation of certain covalent bonds and are likely candidates as the first self-replicative polymers in the evolution of living systems (Joyce and Orgel, 1993). However, it was the development of the unique reaction system of the peptidyl transferase center and decoding of mRNA with tRNA by ribosomes that, during early evolution, marked the point at which self-replication of a nucleic acid polymer was coupled with the synthesis of peptides. This transition was essential for the evolution of catalytically versatile proteins and the generation of "living" systems as represented by modern organisms. Indeed, even though the structural details of ribosomes and tRNA from diverse organisms vary considerably, their overall structure and the fundamentals of the mechanism by which they function appear to be essentially comparable in all ribosomes and may be the most evolutionarily conserved features of living systems.

B. The Peptidyl Transferase Center

Changes in the emission spectrum, fluorescence quantum yield, and fluorescence anisotropy that occur upon binding of N(fluorophore)-aminoacyl-tRNAs into the peptidyl transferase center of ribosomes indicate that the probes are held rigidly in a relatively hydrophobic environment in which they are inaccessible to immuno-globin G (IgG) (Picking et al., 1992). Upon binding of IgG, fluorescence anisotropy increased to nearly the theoretical maximum, 0.40, indicating that the probe and presumably the amino acid to which it was attached was held rigidly in a way that allows very little movement of the probe independent of the ribosome (Odom et al., 1991). After binding to the peptidyl transferase center, a nascent peptide was formed in which the N(fluorophore)-amino acid was in the N-terminal position. Coumarin was used as the fluorescent probe for most of our fluorescence experiments considered below.

C. Homopolymeric Peptides

Coumarin-labeled amino acids were incorporated into the N-terminal positions of various nascent amino acid homopolymers and the coumarin fluorescence parameters were examined as a function of chain length. The first homopolymer to be examined was polyphenylalanine (Odom et al., 1991; Picking et al., 1991b). The emission maximum of coumarin at the N-terminus of polyphenylalanine showed progressive shifts to the blue with increasing chain length, accompanied by corresponding increases in quantum yield. This indicated that the probe encountered an increasingly hydrophobic environment as the peptide was extended. Fluorescence anisotropy remained very high even at peptide lengths approaching 100 phenylalanine residues. Moreover, synthesis of polyphenylalanine was refractory to erythromycin. This antibiotic inhibits the elongation of most peptides beyond a di- or tripeptide but does not inhibit the synthesis of the first peptide bond (Roberts and Rabinowitz, 1989). It binds tightly to 50S subunits near the fluorescent probe of N-coumarin-Phe-tRNA bound in the peptidyl transferase center (see Figure 4). However, it does not bind to ribosomes bearing longer nascent chains of polyphenylalanine or other amino acids. These observations are consistent with the hypothesis proposed by Arevalo et al. (1988) that erythromycin blocks the entrance to the tunnel through which the nascent peptide must normally pass.

Using erythromycylamine labeled with coumarin, it was shown that prebound erythromycin was not displaced during the synthesis of polyphenylalanine (Odom et al., 1991). Rather, the prebound coumarin–erythromycin became increasingly resistant to exchange with unlabeled erythromycin as the polyphenylalanine chain was extended. Fluorescence from the probe underwent a marked increase in intensity and shifted towards the blue indicating that the environment of the probe became increasingly hydrophobic. These results appear to indicate that the coumarin–erythromycin was buried in or under the nascent polyphenylalanine chain. Polyphenylalanine is very insoluble in nearly all aqueous and nonpolar solvents. Considered together, these results indicate that polyphenylalanine forms an insoluble mass near the peptidyl transferase center as it is formed.

N-terminal coumarin-labeled nascent polylysine exhibited greatly different changes in fluorescence properties with increasing chain length when compared to polyphenylalanine. Fluorescence anisotropy dropped from 0.34 to 0.22 as the nascent peptides were extended an average length of four or five lysine residues and fell somewhat more upon further extension of the peptide. The lower value is near that observed for ε(N-coumarin)Lys-tRNA free in solution and indicated the probe was relatively mobile in solution. This decrease in anisotropy was accompanied by a red shift in emission maximum and a decrease in quantum yield, indicating a relatively polar, probably aqueous environment. This rapid increase in rotational flexibility and shift to a hydrophilic environment prompted the suggestion that polylysine may not enter the internal ribosomal channel but rather exit the ribosome directly into the solvent from the peptidyl transferase center. It should be noted that polylysine free in solution should have high inherent flexibility since it exists mainly as a random coil at neutral pH due to charge repulsion.

Polyamino acids, such as polyalanine that have a propensity to form an α-helix, might be used as rulers of designed length to study the environment of the nascent peptides as it is extended through the ribosome. The construction of alanyl-tRNAs and seryl-tRNA having AAA as their anticodons and therefore recognizing poly(U) as a template (Picking et al., 1991a; 1991b) made it possible to investigate by fluorescence the environment encountered by these homopolymers as they are synthesized on the ribosome. Polyalanine forms a stable α-helix in solution (Chou and Fasman, 1974) and is relatively soluble. Polyserine is very soluble and tends to exist as a random coil in aqueous solution, but with a nucleation site at the peptidyl transferase center it was expected to exhibit α-helical formation that is intermediate between polyalanine and polylysine.

Using synthetic CPM-SAcAla-tRNA or CPM-SAcSer-tRNA to initiate either polyalanine synthesis or polyserine synthesis, fluorescence parameters of the N-terminal coumarin were monitored as a function of chain length (Picking et al., 1992), particularly fluorescence anisotropy and accessibility to four soluble molecules of increasing size: methyl viologen, MV^{2+} ($M_r = 257$), proteinase K ($M_r = $

27,000), Fab fragment of anti-coumarin IgG ($M_r = 50,000$), and intact anti-coumarin IgG ($M_r = 150,000$). Initially fluorescence anisotropy was very high, near 0.38, but decreased for both polyalanine and polyserine up to an average chain length of about 80 residues. The rate of decrease was considerably slower for polyalanine than for polyserine. This may reflect a greater tendency of polyserine α-helix to degenerate into a less rigid conformation as elongation continues.

Coumarin at the N-terminus of both polyalanine and polyserine was accessible to quenching by MV^{2+} at all chain lengths. In fact, the quenching was greater at short chain lengths, as described above. This accessibility to MV^{2+} suggests that the interior of the tunnel is accessible to this relatively small organic molecule and presumably to solvent throughout its length.

Probes at the N-terminus of both polyalanine and polyserine were inaccessible to proteinase K at short chain lengths, but became accessible at an average length of about 40 residues with polyalanine and 25–30 residues with polyserine. These lengths are comparable to previously reported protected lengths of natural nascent proteins during limited proteolysis (Malkin and Rich, 1967; Blobel and Sabatini, 1970; Smith et al., 1978; Ryabova et al., 1988).

The N-terminal coumarin on both polyalanine and polyserine was inaccessible to binding of anti-coumarin Fab fragments until chain lengths reached 55 residues and 45 residues, respectively. Finally, accessibility to intact anti-coumarin IgG was only achieved at lengths of about 80 alanine residues or 60 serine residues (Picking et al., 1991b). Binding of anti-coumarin IgG or its Fab to the N-terminal coumarin caused a large increase in quantum yield and a shift in the emission maximum, reflecting the hydrophobic nature of the coumarin binding site in the antibody, as considered above. The difference in accessibility with nascent polyalanine and polyserine apparently reflects differences in the secondary structure of the peptides, whereas the difference between IgG, its Fab fragment, and proteinase K probably reflects differences in their molecular size. If the polyalanine is in the α-helical conformation, 80 residues would correspond to a length of about 120 Å. Yonath and co-workers have shown that the main ribosomal tunnel can accommodate a peptide of this length (Figure 5). In addition, a plausible model for the apparently partial penetration into the tunnel by the Fab fragments was constructed (Eisenstein et al., 1993).

D. Signal Peptides

A number of laboratories have attempted to define the shortest nascent chain that is capable of binding the signal recognition particle (SRP) and translocation of the peptide into a membrane (Ibrahimi et al., 1986; Siegel and Walter, 1988; Okun et al., 1990). The approach has been to prepare a series of synthetic mRNAs that produce truncated proteins of different lengths, then examine the ability of these peptides to interact with SRP and membranes. The results are somewhat variable but indicate that peptides containing 70–75 amino acids were translocated while

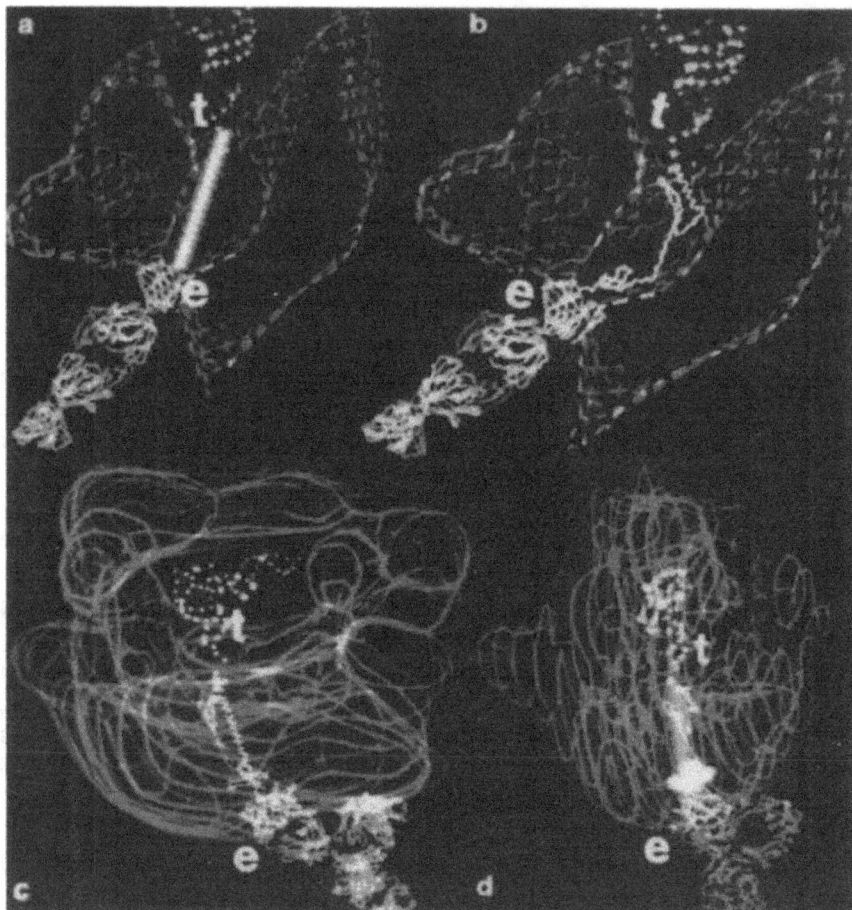

Figure 5. A slice of 50 Å thickness of the 50S subunit. In tunnel t1 (Figure 2) is placed: (a) a theoretical, undeformed stretch of an α-helix of 67 residues; (b) the main chain of the MS2 coat protein in a partially unfolded conformation, maintaining the native fold of its β-stretches and the native (crystallographically determined) conformation of the segment 1–47 (the C-terminus is in the vicinity of the proposed peptidyl transferase center and the N-terminus at the exit domain of the t1 tunnel); (c) and (d) show perpendicular views of the 70S ribosome, with an imaginary composite of (a) and (b) showing in tunnel t1 both the α-helix and the partially folded MS2 protein; (d) the segment 1–47 of the MS2 protein is shown as a space filling model. The partial penetration of the Fab molecule is evident. A molecule of tRNA (*in crosses*) was model built in the intersubunit space, so its anticodon is close to the mRNA binding site and its CCA end is placed in the entrance to the tunnel. A hypothetical IgG molecule (main chain as *solid lines*) was attached at the end of the tunnel, so that it makes the closest possible contacts with the ribosome. **t** shows the entrance to the tunnel and **e**, its exit.

those containing 50–55 amino acids or less were not. It appears likely that the variability of the results with different proteins reflects differences in the folding of these proteins that take place within the ribosomes.

Johnson and co-workers (Crowley et al., 1993) labeled truncated forms of pre-prolactin by incorporation of ϵ-N(NBD)-lysine from the corresponding Lys-tRNA during translation. NBD is a highly fluorescent, environmentally sensitive fluorophore, 6-(7-nitrobenz-2-oxa-1,3-diazol-4-yl)amino hexanoic acid. NBD-lysine was incorporated in place of lysine internally along nascent preprolactin peptides that were blocked at specific points during translation as the ribosomes came to the end of a truncated mRNA that did not contain a termination codon. Analysis of the resulting fluorescence lifetime data was interpreted to indicate that the fluorophore and the nascent peptides to which they were attached were in an aqueous environment as they were extended through the ribosome. Furthermore, the probe was accessible to iodide ions at all peptide lengths as indicated by Stern–Volmer constants for collisional quenching. However, upon addition of an endoplasmic reticulum (ER) membrane fraction to the translational reaction mixture collisional quenching by the iodide ions decreased to very low levels with little or no apparent increased quantum yield. The results indicated that attachment of the ribosomes to the ER membranes blocks access from the aqueous solvent and iodide ions to the nascent chain. They provide strong support to the hypothesis that nascent peptide follows an enclosed tunnel through the ribosomes, rather than an open groove or channel on its surface, to reach the exit domain on the outer surface of the large subunit.

E. MS2 Coat Protein

The homopolymeric peptides considered above are useful to illuminate some basic features of the biosynthetic process, but provide a poor model for following the fate of nascent chains of proteins since they do not fold into the complex tertiary structure of a typical globular protein. MS2 coat protein was chosen for initial experiments involving fluorescence in which the same general strategy described above was used to monitor the extension of nascent homopolymeric peptides. This protein contains 130 amino acids including the N-terminal methionine and is folded into a well-defined globular-like structure containing seven antiparallel β-strand and two short α-helices (Valegard et al., 1990). To label the N-terminus of the nascent protein, a coumarin derivative was covalently attached to the methionine amino group of Met-tRNA$_f$, which was then bound to ribosomes in the presence of MS2 RNA. Synthesis of defined lengths of nascent MS2 coat protein was accomplished by limiting specific tRNAs or amino acids in the polymerization reaction mixture or by truncating the MS2 RNA with ribonuclease H after hybridization with an antisense DNA oligomer. Fluorescence anisotropy was used as a measure of probe mobility as a function of nascent peptide length. The resulting anisotropy profile was more similar to that for polyalanine rather than polyserine, polylysine, or polyphenylalanine. Anisotropy remained very high (near 0.38) until a peptide length of 30–40 residues was reached, followed by a slow decline to values near those for full-length MS2 coat proteins that had been released from the ribosome (Hardesty et al., 1993). MV^{2+} quenching of fluorescence from N-terminal coumarin–methionine was used to characterize the environment of the probe as the nascent peptide was extended. Although quenching for the ribosome-bound MS2 protein initially declined rapidly as the chains were extended, values remained well above those for the corresponding puromycin-released material to lengths of about 125 residues. This result appears to indicate that fluorescence from the N-terminal coumarin even in nearly full-length nascent MS2 peptides is subject to elevated quenching by MV^{2+} due to binding of the quenching agent to RNA within the ribosome. This is a strikingly different result than was observed for nascent polyalanine and appears to indicate that the N-terminus of the nearly full-length nascent MS2 peptides are within the large ribosomal subunit.

Experiments similar to those described above were carried out with coumarin on the N-terminus of nascent MS2 coat protein and IgG or Fab derived from it. N-terminal coumarin becomes accessible to Fab at an average length of about 45 residues. This is approximately the length at which N-terminal coumarin on nascent polyalanine reacts with the Fab. However, in contrast to the results with nascent polyalanine with which N-terminal coumarin became accessible at an average chain length of about 60 residues, the N-terminus of all lengths of ribosome-bound nascent MS2 peptides were largely shielded from IgG. Both IgG and Fab interacted readily with N-terminal coumarin of all lengths of puromycin-released MS2 peptides (Hardesty et al., 1993). The different accessibility for Fab and IgG is consistent with the hypothesis that the N-terminal coumarin of nearly full-length peptides is shielded from IgG by the ribosome in a conformation or position in which it is accessible only to the smaller Fab molecule.

Both MV^{2+} quenching and the antibody accessibility experiments suggest that the nascent MS2 peptides of about 125 residues are in a folded conformation while they are being extended through the ribosome. This conclusion was tested directly by nonradiative energy transfer between coumarin covalently attached to cysteine at position 47, the first cysteine residue in the MS2 sequence, and fluorescein on the N-terminal methionine (Hardesty et al., 1993). The observed level of energy transfer appears to be a minimum value in that a number of factors might reduce the level of transfer that was observed in the experiments. A high level of energy transfer, probably greater than 95%, would be predicted from the distance between the N-terminus and cysteine[47] in the crystal structure of the native protein in the phage (Valegard et al., 1990). Again, the results appear to indicate that the nascent MS2 peptides are folded into some type of compact tertiary structure before they exit from the ribosomes. The relation of this pre-native state to the conformation of folded globular proteins is not known yet. Using fluorescence techniques, we were unable to detect binding of a 26-residue RNA to the nascent MS2 coat protein (unpublished results). The oligonucleotide had the base sequence of the hairpin fragment of the MS2 and R17 RNA to which the coat protein binds (Berzin et al., 1978; Romaniuk et al., 1987). These results—though negative—agree with those described below indicating that even full-length polypeptides have no detectable enzymatic activity as long as they are bound to ribosomes.

· V. *IN VITRO* SYNTHESIS AND FOLDING OF PROTEINS INTO BIOLOGICALLY ACTIVE ENZYMES

Specific binding of a ligand to an enzyme protein generally requires at least partial folding of the peptide into the native conformation. Therefore this might be used to monitor folding of nascent proteins into their native form. Similarly, the ultimate test of whether or not a nascent enzyme is folded into its native conformation is expression of its enzymatic activity. Covalent attachment of heme to cytochrome *c* was shown to occur posttranslationally (Dumont et al., 1988); however, other studies suggest that chlorophyll is bound to protein D1 of the membrane-bound chloroplast reaction center as a cotranslational event (Mullet et al., 1990; Kim et al., 1991). Spirin and his co-workers have reported that [^{3}H]-labeled hemin binds to nascent globin chains that are still attached to ribosomes (Komar et al., 1993).

Nascent peptides attached to the ribosome were reported in early publications to acquire the characteristics of the completed protein with a native tertiary structure. β-galactosidase provides a well-known example. In addition to folding of the polypeptide into the correct tertiary structure, its enzymatic activity requires that the four subunits of the native enzyme be united into a defined quaternary structure. It was shown that the nascent β-glactosidase peptide, prior to its completion and when it was still attached to the ribosome, associated with free subunits of the same protein and that this complex exhibited enzymatic activity while it was associated with the ribosome (Zipser and Perrin, 1963; Kiho and Rich, 1964). From these experiments it is not clear whether or not the nascent peptide itself was enzymatically active, but it can be assumed that substantial proper folding of the nascent peptide must have taken place to provide a basis for its interaction with free subunits of the enzyme.

Other results indicate that nascent chains of β-galactosidase acquire a conformation in which they can react with antibodies that recognize specific regions on the surface of the mature enzyme before translation is completed and the protein is released from the ribosome (Hamlin and Zabin, 1972). The antibodies used were concluded to be specific for the tertiary structure of the protein in that denaturation of the ribosome-bound material by heating destroyed its capacity to react with these antibodies. Much of the immunologically reactive peptide was released from the ribosomes by incubation with puromycin.

The results of these and other similar experiments were interpreted to indicate that the polyclonal antibodies that were used did indeed recognize and bind to the nascent peptides. However, the specificity of the antibodies for tertiary conformation is limited to small surface patches so that some questions remain concerning whether or not the nascent peptides were folded into their native conformation. Goldberg and his co-workers (Fedorov et al., 1992) carried out rather similar experiments with monoclonal antibodies that were shown to be highly specific for determinants formed by the tertiary structure of the β subunit of tryptophan synthetase (M_r~44,000). One monoclonal antibody isolate was found to react with ribosome-bound nascent peptides provided that their size was above 11.5 kDa. Further, it was shown by gel filtration chromatography that the 11.5-kDa fragment formed a condensed structure characteristic of a folded protein. The results were interpreted to indicate that a nascent peptide starts to fold into its native configuration while it is being formed on a ribosome. It is unfortunate that the nascent chains were not shown to be released from the ribosomes by puromycin or to be extended to give enzymatically active full-length polypeptides.

Experiments in our laboratory thus far have failed to demonstrate enzymatic activity of nascent chains of either *E. coli* DHFR (dihydrofolate reductase) or mammalian rhodanese on ribosomes (Kudlicki et al., 1994). The ribosomes were isolated from a bacterial coupled transcription/translation reaction system in which these enzymes were expressed from the respective plasmid. The folding of these enzymes after *in vitro* synthesis is considered in the subsequent sections.

E. coli DHFR cloned into a plasmid from the *fol A* gene was expressed in the *in vitro* coupled transcription/translation system and its activity tested. Mostly full-length DHFR was synthesized (Ma et al., 1993) and over 80% of newly synthesized DHFR was released into the soluble phase after 30 min synthesis at 37 °C (unpublished results). Enzymatic activity was followed by the decrease in absorbance at 340 nm over time due to oxidation of NADPH. One unit of DHFR activity is the amount of enzyme required to reduce 1 μmol DHF/min based on a molar extinction coefficient for NADPH of 12.3×10^3 at 340 nm. *In vitro* synthesized DHFR was calculated to have a specific activity of about 58 units/mg protein (Ma et al., 1993). Isolated native DHFR was reported to have 19 units/mg (Baccanari et al., 1975). The basis for this apparent difference between the specific enzymatic activity of newly formed and isolated DHFR is unknown.

Rhodanese deserves special attention as a model enzyme for studies of protein folding and membrane transport. Mammalian rhodanese is a monomeric 33-kDa sulfur transferase which is coded for by a nuclear gene and synthesized on cytoplasmic ribosomes and then transported into mitochondria where it is found as a soluble enzyme in the matrix. The crystal structure shows the enzyme to be composed of two domains of approximately equal size (Ploegman et al., 1978).

It has been shown that all the information required for transport, targeting, and folding into an enzymatically active conformation is contained in the primary sequence of the enzyme. Rhodanese contains four cysteine residues, yet has no disulfide bonds (Ploegman et al., 1978). Its only posttranslational processing is cleavage of its N-terminal methionine residue (Miller et al., 1978). Its enzymatic activity is easily monitored by a relatively simple and sensitive calorimetric assay

(Sorbo, 1953). This activity is lost in the first stage of denaturation (Tandon and Horowitz, 1986). Refolding from the denatured state has been extensively studied by Horowitz and his co-workers. Under proper conditions the denatured enzyme can refold spontaneously (Mendoza et al., 1991a), but the efficiency of refolding can be enhanced and greatly facilitated by the chaperonins, GroEL plus GroES (Mendoza et al., 1991b), or lipid micelles (Zardeneta and Horowitz, 1993).

Enzymatically active rhodanese has been produced from plasmid DNA containing its coding sequence in cell-free coupled transcription/translation systems (Tsalkova et al., 1993). Two additions when present during its synthesis were found to stabilize the newly synthesized enzyme and increase its specific activity. One component is thiosulfate, one of the enzyme's substrates. Though thiosulfate above 15 mM strongly inhibits protein synthesis (with all plasmids tested), it increased the specific enzymatic activity of rhodanese formed *in vitro* apparently by stabilizing the protein. Secondly, the chaperonins GroEL/ES increased the specific enzymatic activity of *in vitro* synthesized rhodanese, but only when they were included during protein synthesis (Tsalkova et al., 1993).

VI. CHAPERONES

A. Association with Ribosomes

In vitro renaturation of many proteins from their denatured state is facilitated by chaperones as is considered in other chapters. A central issue involves the question of whether or not chaperones play a role in folding of nascent peptides while they are attached to ribosomes. Several recent reports indicate this may be the case. Beckmann et al. (1990) observed that in HeLa cell extracts newly synthesized peptides ranging in size from 20 to 200 kDa were precipitated by monoclonal antibodies that were specific for the 70-kDa heat-shock protein, Hsp 72,73, in HeLa cells. The peptides were associated with polysomes, but could be released by reaction with puromycin. The results indicated that the Hsp 70 was transiently associated with the nascent peptide on ribosomes as they were being extended.

Craig and her co-workers (Nelson et al., 1992) reported intriguing results from a series of experiments with yeast which indicate that two members of the Hsp70 family of heat-shock proteins, Ssb1 and Ssb2, bind directly to nascent peptides as they are formed on polysomes. Puromycin caused the release of Ssb1 and Ssb2 with the nascent peptides. "Knock out" alleles of both Ssb proteins were prepared and studied with respect to their sensitivity to antibiotics that inhibit peptide elongation, but at different points in the peptide elongation cycle. Growth of the mutant strains was hypersensitive to puromycin, hygromycin B, and G418, all of which are amino glycoside antibiotics that act on the 40S ribosomal subunit. However, the mutant strains did not exhibit hypersensitivity to cycloheximide or anisomycin, both of which bind to the 60S subunit and block peptide bond formation. Hypersensitivity suggests that the mutants affect a component of a reaction step that is inhibited by the antibiotic. The slow-growth phenotype of the Ssb1 and Ssb2 mutant strains could be partially overcome by increasing the copy number the HBS1 gene which encodes a protein that appears to be homologous to peptide elongation factor eEF1-α.

These provocative results were interpreted to indicate that the Hsp70 proteins, Ssb1 and Ssb2, bind to the nascent peptide and cause a direct inhibition of peptide elongation. It was suggested that the absence of these proteins causes the nascent peptide to back up in the channel or tunnel through the large subunit, thereby slowing the rate of peptide elongation.

A corollary of this hypothesis is that movement of the nascent peptide through the large ribosomal subunit to the exit domain is dependent on binding of Hsp70 to a region near its N-terminus. It is difficult to imagine the mechanism by which this might occur, and certainly there is no indication that Hsp70 is required for translation of most peptides. The effect of overcoming such an inhibition by increasing the concentration of eEF1-α•amino acyl–tRNA complex in the cell is even more puzzling. Clearly these provocative observations require further investigation.

The inhibition of peptide elongation described above is reminiscent of inhibition of peptide elongation that is caused by the SRP that binds to the N-terminal signal sequence of proteins destined to be transported into or through membranes (Walter and Blobel, 1981; Wolin and Walter, 1989). The SRP facilitates establishment of a ribosome membrane junction. In the case of SRP, however, inhibition of peptide elongation is inhibited by binding of the SRP, not its absence. Although the mechanism by which the SRP inhibits peptide elongation is not understood, it is likely that it somehow blocks the tunnel or exit site so that the N-terminal domain of the nascent peptide cannot leave the ribosome.

B. Chaperone-Deficient *In Vitro* Translation Systems

At least 30 different protein components are required for translation. The list includes at least 20 aminoacyl·tRNA synthetases as well as the individual factors that are required for the ribosomal steps of peptide initiation, elongation, and termination. Weissbach and his co-workers (Weissbach et al., 1981) carried out the monumental task of isolating each of these components from prokaryotic sources and then reconstituting them into a completely defined translation system in which enzymatically active β-glactosidase was formed on bacterial ribosomes. These are important observations in that they provide very strong support for the hypothesis put forth by Anfinsen in which he proposed that folding of a protein into its tertiary conformation is dependent on the sequence of amino acids that constitute its primary structure, and thus is inherent in the base sequence of the DNA from which it was coded (Anfinsen and Scheraga, 1975). However, the amount of protein product that could be formed in a completely defined system was low and isolation of all of the required components is far beyond the resources that are available to most laboratories. Furthermore, a number of more recent studies indicate that chaperone proteins may facilitate folding of newly formed protein as well as function by promoting refolding from the denatured state. Studies designed to elucidate the role of chaperones in folding of nascent peptides on ribosomes have been severely hampered by the lack of an adequate translation system in which the effect of the chaperones on folding could be studied directly.

Recently we developed a coupled transcription/translation system from *E. coli* that is deficient in certain chaperones, at least in GroEL and DnaK which we could quantitate by antibodies (Kudlicki et al., 1994). DnaK and GroEL are the representatives of two "classes of molecular chaperone machines" (Georgopoulos, 1992) that have been hypothesized to act sequentially in the folding of newly synthesized proteins (Langer et al., 1992). Even though the amounts of GroEL and DnaK were reduced to 17 and 3%, respectively, of the level determined for the S30 extract, we were able to synthesize DHFR and rhodanese in high yield and at nearly the same specific enzymatic activity as seen in the cell-free system using S30 or crude ribosomes (cf. Section V). Thus it appears that enzymes can be synthesized and folded on ribosomes without the requirement of chaperones.

VII. CONCLUSIONS, PROBLEMS, AND PROSPECTS

Images of ribosomes reconstructed from electron micrographs of crystalline arrays viewed at different tilt angles show an empty space between the ribosomal subunits and a cavity within the large subunit that extends to the exit domain on its outer surface. Although polyphenylalanine and polylysine may be exceptions, most nascent peptides including polyalanine and MS2 coat protein appear to be extended from the peptidyl transferase center into a cavity. Nascent MS2 coat protein and probably most other globular proteins fold into some type of tertiary conformation within this cavity as they are elongated by addition of new amino acids to their C-terminal end. A nascent protein eventually emerges from the cavity at a point on the outer surface of the large subunit called the exit domain that is distal to the peptidyl transferase center. It is this region of the large ribosomal subunit that provides the docking site for ribosomes on the endoplasmic reticulum. The cavity within the large subunit is accessible to the solvent in that small molecules of the size of methyl viologen and iodide ions penetrate into the subunit. However, binding of the ribosome to a membrane appears to cap or block the outer entrance so that access by iodide ions is prevented. These observations led to the conclusion that the cavity is closed in the form of a tunnel rather than an open channel. The cavity in the large subunit of bacterial ribosomes is adequate to accommodate nearly full-length MS2 coat protein and can be penetrated to some depth by relatively small proteins such as trypsin, proteinase K, or a Fab fragment of IgG. However, IgG itself is excluded. These considerations led to the hypothesis that the N-terminal region of a nascent peptide begins to fold into a tertiary conformation within the ribosomal subunit as the peptide is extended. It follows that the cavity may provide a sheltered environment in which the nascent peptide is at least partially protected from proteolysis and other potentially destructive agents in the cytoplasm surrounding the ribosome. It should be emphasized that the relation of the conformation of the nascent peptide within the large ribosomal subunit to native conformation of the active protein is not known. Thus far, attempts to demonstrate functional activity of nascent peptides bound to ribosomes as peptidyl–tRNA have been negative or inconclusive.

Chaperones facilitate folding of at least some species of newly formed protein into enzymatically active conformation, but in most or all of the cases examined thus far some proportion of newly formed enzyme protein is enzymatically active in their absence, providing strong support for the hypothesis that folding of a protein during its synthesis is an inherent property of its primary structure. The proportion of enzyme protein that is folded into an active conformation in the absence of chaperones varied widely with the protein that is being considered and the conditions of its synthesis. Whether or not vectorial folding from the N-terminus as the protein is formed is a requirement for most proteins to reach their native conformation remains to be established. The requirement for vectorial folding to achieve the native state is likely to be dependent on the protein species and conditions.

A number of lines of evidence strongly suggest but do not conclusively prove that some of the chaperones and other factors can facilitate folding of a nascent peptide while it exists as peptidyl–tRNA on a ribosome. Clearly the signal recognition particle can recognize the signal sequence of nascent peptide that is destined for transport out of the cytoplasm. The mechanism by which elongation of such peptides is blocked in the absence of this interaction is unknown. It appears likely that some newly formed peptides interact with the GroEL/ES complex after they are released from the ribosome but that DnaK or DnaJ may interact with nascent chains. More detailed information about the structure of ribosomes and their function is required to resolve these questions. Generation of a high resolution crystal structure by X-ray diffraction of crystallized ribosomes holds great promise for the former. Fluorescence techniques coupled with procedures by which fluorophores can be incorporated at specific sites in a nascent peptide provide a powerful approach to the latter.

ACKNOWLEDGMENTS

We would like to thank Willa Mae Hardesty and Ronda Barnett for typing of the manuscript. Support was provided by the National Science Foundation (DMB 9018260) and The Foundation for Research (to B.H.) and by the National Institute of Health (NIH GM 34360), the Federal Ministry for Research and Technology (BMFT 05 180 MP BO), the France-Israel Binational Foundation (NRCD-334190) and the Kimmelman Center for Macromolecular

Assembly at the Weizmann Institute (to A.Y.). A.Y. holds the Martin S. Kimmel Professional chair.

REFERENCES

Abola, E., Bernstein, F.C., Bryant, S.H., Koetzla, T.F., & Weng, J. (1987). Protein Data Bank. In: Crystallographic Database (Allen, F.H., Berggerhoff, F., & Sievers, R., Eds.), pp. 107–132, Bonn/Cambridge/Chester.

Anfinsen, C.B. & Scheraga, H.A. (1975). Experimental and theoretical aspects of protein folding. Adv. Prot. Chem. 29, 205–300.

Arad, T., Piefke, J., Weinstein, S., Gewitz, H.S., Yonath, A., & Wittmann, H.G. (1987). Three-dimensional image reconstruction from ordered arrays of 70S ribosomes. Biochimie 69, 1001–1005.

Arevalo, M.A., Tejedor, R., Polo, R., & Ballesta, J.P. (1988). Protein components of the erythromycin binding site in bacterial ribosomes. J. Biol. Chem. 263, 58–63.

Avila-Sakar, A.J., Guan, T.L., Schmid, M.F., Loke, T.L., Arad, T., Yonath, A., Piefke, J., Franceschi, F., & Chiu, W. (1994). Electron Cryomicroscopy of 50S Ribosomal Subunits from *B. stearothermophilus* Crystallized on Phospholipid Monolayers. J. Mol. Biol. 239, 689–697.

Baccanari, D., Phillips, A., Smith, S., Sinski, D., & Burchall, J. (1975). Purification and properties of *Escherichia coli* dihydrofolate reductase. Biochemistry 14, 5267–5273.

Beckmann, R.P., Mizzen, L.A., & Welch, W.J. (1990). Interaction of Hsp 70 with Newly Synthesized Proteins: Implications for Protein Folding and Assembly. Science 248, 850–854.

Berkovitch-Yellin, Z., Wittmann, H.G., & Yonath, A. (1990). Low resolution models for ribosomal particles reconstructed from electron micrographs of tilted two-dimensional sheets: Tentative assignments of functional sites. Acta Cryst. B46 637, 643.

Berkovitch-Yellin, Z., Bennett, W.S., & Yonath, A. (1992). Aspects in Structural Studies on Ribosomes. CRC Rev. Biochem. & Mol. Biol. 27, 403–444.

Berzin, V., Borisova, G., Cielens, I., Gribanov, V., Jansone, I., Rosenthal, G., & Gren E. (1978). The Regulatory Region of MS2 Phage RNA Replicase Cistron. J. Mol. Biol. 119, 101–131.

Blobel, G. & Sabatini, D. (1970). Controlled proteolysis of nascent polypeptides in rat liver cell fractions. I. Location of the Peptides within ribosomes. J. Cell Biol. 45, 130–145.

von Boehlen, K., Makowski, I., Hansen, H.A.S., Bartels, H., Berkovitch-Yellin, Z., Zaytzev-Bashan, A., Meyer, S., Paulke, C., Franceschi, F., & Yonath, A. (1991). Characterization and preliminary attempts for derivation of crystals of large ribosomal subunits from H. marismortui diffracting to 3 Å resolution. J. Mol. Biol. 222, 11–15.

Chantrenne, H. (1961). The Biosynthesis of Proteins In: Modern Trends in Physiological Science (Alexander, P. & Bacq, Z., Eds.), p. 122. Pergamon Press, London.

Chou, P.Y. & Fasman, G.D. (1974). Conformational parameters for amino acids in helical, β-sheet and random coil regions calculated from proteins. Biochemistry 13, 211–245.

Crowley, K.S., Reinhart, G.D., & Johnson, A.E. (1993). The signal sequence moves through a ribosomal tunnel into a noncytoplasmic aqueous environment at the ER membrane early in translocation. Cell 73, 1101–1115.

Dumont, M.E., Ernst, J.F., & Sherman, F. (1988). Coupling of heme attachment to import cytochrome *c* into yeast mitochondria. J. Biol. Chem. 263, 15928–15937.

Eisenstein, M., Sharon, R., Berkovitch-Yellin, Z., Gewitz, H.S., Weinstein, S., Pebay-Peyroula, E., Roth, M., & Yonath, A. (1991). The interplay between X-ray crystallography, neutron diffraction, image reconstruction, organo-metallic chemistry and biochemistry in structural studies of ribosomes. Biochimie 73, 879–886.

Eisenstein, M., Hardesty, B., Odom, O.W., Kudlicki, W., Kramer, G., Arad, T., Franceschi, F., & Yonath, A. (1994). Modelling and the experimental study of progression of nascent protein in ribosomes. In: Biophysical Methods in Molecular Biology (Pifat, G., ed.), pp. 213–246, Balaban Publ.

Fahnestock, S., Neumann, H., Shashoua, V., & Rich, A. (1970). Ribosome-catalyzed ester formation. Biochemistry 9, 2477–2483.

Frank, J., Penszek, P., Grassucci, R., & Srivastava, S. (1991). Three-dimensional reconstruction of the 70S E. coli ribosome in ice: the distribution of ribosomal RNA. J. Cell. Biol. 115, 597–605.

Fedorov, A.N., Friguet, B., Djavadi-Ohaniance, J., Alakhov, Y.B., & Goldberg, M.E. (1992). Folding on the ribosome of Escherichia coli tryptophan synthase β subunit nascent chains probed with a conformation-dependent monoclonal antibody. J. Mol. Biol. 228, 351–358.

Franceschi, F., Weinstein, S., Evers, U., Arndt, E., Jahn, W., Hansen, H.A.S., von Boehlen, K., Berkovitch-Yellin, Z., Eisenstein, M., Zaytzev-Bashan, A., Sharon, R., Levin, I., Dribin, A., Sagi, I., Choli-Papadopoulou, T., Tsiboly, P., Kryger, G., Bennett, W.S., & Yonath, A. (1993). Towards atomic resolution of prokaryotic ribosomes: crystallographic, genetic and biochemical studies. In: The Translational Apparatus (Nierhaus, K.H., Subramanian, A.R., Erdmann, V.A., Franceschi, F., & Wittmann-Liebold, B., Eds.). Plenum Press, pp. 397–440.

Georgopoulos, C. (1992). The emergence of the chaperone machines. TIBS 17, 295–299.

Gooch, J. & Hawtrey, A.O. (1975). Synthesis of thiol-containing analogues of puromycin and a study of their interaction with N-acetylphenylalanyl-transfer ribonucleic acid on ribosomes to form thioesters. Biochem. J. 149, 209–220.

Hamlin, J. & Zabin, I. (1972). β-Galactosidase: Immunological activity of ribosome-bound, growing polypeptide chains. Proc. Natl. Acad. Sci. USA 69, 412–416.

Hardesty, B., Odom, O.W., and Deng H.-Y. (1986). The movement of tRNA through ribosomes during peptide elongation: The displacement reaction. In: Structure, Function and Genetics of Ribosomes (Hardesty, B. & Kramer, G., Eds.), pp. 495–508. Springer-Verlag, New York.

Hardesty, B., Odom, O.W., & Czworkowski, J. (1990). Movement of tRNA through ribosomes during peptide elongation. In: The Ribosomes: Structure, Function and Evolution (Hill, E.W., Dahlberg, A., Garrett, R.A., Moore, P.B., Schlesinger, D., & Warner, J.R., Eds.), pp. 366–372. American Society for Microbiology, Washington, DC.

Hardesty, B., Odom, O.W., & Picking. W. (1992). Ribosome function determined by fluorescence. Biochimie 74, 391–401.

Hardesty, B., Odom, O.W., Kudlicki, W. & Kramer, G. (1993). Extension and folding of nascent peptides on ribosomes. In: The Translational Apparatus (Nierhaus, K.H., Subramanian, A.R., Erdmann, V.A., Franceschi, F., & Wittmann-Liebold, B., Eds.). Plenum Press, pp. 347–358.

Huber, R., Deisenhofer, J., Colman, P.M., & Matsushina, M. (1976). Crystallographic structure studies of an IgG molecule and an Fc fragment. Nature 264, 415–420.

Huettenhofer, A. & Noller, H.F. (1992). Hydroxyl radical cleavage of tRNA in the ribosomal P site. Proc. Natl. Acad. Sci. USA 89, 7851–7855.

Ibrahimi, I., Cutler, D., Steuber, D., & Bujard, H. (1986). Determinants for protein translocation across mammalian endoplasmic reticulum. Membrane insertion of truncated and full-length prolysozyme molecules. Eur. J. Biochem. 155, 571–576.

Johnson, A.E., Woodward, W.R., Herbert, E., & Menninger, J.R. (1976). N-ε-acetyl lysine transfer ribonucleic acid: A biologically active analogue of aminoacyl transfer ribonucleic acids. Biochem. 15, 569–575.

Jones, T.A. (1978). A graphics model building and refinement system for macromolecules. J. Appl. Crystallog. 11, 268–272.

Joyce, G.F. & Orgel, L.E. (1993). Prospects for understanding the origins of the RNA world. In: The RNA World (Gesteland, R.F. & Atkins, J.F., Eds.), pp. 1–25. Cold Spring Harbor Laboratory Press.

Kang, C. & Cantor, C.R. (1985). Structure of ribosome bound mRNA as revealed by enzymatic accessibility studies. J. Mol. Biol. 210, 659–663.

Kiho, Y. & Rich, A. (1964). Induced enzyme formed on bacterial polyribosomes. Proc. Natl. Acad. Sci. USA 51, 111–118.

Kim, J., Klein, P.G., & Mullet, J.E. (1991). Ribosomes pause at specific sites during synthesis of membrane-bound chloroplast reaction center protein D1. J. Biol. Chem. 266, 14931–14938.

Kim, S.H., Suddath, F.L., Quigley, G.J., McPherson, J.L., Sussman, A., Wang. H.-J., Seeman, N.C., & Rich, A. (1974). Three dimensional tertiary structure of yeast phenylalanine transfer RNA. Science 185, 435–440.

Komar, A.A., Kommer, A, Krasheninnikov, I.A., & Spirin, A.S. (1993). Cotranslational heme binding to nascent globin chains. FEBS Lett. 326, 261–263.

Kudlicki, W., Kramer, G., & Hardesty, B. (1992). High efficiency cell-free synthesis of proteins: Refinement of the coupled transcription/translation system. Anal. Biochem. 206, 389–393.

Kudlicki, W., Mouat, M., Walterscheid, J.P., Kramer, G., & Hardesty, B. (1994). Development of a chaperone-deficient system by fractionation of a prokaryotic coupled transcription/translation system. Anal. Biochem. 217, 12–19.

Langer, T., Lu, C., Echols, H., Flanagan, J., Hayer, M., & Hartl, F. (1992). Successive action of DnaK, DnaJ and GroEL along the pathway of chaperone-mediated protein folding. Nature 356, 683–689.

Lim, V.I. & Spirin, A.S. (1986). Stereochemical analysis of ribosomal transpeptidation. Conformation of nascent peptide. J. Mol. Biol. 188, 565–577.

Ma, C., Kudlicki, W., Odom, O.W., Kramer, G., & Hardesty, B. (1993). *In vitro* protein engineering using synthetic tRNA^{Ala} with different anticodons. Biochemistry 32, 7939–7945.

Malkin, L.I. & Rich, A. (1967). Partial resistance of nascent polypeptide chains to proteolytic digestion due to ribosomal shielding. J. Mol. Biol. 26, 329–346.

Matthews, R. (1993). Pathways of protein folding. Annu. Rev. Biochem. 62, 653–683.

Mendoza, J.A., Rogers, E., Lorimer, G.H., & Horowitz, P.M. (1991a). Unassisted refolding of urea unfolded rhodanese. J. Biol. Chem. 266, 13587–13591.

Mendoza, J.A., Rogers, E., Lorimer, G.H., & Horowitz, P.M. (1991b). Chaperonins facilitate *in vitro* folding of monomeric mitochondrial rhodanese. J. Biol. Chem. 266, 13044–13049.

Miller, D.M., Delgado, R., Chirgwin, J.M., Hardies, S.C., & Horowitz, P.M. (1991). Expression of cloned bovine adrenal rhodanese. J. Biol. Chem. 266, 4686–4691.

Milligan, R.A. & Unwin, P.N.T. (1986). Location of the exit channel for nascent proteins in 80S ribosomes. Nature 319, 693–696.

Mitchell, P., Osswald, M., & Brimacombe, R. (1992). Identification of intermolecular RNA cross-links at the subunit interface of the *E. coli* ribosome. Biochemistry 31, 3004–3011.

Moazed, D. & Noller, H.F. (1989a). Interaction of tRNA with 23S rRNA in the ribosomal A, P, and E sites. Cell 57, 585–597.

Moazed, D. & Noller, H.F. (1989b). Intermediate states in the movement of transfer RNA in the ribosome. Nature 342, 142–148.

Moras, D. (1989). Crystal structure of tRNAs, In: Landolt-Borstein New Series 1b. Nucleic Acids (Saenger, W., Ed.), pp. 1–30. Springer Verlag, Berlin/New York.

Mullet, J.E., Klein, P.G., & Klein, R.R. (1990). Chlorophyll regulates accumulation of the plastid-encoded chlorophyll apoproteins CP43 and D1 by increasing apoprotein stability. Proc. Natl. Acad. Sci. USA 87, 4038–4042.

Nelson, R.J., Ziegelhoffer, T., Nicolet, C., Werner-Washburne, M., & Craig, E.A. (1992). The translation machinery and 70 Kda heat shock protein cooperate in protein synthesis. Cell 71, 97–105.

Noller, H.F. (1993). tRNA-rRNA Interactions and peptidyl transferase. FASEB J. 7, 87–93.

Odom, O.W., Picking, W.D., & Hardesty, B. (1990). Movement of tRNA but not the nascent peptide during peptide bond formation on ribosomes. Biochemistry 29, 10734–10744.

Odom, O.W., Picking, W.D., Tsalkova, T., & Hardesty, B. (1991). The synthesis of polyphenylalanine on ribosomes to which erythromycin is bound. Eur. J. Biochem. 198, 713–722.

Okun, M.M., Eskridge E.M., & Shields, D. (1990). Truncations of a secretory protein define minimum lengths required for binding to signal recognition particle and translocation across the endoplasmic reticulum membrane. J. Biol. Chem. 265, 7478–7484.

Picking, W.L., Picking, W.D., & Hardesty, B. (1991a). The use of synthetic tRNAs as probes for examining nascent peptides on *Escherichia coli* ribosomes. Biochimie 73, 1101–1107.

Picking, W.D., Odom, O.W., Tsalkova, T., Serdyuk, I., & Hardesty, B. (1991b). The conformation of nascent polylysine and polyphenylalanine peptides on ribosomes. J. Biol. Chem. 266, 1534–1542.

Picking, W.D., Odom, O.W., & Hardesty, B. (1992). Evidence for RNA in the peptidyl transferase center of *Escherichia coli* ribosomes as indicated by fluorescence. Biochemistry 31, 12565–12570.

Ploegman, J.H., Drent, G., Kalk, K.H., Hol, W.G.J., Heinrikson, R.L., Keim, P., Weng, L., & Russell, J. (1978). The covalent and tertiary structure of bovine liver rhodanese. Nature 273, 124–129.

Radermacher, M., Wagenknecht, T., Verschoor, A., & Frank, J. (1987). Three-dimensional structure of the large ribosomal subunit from *E. coli*. EMBO J. 6, 1107–1114.

Radermacher, M., Srivastava, S., & Frank, J. (1992). Abstract 19, European EM Conference, Granada.

Roberts, M.W. & Rabinowitz, J.C. (1989). The effect of *Escherichia coli* ribosomal protein S1 on the translational specificity of bacterial ribosomes. J. Biol. Chem. 264, 2228–2235.

Robertus, J., Ladner, J., Finch, J., Rhodes, D., Brown, R., Clark, B., & Klug, A. (1974). Structure of yeast phenylalanine tRNA at 3 Å resolution. Nature 250, 546–551.

Romaniuk, P.J., Lowary, P., Wu, H., Stormo, G., & Uhlenbeck, O. (1987). RNA binding site of R17 coat protein. Biochemistry 26, 1563–1568.

Ryabova, L.A., Selivanova, O.M., Baranov, V.I., Vasiliev, V.D., & Spirin, A.S. (1988). Does the channel for nascent peptide exist inside the ribosome? FEBS Lett. 226, 255–260.

Siegel, V., & Walter, P. (1988). The affinity of signal recognition particle for presecretory proteins is dependent on nascent chain length. EMBO J. 7, 1769–1775.

Silverston, E.W., Navia, M.A., & Davies, D.R. (1977). Three-dimensional structure of an intact human immunoglobulin. Proc. Natl. Acad. Sci. USA 74, 5140–5144.

Smith, W.P., Tai, P.-C., & Davis, B.D. (1978). Nascent peptide as sole attachment of polysomes to membranes in bacteria. Proc. Natl. Acad. Sci. USA 75, 814–817.

Sörbo, B.H. (1953). Crystalline rhodanese. Purification and physicochemical examination. Acta Chem. Scand. 7, 1129–1136.

Spirin, A.S., Baranov, V.I., Rybova, L.A., & Ovodov, S.Y. (1988). A continuous cell-free translation system capable of producing polypeptides in high yield. Science 242, 1162–1164.

Spirin, A.S., Lim, V.J., & Brimacombe, R. (1993). The arrangement of tRNA in the ribosome. In: The Translational Apparatus (Nierhaus, K.H., Subramanian, A.R., Erdmann, V.A., Franceschi, F., & Wittmann-Liebold, B., Eds.), pp. 445–454. Plenum Press, New York.

Stöffler, G. & Stöffler-Meilicke, M. (1986). Immuno electron microscopy on *Escherichia coli* ribosomes. In: Structure, Function and Genetics of Ribosomes (Hardesty, B. & Kramer, G., Eds.), pp. 28–46. Springer-Verlag, New York.

Stryer, L. & Hoagland, R.P. (1967). Energy transfer: a spectroscopic ruler. Proc. Natl. Acad. Sci. USA 58, 719–726.

Sussman, J.L., Holbrook, S.R., Warrent, W.R., Church, G.M., & Kim, S.H. (1978). Crystal structure of yeast phenylalanine transfer RNA I. Crystallographic refinement. J. Mol. Biol. 123, 607–613.

Tandon, S. & Horowitz, P. (1986). Detergent-assisted refolding of guanidinium chloride-denatured rhodanese. J. Biol. Chem. 261, 15615–15618.

Taressova, N.B., Jacovleva, G.M., Victorova, L.S., Kukhanova, M.K., & Khomutov, R.M. (1981). FEBS Letters 130, 85–87.

Tsalkova, T., Zardeneta, G., Kudlicki, W., Kramer, G., Horowitz, P., & Hardesty, B. (1993). GroEL and GroES increase the specific enzymatic activity of newly-synthesized rhodanese if present during *in vitro* transcription/translation. Biochemistry 32, 3377–3380.

Valegard, K., Liljas, L., Fridborg., K., & Unge, T. (1990). The three-dimensional structure of the bacterial virus MS2. Nature 345, 36–41.

Victorova, L.S., Kotusov, L.S., Azhayev, A.V., Krayevsky, A.A., Kukhanova, M.K., & Gottikh, B.P. (1976). Synthesis of thioamide bond catalyzed by *E. coli* ribosomes. FEBS Letters 68, 215–218.

Vogel, R.H. (1983). Three-dimensional reconstruction from electron micrographs of disordered specimens. EMBL Annual Report, pp. 23.

Walter, P. & Blobel, G. (1981). Translocation of proteins across the endoplasmic reticulum III. Signal recognition protein (SRP) causes signal sequence-dependent and site-specific arrest of chain elongation that is released by microsomal membranes. J. Cell Biol. 91, 557–561.

Watson, J.D. (1964). The synthesis of proteins upon ribosomes. Bull. Soc. Chem. Biol. 46, 1399–1425.

Weissbach, H., Zarucki-Schulz, T., Kung, H.-F., Spears, C., Redfield, B., Caldwell, P., & Brot, N. (1981). Use of DNA-directed *in vitro* systems to study bacterial gene expression. In: Molecular Approaches to Gene Expression and Protein Structure (Siddigui, M., Krauskopf, M., & Weissbach, H., Eds.), pp. 215–243. Academic Press, New York.

Wolin. S.L. & Walter, P. (1989). Signal recognition particle mediates a transient elongation arrest of preprolactin in reticulocyte lysate. J. Cell Biol. 109, 2617–2622.

Yonath, A., Leonard, K.R., & Wittmann, H.G. (1987a). A tunnel in the large ribosomal subunit revealed by three-dimensional image reconstruction. Science 236, 813–816.

Yonath, A., Leonard, K.R., Weinstein, S., & Wittmann, H.G. (1987b). Approaches to the determination of the three-dimensional architecture of ribosomal particles. In: Cold Spring Harbor Symposium on Quantum Molecular Biology (Watson, J., Ed.), pp. 729–742.

Yonath, A. & Wittmann, H.G. (1989). Challenging the three-dimensional structure of ribosomes. TIBS 14, 329–335.

Yonath, A., Bennett, W.S., Weinstein, W., & Wittmann, H.G. (1990). Crystallography and image reconstruction of ribosomes. In: Ribosomes: Structure, Function and Evolution. (Hill, E.W., Dahlberg, A., Garrett, R.A., Moore, P.B., Schlesinger, D., & Warner, J.R., Eds.), pp. 134–147. Washington, DC.

Yonath, A. & Berkovitch-Yellin, Z. (1993). Hollows, voids, gaps and tunnels in the ribosome. Current Opin. Struct. Biol. 3, 175–181.

Zardeneta, G. & Horowitz, P.M. (1992). Micelle-assisted Protein Folding. J. Biol. Chem. 267, 5811–5816.

Zipser, D. & Perrin, D. (1963). Complementation on ribosomes. Cold Spring Harbor Symp. Quant. Biol. 28, 533–537.

Zubay, G.(1973). *In vitro* synthesis of protein in microbial systems. Annu. Rev. Genet. 7, 267–287.

PROTEIN IMPORT INTO MITOCHONDRIA

Martin Horst and Nafsika G. Kronidou

Membrane Protein Transport
Volume 1, pages 109–143.
ISBN: 1-55938-907-9

ABSTRACT

Most mitochondrial proteins are encoded by the nucleus and synthesized in the cytoplasm. Protein sorting within mitochondria depends on the precise coordination of several components of the import machinery. In this review we will discuss how precursor proteins are targeted to mitochondria, the energetics of import, intramitochondrial sorting, the requirement of cytosolic and matrix localized chaperones, as well as the import channels themselves.

I. INTRODUCTION

Mitochondria, the power stations of eukaryotic cells, were first described by Altmann in 1890 (for historic details see: Slonimski, 1953; Roodyn and Wilkie, 1968; Sager, 1972; Ernster and Schatz, 1981). In a metabolically active cell, mitochondria may contain up to 20% of the intracellular protein (for review see: Tzagaloff, 1982). Mitochondria are enveloped by two membranes which surround two aqueous compartments: the intermembrane space and the matrix space. Each of these four compartments has a unique protein composition and a particular set of functions. Although mitochondria have their own DNA and have probably evolved from endosymbionts (Pon and Schatz, 1991), they are bona fide organelles which communicate with the nucleus (Glover and Lindsay, 1992). The mitochondrial genome encodes most or all mitochondrial RNAs and a few proteins of the oxidative phosphorylation machinery. In humans the mitochondrial genome encodes only 13 proteins, all of them components of the inner membrane (Attardi and Schatz, 1988). Most mitochondrial proteins are therefore encoded by the nuclear genome.

Nuclear-encoded mitochondrial proteins are synthesized on cytoplasmic ribosomes as precursor proteins, and must be co- or posttranslationally imported and sorted to their final mitochondrial destination. The precise mechanism by which precursor proteins are recognized by the mitochondrial import machinery is still unclear. It is only known that a precursor binds to one or more import receptors on the mitochondrial surface before being delivered to the translocation channel (Glick and Schatz, 1991). Translocation of a precursor into the matrix requires an unfolded or loosely folded conformation of the precursor (Eilers and Schatz, 1986), the presence of an electrochemical potential across the mitochondrial inner membrane (Gasser et al., 1982; Schleyer et al., 1982), and ATP in the matrix (Hwang and Schatz, 1989). Translocation occurs through proteinaceous channels in the outer and inner membranes (Ohba and Schatz, 1987a,b). These two channels can reversibly associate, and each channel is independently capable of translocating proteins (Glick and Schatz, 1991; Pfanner et al., 1992; Horst et al., 1993b; Mayer et al., 1993). The information for targeting a precursor to mitochondria and for intramitochondrial sorting is often located in a transient N-terminal presequence (Hurt and van Loon, 1986). Following import, this presequence is removed by one or more specific processing proteases. A matrix-localized chaperone is believed to generate the "pulling force" for translocating precursor proteins into the matrix (Kang et al., 1990; Manning-Krieg et al., 1991). Finally, another class of chaperones seems to be involved in the folding and assembly of imported proteins into oligomeric complexes (Cheng et al., 1989).

The last decade has witnessed rapid progress in unraveling the mechanisms by which proteins traverse biological membranes and in characterizing the components that are involved in this complex process. The purpose of this article is to summarize the latest results in the field of protein import into mitochondria. We shall focus on the identification of new components of the mitochondrial import machinery and on the molecular mechanisms of targeting.

II. SIGNALS FOR MITOCHONDRIAL IMPORT

Proteins destined for a cellular subcompartment other than the cytoplasm must contain targeting information. This targeting information can reside in a sugar side chain of a polypeptide, as in lysosomal enzymes (von Figura and Hasilik, 1986; Kornfeld and Mellman, 1989); in a short C-terminal amino acid sequence; as in proteins that are retained in the ER (Pelham, 1989) or targeted to peroxisomes (van den Bosch et al., 1992); or in an N-terminal "signal" sequence, as in proteins targeted across various eukaryotic or prokaryotic membranes (Verner and Schatz, 1988).

The information for targeting a precursor protein to mitochondria is usually located close to its N-terminus (Hurt and van Loon, 1986; Schatz, 1987; Hartl et al., 1989). In the absence of additional sorting signals, a precursor will follow the default pathway into the matrix. Some precursors contain signals that target them to one of the mitochondrial subcompartments, such as the intermembrane space (Glick et al., 1992a).

Precursors destined for the matrix usually contain positively charged amphiphilic α-helices near their N-termini. These "matrix-targeting presequences" are usually 12–30 amino acids long, but the actual matrix-targeting signal is often much shorter (Verner and Lemire, 1989). If matrix-targeting signals are fused to non-mitochondrial passenger proteins such as mouse dihydrofolate reductase (DHFR), the resulting fusion proteins are imported into the matrix (Hurt et al., 1984a,b, 1985a). Matrix-targeting signals show no significant homology to each other. In general, they contain few if any acidic amino acids, and are rich in basic and hydroxylated amino acids. If random peptide sequences are fused in front of a passenger protein, up to 25% of these constructs are targeted to the matrix (Hurt et al., 1986; Vassarotti et al., 1987; Bibus et al., 1988; Lemire et al., 1989). The targeting efficiency of these random sequences is related to their amphiphilicity (Bedwell et al., 1989; Lemire et al., 1989). Since a great variety of sequences can function as mitochondrial targeting signals, the question arises how targeting specificity is maintained. Most likely evolution has not only selected for the addition of mitochondrial targeting sequences to proteins destined for mitochondria, but also for the absence of such signals in proteins destined for other organelles (Schatz, 1993).

Many matrix-targeting presequences are able to form amphiphilic α-helices if they come into contact with a lipid bilayer (von Heijne, 1986; Roise et al., 1986; for a review see: Hartl et al., 1989; Baker and Schatz, 1991). It is still not clear whether these amphiphilic helices interact with the phospholipid bilayer, with import receptors, or with both. There is experimental evidence supporting each of these possibilities. A chemically synthesized mitochondrial presequence peptide can translocate across a protein-free phospholipid bilayer (Maduke and Roise, 1993). Karslake et al. (1990) proposed a model for the interaction of mitochondrial targeting signals with the mitochondrial outer membrane. Using a slightly modified rat liver aldehyde dehydrogenase presequence which contained two amphiphilic helices joined by a flexible hinge region, they could show that during the initial binding steps the more stable C-terminal helix interacted with the lipid bilayer and formed a membrane anchor. They proposed that lateral diffusion in the bilayer allowed the N-terminal helix to bind to the appropriate receptor. In this model, targeting is a two step event: binding to the mitochondrial outer membrane through an interaction of a part of the presequence with the lipid bilayer, followed by a more specific interaction of another part of the presequence with an import receptor. This two-step targeting mechanism may enhance targeting specificity and explain the low frequency of precursor mistargeting to different organelles.

The ability of targeting sequences to interact with phospholipid bilayers may be one of the reasons why the targeting sequence is proteolytically removed upon import. This may also be the reason that the protease catalyzing the removal of matrix-targeting signals is essential for life (Yaffe and Schatz, 1984). Presequences are probably degraded following cleavage from the precursor protein, as accumulation could disrupt the mitochondrial membranes. Cleavage of the presequences is not necessary for the actual translocation across the membranes, as translocation occurs even if cleavage is blocked (Reid and Schatz, 1982b; Zwizinski and Neupert, 1983; Hurt et al., 1985a; Vassarotti et al., 1987).

It seems that all proteins destined for the mitochondrial matrix use the same import pathway. The transport routes into the other mitochondrial subcompartments are less well-defined. However, some of these pathways seem to share at least parts of the matrix import pathway. For example, matrix-targeting signals are also present on precursors of some mitochondrial inner membrane proteins. These proteins are first transported into the matrix space and then inserted into the inner membrane. Examples for precursor proteins using this pathway include cytochrome oxidase subunit IV, the iron–sulfur protein of cytochrome c reductase and subunit 9 of the F_1F_0-ATPase (Hurt et al., 1984a; Hartl et al., 1986;, Mahlke et al., 1990). The iron–sulfur protein belongs to a class of inner membrane proteins whose presequences are cleaved twice in the matrix by two different enzymes (Hendrick et al.,

1989); these presequences contain only matrix-targeting information. The second cleavage is perhaps necessary for the correct processing and assembly into the heterooligomeric cytochrome *c* reductase complex. The information for the membrane association is most likely localized in the mature part, as with the bacterial homolog of this protein (Hartl and Neupert, 1990).

The ATP/ADP translocator, a homodimeric integral inner membrane protein, is synthesized without a cleavable presequence. This molecule contains internal targeting information (Pfanner et al., 1987a; Smagula and Douglas, 1988). Another example of an unusual inner membrane targeting signal is that of cytochrome oxidase subunit Va. This protein is synthesized with a cleavable matrix targeting signal and with a hydrophobic stretch near the C-terminus that is necessary for insertion into the inner membrane (Miller and Cumsky, 1991). This hydrophobic stretch is thought to act as a stop-transfer signal (Miller and Cumsky, 1993) (see below).

Sorting to the outer membrane appears to utilize the most straightforward pathway: proteins insert directly into the outer membrane without requiring an electrochemical potential across the inner membrane or ATP in the matrix. Nevertheless, the mechanism of insertion is poorly understood. So far only few outer membrane proteins have been investigated in detail: for example porin (Mihara et al., 1982; Freitag et al., 1982; Gasser and Schatz, 1983; Pfaller and Neupert, 1987; Mayer et al., 1993) and MAS70 (Riezman et al., 1983; Hase et al., 1984, 1986). Outer membrane proteins are not synthesized with cleavable presequences. They either contain N-terminal targeting information or internal targeting information (Schatz, 1987; Hartl and Neupert, 1990; Glick and Schatz, 1991; Glover and Lindsay, 1992). The N-terminal 12 amino acids of yeast MAS70 function as a mitochondrial targeting signal (Hurt et al., 1985b; Nakai et al., 1989). Downstream of this sequence there is a stretch of hydrophobic amino acids that may represent a stop-transfer sequence that anchors the protein in the outer membrane. Interestingly, some proteins destined for the intermembrane space also display a hydrophobic stretch immediately following a mitochondrial targeting sequence, yet these precursors do not get stuck across the outer membrane (Kaput et al., 1982; Reid et al., 1982; van Loon and Schatz, 1987; van Loon et al., 1987). The orientation of the hydrophobic stretch in the MAS70 targeting signal and thereby the orientation of an attached passenger protein depend on the positive charge and the length of the upstream sequence (Li and Shore, 1992b). Monoamine oxidase B contains a very unusual sorting signal which is located near its C-terminus (Mitoma and Ito, 1992). Furthermore, *in vitro* insertion of this protein into the outer membrane requires ubiquitin (Zhuang and McCauley, 1989).

Many precursor proteins destined for the intermembrane space, including cytochrome b_2, cytochrome *c* peroxidase, and cytochrome c_1 contain a bipartite N-terminal presequence consisting of a typical mitochondrial targeting sequence followed by a stretch of hydrophobic amino acids (Hurt and van Loon, 1986). Deletion or inactivation of the hydrophobic stretch causes mistargeting to the matrix. Recent results have shown that the sorting signal for cytochrome b_2 is even more complicated (Beasley et al., 1993; Schwarz et al., 1993). It contains at least three distinct regions including a cluster of three basic amino acids directly upstream of the hydrophobic stretch, the hydrophobic stretch itself, and the first residue of the mature protein. Amino acid substitutions in each of these regions cause missorting of cytochrome b_2 into the matrix. Interestingly, the amount of missorting correlates with the nature of the amino acid substitution. Proline substitutions, which are likely to change the conformation of the sorting signal cause more extensive missorting than those which only decrease the overall hydrophobicity. Thus, protein–protein interactions between the precursor and the inner membrane import channel appear to be important for proper sorting to the intermembrane space (Beasley et al., 1993).

Proteins with bipartite presequences are cleaved twice: the N-terminal mitochon-drial-targeting signal is removed by the matrix protease, and the sorting signal is removed by a second protease which is located in the inner membrane facing the intermembrane space (Behrens et al., 1991; Schneider, A. et al., 1991). The hydrophobic stretch acts as a stop-transfer signal for the inner membrane translo-cation machinery (van Loon and Schatz, 1987; Nguyen et al., 1988; Glick et al., 1992b). In an alternative model the precursor is first translocated into the matrix and then reexported to the intermembrane space by a "conservative sorting" pathway (Hartl et al., 1987; Hartl and Neupert, 1990). As will be discussed below, current evidence favors the stop-transfer model.

Other intermembrane space proteins use different sorting mechanisms. Cyto-chrome c has a quite simple import pathway. Cytochrome c is synthesized as a heme-free apoprotein (Stuart et al., 1990) which is amphiphilic and spontaneously inserts across the lipid bilayer of the mitochondrial outer membrane. Covalent attachment of the heme group to apocytochrome c by cytochrome c heme lyase generates the mature cytochrome c. Heme attachment is probably necessary for completing the translocation and keeping cytochrome c in the intermembrane space. Cytochrome c heme lyase itself also lacks a cleavable presequence (Lill et al., 1992).

III. EARLY EVENTS IN IMPORT: CYTOSOLIC CHAPERONES AND MITOCHONDRIAL IMPORT RECEPTORS

It was initially proposed that import of nuclear-encoded proteins into mitochondria occurs cotranslationally, because mitochondria isolated in the presence of cyclo-heximide contain cytoplasmic ribosomes anchored to the mitochondrial outer membrane by nascent chains of mitochondrial precursor proteins (Kellems and Butow, 1972, 1974; Kellems et al., 1974; Ades and Butow, 1980a,b). These bound ribosomes are enriched in so called "contact-sites" where the outer and inner mitochondrial membranes are in close contact. *In vitro* import studies using radiolabeled precursor proteins and isolated mitochondria have shown that import can occur posttranslationally (Maccecchini et al., 1979; Eilers and Schatz, 1986). Nevertheless, import of at least some precursors in a cell-free yeast system is strongly enhanced when translation of the precursor occurs in the presence of mitochondria (Fujiki and Verner, 1991), suggesting that cotranslational import can also take place *in vitro*. Recently, cotranslational import has also been shown in *in vivo* experiments (Fujiki and Verner, 1993). It is still unknown what determines whether import into mitochondria is co- or posttranslational. The ratio of co- to posttranslational import is probably dependent on the relative rates of protein synthesis and import (Reid and Schatz, 1982b; Schatz and Butow, 1983; Smith and Yaffee, 1991). Under normal growth conditions, import is believed to occur mainly posttranslationally (Schatz and Butow, 1983).

It is well established that proteins cannot cross biological membranes in their native, tightly folded conformations. This was first demonstrated with an artificial precursor protein consisting of the presequence of cytochrome oxidase subunit IV fused to mouse dihydrofolate reductase (Eilers and Schatz, 1986). When methotrexate which stabilizes the DHFR domain was added to the precursor prior to import, import was blocked (Eilers and Schatz, 1986). Adriamycin, a drug binding to acidic phospholipids, was also shown to block import of this fusion protein, suggesting that acidic phospholipids in the outer membrane may facilitate unfolding and therefore import (Eilers et al., 1989).

A. Cytosolic Chaperones

The first hint that cytosolic factors are involved in import came from observations that cytosol stimulates import of precursor proteins into isolated mitochondria (Argant et al., 1983; Miura et al., 1983; Ohta and Schatz, 1984; Randall and Shore, 1989). Several proteins in the cytosol are believed to be responsible for keeping precursor proteins in a translocation-competent conformation. These so-called chaperones probably bind to the nascent chain as it emerges from the ribosome and prevent tight folding of the precursor. In addition to keeping precursors in an import-competent conformation, chaperones may also prevent aggregation of these proteins. How chaperones deliver the bound precursor to the mitochondria is not fully understood. However, this step most likely involves ATP hydrolysis (Pelham, 1986). This ATP requirement in the cytosol can be bypassed by destabilizing the precursor: for example, by denaturation (as proposed by: Pfanner et al., 1988; Fujiki and Verner, 1991), or by using mutant precursors that are less tightly folded (Chen and Douglas, 1987a,b). Import of some precursors—for example, apocytochrome *c* (Stuart and Neupert, 1990; Stuart et al., 1990), cytochrome *c* heme lyase (Lill et al., 1992), cytochrome oxidase subunit Va (Miller and Cumsky, 1991) and cytochrome b_2 (Glick et al., 1993)—may not require antifolding proteins.

So far, three proteins which stimulate import into isolated mitochondria have been isolated and characterized. A protein called PBF (presequence binding factor), which is a homooligomer composed of 50-kDa subunits, prevents folding and aggregation of the ornithine transcarbamylase precursor by binding to the presequence (Murakami et al., 1990; Murakami and Mori, 1990). MSF (mitochondrial import stimulation factor), another cytosolic import stimulating factor, recognizes the presequences of mitochondrial precursor proteins and catalyzes the unfolding of *in vitro* synthesized precursor molecules in an ATP-dependent fashion (Hachiya et al., 1993). The active MSF enzyme is a heterooligomeric complex composed of two subunits of 30 and 32 kDa. It was furthermore shown that a precursor–MSF complex was specifically bound to mitochondrial outer membranes (Hachiya et al., 1994) probably via interaction with the MAS37p/MMas70p receptor (Gratzer et al., 1994; Hachiya and Lithgow, in preparation). The molecular mass of the small subunit of MSF is similar to that of another protein that has been described as a targeting factor and that has a similar function as PBF (Ono and Tuboi, 1990a,b). So far it is not clear whether the two proteins are identical. PBF and MSF have similar functions but they are clearly different in their ATP requirements. The ATP-independent PBF is thought to cooperate with cytoplasmic hsp70 (see below) in keeping precursor proteins in an import-competent conformation, whereas MSF may represent the NEM-sensitive factor that is involved in stimulation of mitochondrial import by hsp70 (Murakami et al., 1988). MSF on the other hand hydrolyzes ATP and seems to be a bona fide ATP-dependent unfoldase (Rothman and Kornberg, 1986; Pfanner et al., 1987b). A yeast homolog of the bacterial protein DnaJ, Ydj1p [yeast DnaJ homolog; also called MAS5p; (Caplan et al., 1992a; Atencio and Yaffe, 1992)] is also involved in early events of mitochondrial import. This protein stimulates the ATPase activity of hsp70 and facilitates the dissociation of hsp70–precursor complexes (Cyr et al., 1992). Interestingly the C-terminus of YDJ1p is farnesylated (Caplan et al., 1992b), suggesting a possible attachment to the outer membrane.

Another group of chaperones involved in import of precursor proteins into mitochondria are the 70-kDa heat-shock proteins (hsp70s; Deshaies et al., 1988a,b; Chirico et al., 1988; Murakami et al., 1988). This group of proteins are the eukaryotic homologues of the prokaryotic DnaK chaperones (Hemmingsen et al., 1988). Hsp70s participate in different aspects of intracellular transport and protein folding (Chirico et al., 1988) and are constitutively expressed in all eukaryotic cells. The expression of some members of this family is dramatically increased during stress. The *S. cerevisiae* genome encodes nine different hsp70s. Four of them (SSA1p, SSA2p, SSA3p, and SSA4p) function in maintaining precursor proteins in a translocation-competent conformation in the cytoplasm (for a review see: Lindquist and Craig, 1988). Depletion of three of them leads to an accumulation of precursor proteins in the cytoplasm and eventually causes cell death (Deshaies et al., 1988a,b). Although these chaperones are collectively essential, depletion of any one of them is not lethal, presumably because of functional redundancy and overlapping specificities of the four proteins. Release of bound precursors probably requires ATP hydrolysis by hsp70 (Pelham, 1986; Rothman and Schmidt, 1986). So far, however, this has not been shown directly.

It is not clear whether cytosolic chaperones distinguish between newly synthesized cytosolic proteins and precursors destined for transport into a cellular subcompartment such as mitochondria. If they do, they may recognize the targeting signal directly (Park et al., 1988; Murakami et al., 1990; Murakami and Mori, 1990; Schmid et al., 1992). Alternatively, the presequences may slow down folding of the precursor and thereby enhance the probability of binding to the chaperones. The precursors are then delivered to the translocation machinery (Schatz, 1993). It is still open whether the antifolding proteins or one of their proposed associated proteins actively function in targeting the bound precursor to the subcellular compartment or only in keeping the bound precursor in a translocation-competent state. Specialized receptors on the surface of the different cellular subcompartments could then recognize not only the precursor itself, but the precursor–chaperone complex. Current evidence suggests that the cytoplasm contains different chaperones and that different precursors may use different strategies to maintain their translocation-competent state.

B. Mitochondrial Import Receptors

Mitochondria contain several proteins on their outer membrane that facilitate import of cytoplasmically synthesized precursor proteins (for a review see: Pfanner et al., 1991). The first hints of the existence of these import receptors came from experiments showing that import into isolated mitochondria is inhibited when mitochondria are pretreated either with proteases (Riezman et al., 1983a; Zwizinski et al., 1984) or with antibodies against total outer membrane proteins (Ohba and Schatz, 1987b). Different degrees of import inhibition for various precursors after protease treatment of mitochondria led to the hypothesis that there are several import receptors with overlapping specificities (Zwizinski et al., 1984; Pfaller et al., 1988).

So far, two receptors have been identified: Mas70p (70-kDa mitochondrial assembly protein; Hines et al., 1993), and Mas20p (Ramage et al., 1993), and Mas37p (Gratzer et al., 1994) in the yeast *S. cerevisiae*; the homologous proteins in *N. crassa* are MOM72 (72-kDa mitochondrial outer membrane protein; Söllner et al., 1990) and MOM19 (Söllner et al., 1989). Another *N. crassa* gene MOM22 has been reported to function at a step after binding of the precursor to the receptors (Kiebler et al., 1994). MOM72/Mas70p was originally proposed to be involved only in the import of the ADP/ATP translocator (Söllner et al., 1990), whereas MOM19/Mas20p was thought to be the "master receptor" (Schneider et al., 1991). Evidence from our laboratory has, however, suggested that these two receptors have overlapping specificities at least in yeast (Hines et al., 1990; Ramage et al., 1993). Pulse-chase studies with MAS70-deficient strains showed that the import efficiency of several mitochondrial precursors was strongly reduced. Moreover, *in vitro* experiments with αMas70p antibodies and with mitochondria isolated from a MAS70-deficient yeast strain have shown that import of most precursors tested is accelerated by Mas70p (Hines and Schatz, 1993). However, Mas70p does not accelerate the import of artificial precursor proteins containing a mitochondrial presequence fused to DHFR. This may suggest that the presequence is not the only determinant for productive binding to the Mas70p receptor (Hines and Schatz, 1993).

Mas37p is the third receptor in yeast. Antibodies against Mas37p or disruption of the gene inhibit binding and import of the same subset of precursor proteins that are affected by the inactivation of Mas70p. Biochemical and genetic studies could show that Mas37p and Mas70p form a heterooligomeric receptor complex (Gratzer et al., 1994).

MOM19 and Mas20p were identified using similar approaches: antibodies against various outer membrane proteins were screened for their ability to inhibit import of a mitochondrial precursor into isolated mitochondria (Söllner et al., 1990; Ramage et al., 1993). *In vitro* studies have shown that MOM19 specifically interacts with the presequence of several precursors (Becker et al., 1992). Nevertheless, inactivation of MOM19 or Mas20p with specific antibodies does not completely block import into isolated mitochondria and deletion of MAS20 from yeast cells only partially inhibits import of precursor proteins. However, deletion of MAS20 renders the yeast cells respiratory-deficient (Ramage et al., 1993). Surprisingly the cells recover from these defects after some days and grow as wild-type cells do. This is probably due to an overexpression of a putative partner protein of Mas20p which can restore protein import to cells without Mas20p (Lithgow et al., 1994). As expected, a deletion of both MAS70 and MAS20 is lethal (Ramage et al., 1993). The best evidence that the two receptors are functionally redundant came from experiments which showed that overexpression of Mas70p can suppress the growth defect caused by a MAS20 deletion (Ramage et al., 1993).

Experiments using anti-idiotypic antibodies against a purified presequence peptide have been interpreted as evidence for another mitochondrial protein import receptor in yeast (Pain et al., 1990). One of the proteins recognized by these antibodies is a 32 kDa integral membrane protein called p32. Polyclonal antibodies against p32 inhibit import of some precursor proteins into mitochondria. Furthermore, precursor proteins were found in a complex with p32 in the outer membrane. p32 is not essential for cell viability but disruption of the corresponding gene causes respiratory deficiency (Murakami et al., 1990). It has now been shown that this protein is identical to the mitochondrial phosphate carrier protein, a protein located in the mitochondrial inner membrane (Phelps et al., 1991). It is presently unclear whether this protein has a dual function or whether its role in mitochondrial import must be reevaluated (Meyer, 1990).

IV. PROTEIN TRANSLOCATION CHANNELS IN MITOCHONDRIA

Once a precursor has bound to a mitochondrial outer membrane receptor it has to be delivered to the import site. This transfer most likely occurs through lateral diffusion of the receptor with the bound precursor to the import site where it is delivered to the translocation machinery. If import of a precursor into mitochondria is arrested—for example, by lowering the temperature (Schleyer and Neupert, 1985), or by using precursor proteins whose C-terminal domain cannot be unfolded (Schwaiger et al., 1987; Vestweber and Schatz, 1988)—the precursor protein remains stuck in the import channel. Such stuck precursors are found exclusively in contact sites. The N-terminus of the stuck precursor faces the matrix while the C-terminus is still in the cytosol. Such import intermediates can be extracted from the import sites by salt, urea, or high pH, suggesting that they are in a proteinaceous environment and are not simply stuck in the lipid bilayer (Pfanner et al., 1987c; Sztul et al., 1989). It seems that many precursors imported into mitochondria utilize a common import site because a stuck precursor can block import of several other precursors (Vestweber and Schatz, 1988). These general protein import channels are most likely flexible structures, which allow passage of proteins while at the same time preventing leakage of ions across the inner membrane. The largest structure that could be imported through these channels in experimental tests was a stretch of double-stranded DNA (diameter = 20 Å) that was chemically linked to a precursor protein (Vestweber and Schatz, 1989).

The first component of the outer mitochondrial membrane translocation machinery was detected by photo-cross-linking it to a precursor stuck in the import channel (Vestweber et al., 1989). This protein, termed Isp42p (42-kDa import site protein), lacks both an obvious mitochondrial targeting signal and typical transmembrane domains (Baker et al., 1990). Nevertheless, Isp42p is an integral membrane protein of the mitochondrial outer membrane whose depletion leads to an accumulation of precursor proteins in the cytosol and to cell death. If Isp42p is overexpressed in yeast, the protein fails to assemble correctly in the outer membrane, suggesting the existence of partner proteins. Isp42p is most likely a member of a heterooligomeric outer membrane complex (Baker et al., 1990). In *N. crassa,* the Isp42p homolog, MOM38 (38-kDa mitochondrial outer membrane protein) (Kiebler et al., 1990) is found in a complex containing at least six other components that have been characterized primarily by coimmunoprecipitation and cross-linking experiments (Söllner et al., 1992). These components are the receptors MOM72 and MOM19, and the GIP- (General Insertion Pore) forming proteins MOM30, MOM7, and MOM8 (Söllner et al., 1992). Yet another protein, MOM22, is probably involved in the transfer of the precursors from the receptors to GIP (Kiebler et al., 1993). Furthermore a complex with a very similar protein composition (Mas70p, Mas20p, Isp42p, and homologs of MOM30, MOM22, MOM8 and MOM7) was isolated in *S. cerevisiae* (Moczko et al., 1992). It should now be possible to investigate the function and the interaction of the different components using yeast molecular biology techniques such as gene disruptions and synthetic lethality screens. Recently another component, Isp6p (6-kDa import site protein), was found to be an integral membrane protein of the outer membrane translocation machinery in *S. cerevisiae*. The ISP6 gene was isolated as a high-copy suppressor of a temperature-sensitive allele of Isp42p (Kassenbrock, 1993). Isp6p can be coimmunoprecipitated with Isp42p, indicating a close association of these two proteins.

There is now general agreement that the mitochondrial inner membrane contains a functional translocation channel that can act independently from the one in the outer membrane (Glick and Schatz, 1991; Pfanner et al., 1992; Horst et al., 1993b). This "dynamic model" of import will be discussed in Section VIII. A combination of approaches has now identified the first two components of the inner membrane system in yeast. Our laboratory identified one of these proteins using two complementary approaches (Scherer et al., 1992). In one approach, we cross-linked putative subunits of the inner membrane transport system to a radioactive precursor protein that had been arrested during translocation through the inner membrane (the "ATP-depletion intermediate" mentioned earlier). Cross-links were obtained to four mitochondrial polypeptides of apparent molecular masses of 55, 45, 20, and 5 kDa. As the radiolabeled cross-linked product containing the 45-kDa protein could be immunoprecipitated with a complex antiserum, we could purify the 45-kDa protein on the basis of its ability to quench the immunoprecipitation. The second approach used anti-inner membrane antisera that inhibited protein import into inner membrane vesicles, but not into intact mitochondria. The antigens responsible for the inhibition were then identified by affinity-purifying the antibodies against different subpopulations of mitochondrial proteins. Both approaches identified the same 45-kDa inner membrane protein, which we termed Isp45p (45-kDa import site protein; Scherer et al., 1992). When we purified the protein to homogeneity and determined its partial amino acid sequence (Horst et al., 1993a), it proved to be the product of the recently described *MPI1* gene (Maarse et al., 1992). This gene had been discovered by an elegant genetic screen which selected for yeast mutants that were partially defective in the import of a matrix-targeted fusion protein. *MPI1* encodes a 48.8-kDa membrane protein with a cleavable presequence. Mpi1p is essential for viability, and its depletion leads to the accumulation of uncleaved mitochondrial precursor proteins in the cytoplasm. Taken together, the genetic and biochemical results clearly show that Isp45p is a component of the protein transport system of the inner membrane.

Another component of the inner membrane translocation system in yeast is encoded by the *MAS6* gene. This gene was originally isolated as a temperature-sensitive allele causing accumulation of mitochondrial precursor proteins in the cytosol (Yaffe and Schatz, 1984). Mas6p is a 23-kDa integral inner membrane protein that is essential for life (Emtage and Jensen, 1993). The same gene was independently isolated by Dekker and co-workers as one of the complementation groups (*MPI3*) of the genetic screen that also yield *MPI1* (Dekker et al., 1993).

Antibodies raised against this protein block import of various precursor proteins into mitoplasts but not into intact mitochondria. Mas6p is also the 20-kDa protein cross-linked to the already described (Scherer et al., 1992) ATP-depletion intermediate (Ryan and Jensen, 1993; Kübrich et al., 1994).

The third known component of the mitochondrial inner membrane translocation machinery is encoded by the essential *MP12* gene (Maarse et al., 1994). The gene was independently isolated as a high copy number suppressor of *MAS6* called *SMS1* (Ryan et al., 1994). The gene encodes a 16.5-kDa integral inner membrane protein which is homologous to the C-terminal domain of Mas6p/MIM23. Although this has not been directly shown, Mas6p and MIM17 are probably subunits of a heterooligomeric complex that also includes Isp45p and other as yet unknown proteins (Kübrich et al., 1994).

Genetic and biochemical evidence implicate mhsp70 as a component of the inner membrane import machinery. Mitochondrial hsp70 is an essential protein (Craig et al., 1987). *In vitro* import studies of mitochondria isolated from a temperature-sensitive mhsp70 mutant show compromised protein import and an accumulation of precursor proteins at contact sites (Kang et al., 1990). Furthermore, mhsp70 can be cross-linked to or coimmunoprecipitated with precursor proteins on their way to the matrix (Kang et al., 1990; Scherer et al., 1990). Subsequent experiments have shown that mhsp70 forms a transient, ATP-dependent interaction with newly imported precursor proteins (Manning-Krieg et al., 1991). These results have led to the hypothesis that mhsp70 acts as an ATP-dependent motor which pulls precursor proteins into the matrix. This model is consistent with the observation that ATP-depletion in the matrix arrests translocation of precursor proteins across the inner membrane (Hwang and Schatz, 1989; Hwang et al., 1991). In addition , it has recently been shown that a temperature-sensitive mhsp70 allele with a mutated ATPase domain fails to bind and complete import of a partially translocated precursor protein (Gambill et al., 1993). Interestingly, import and assembly of precursors into the yeast endoplasmic reticulum also requires a luminal hsp70 homolog, BiP (Munro and Pelham, 1986; Vogel et al., 1990; Sanders et al., 1992). How hsp70 mediates translocation in either system is not yet understood. In one possible mechanism, mhsp70 exerts force on the translocating polypeptide and pulls the protein into the matrix by virtue of being anchored to the inner membrane translocation machinery. Consistent with this model is the observation that mhsp70 and its partner protein GrpEp (Bollinger et al., 1994) dynamically interact with Isp45p/Mpi1p (Kronidou et al., 1994). In an alternative model for mhsp70 action, the so-called "Brownian ratchet" model (Neupert et al., 1990; Simon et al., 1992), the matrix-targeting presequence of a mitochondrial precursor would insert across the inner membrane and the matrix-targeting signal would be cleaved. In the absence of the matrix-targeting signal the precursor could diffuse within the import channel in either direction. However, the binding of a mhsp70 molecule to the precursor would prevent diffusion out of the import channel and ensure that movement is unidirectional. Binding of additional mhsp70 molecules to the precursor could then pull the precursor completely into the matrix. According to this model, no membrane-associated mhsp70 would be required. Obviously, more experiments are needed to determine the mechanism of mhsp70 action.

V. PROCESSING AND ASSEMBLY OF MITOCHONDRIAL PRECURSOR PROTEINS IN THE MATRIX

At least four mitochondrial matrix proteins have been shown to be involved in the maturation of an incoming precursor and its subsequent assembly: the Mas1p and Mas2p subunits of the matrix protease (Yang et al., 1988), hsp60 (McMullin and Hallberg, 1987, 1988; Cheng et al., 1989; Reading et al., 1989), and the DnaK-like chaperone mhsp70 (Kang et al., 1990; Scherer et al., 1990). Each of these four proteins is essential for life.

Once the N-terminus of a matrix-targeted protein crosses the inner membrane it transiently binds to mhsp70 (mitochondrial heat-shock protein of 70 kDa). This was first demonstrated by cross-linking a stuck precursor on its way into the matrix to mhsp70 followed by coimmunoprecipitation experiments (Ostermann et al., 1990; Scherer et al., 1990). The rate of association of a precursor with mhsp70 correlates with the import rate, suggesting that mhsp70 directly catalyzes import (Manning-Krieg et al., 1991). Mitochondrial hsp70 is thought to "pull" the precursor into the matrix by multiple cycles of binding and release. Another model favors the binding of more than one molecule of mhsp70 per precursor molecule which would step-wise pull the precursor into the matrix (Neupert et al., 1990). Since mhsp70 is believed to require ATP hydrolysis for its function, this could explain why import of precursors into the matrix always requires ATP in the matrix (Hwang and Schatz, 1989; Wachter et al., 1992).

Mitochondrial hsp70 is encoded by the nuclear SSC1 gene (Craig et al., 1987), and is synthesized with a cleavable matrix-targeting presequence. Import of mhsp70 into the matrix is apparently autocatalyzed. Incubation of isolated mitochondria containing a temperature-sensitive mhsp70 protein at the nonpermissive temperature leads to accumulation of precursor proteins stuck in the import channel. The resulting inhibition of protein import is probably the reason that cells containing this temperature-sensitive allele are inviable at high temperatures (Kang et al., 1990). The *E. coli* homolog of hsp70, DnaK, functions in DNA replication, raising the question whether the mitochondrial protein performs a similar function. DnaK associates with another protein called DnaJ and grpE which modulates the activity of DnaK (Georgopoulos, 1992). Recently DnaJ and grpE homologs have also been found in mitochondria (Rowley et al., 1994; Bolliger et al., 1994).

Precursors that have been translocated into the matrix are released from mhsp70 in an incompletely folded form (Manning-Krieg et al., 1991). Some imported proteins fold and assemble with the assistance of the ATP-dependant chaperone hsp60 (Cheng et al., 1989; Reading et al., 1989). A sequential action very similar to the one of mhsp70 and hsp60 has been demonstrated in an *in vitro* system with the *E. coli* homologs (Langer et al., 1992a,b). However, it is not yet clear whether all proteins need hsp60 for folding (Cheng et al., 1989; Ostermann et al., 1990; Manning-Krieg et al., 1991).

Mitochondrial hsp60 is a homolog of groEL from *E. coli* (McMullin and Hallberg, 1988) and of the Rubisco (ribulose-bisphosphate-carboxylase) binding protein of chloroplasts (Hemmingsen et al., 1988). The hsp60 proteins form homooligomeric complexes consisting of 14 subunits arranged into two 7mer rings (Cheng et al., 1989; Prasad et al., 1990). Yeast strains with hsp60 temperature sensitive mutations are inviable at the nonpermissive temperature. They are still able to import cytoplasmically-made precursors at the nonpermissive temperature, but they fail to assemble these proteins into functional complexes (Cheng et al., 1989; Glick et al., 1992b). In *E. coli*, chloroplasts, and mitochondria, hsp60 has a small partner protein of about 10 kDa, termed groES in *E. coli* (Chandrasekhar et al., 1986) and cpn10 in mitochondria and chloroplasts (Lubben et al., 1989; Rospert et al., 1993). The function of this mitochondrial co-chaperonin is currently under investigation. The nuclear-encoded hsp60 protein is synthesized with a cleavable matrix-targeting signal, and needs both mhsp70 and preexisting hsp60 for its import and assembly (Cheng et al., 1990; Manning-Krieg et al., 1991).

The N-terminal matrix-targeting signal of most precursors is removed by a matrix-localized metallo-endoprotease. This enzyme is sensitive to divalent metal chelators such as EDTA. So far no consensus cleavage site has been found. It is possible that the enzyme recognizes a three-dimensional structure such as an amphiphilic helix which is a structure found in many mitochondrial presequences.

The protease has been purified to homogeneity from several different organisms including *N. crassa* (Hawlitschek et al., 1988), *S. cerevisiae* (Yang et al., 1988), rat liver (Ou et al., 1989; Kleiber et al., 1990), and potato (Braun et al., 1992). It consists of two nonidentical but very similar nuclear-encoded subunits which probably arose from a common ancestor. These subunits are termed Mas1p/Mas2p (mitochondrial assembly protein) in *S. cerevisiae*, PEP (processing enhancing protein)/MPP (matrix processing peptidase) in *N. crassa*, and p52/p55 in rat liver. Interestingly, one subunit in *N. crassa* (PEP) (Schulte et al., 1989) and at least one of the subunits in potato (MPP) (Braun et al., 1992) is associated with cytochrome *c* reductase. In fact, PEP is identical to subunit I of the cytochrome *c* reductase complex in *N. crassa* (Schulte et al., 1989) and the MPP-like subunit of the potato enzyme is identical to subunit III of the cytochrome bc_1 complex (Braun et al., 1992). In yeast, both subunits of the matrix protease are synthesized with N-terminal matrix-targeting sequences which are cleaved upon import showing that the protease is required for processing of its own subunits. Neither of these subunits is by itself active (Geli et al., 1990). The subunits are now called MPPα and MPPβ (instead of Mas2p/MPP and Mas1p/PEP; Kalousek et al., 1993).

As mentioned earlier, cleavage of the presequence is not necessary for import of precursors, but inactivation or depletion of the protease *in vivo* is lethal. In this case import into mitochondria is completely inhibited, either because the uncleaved precursors block the import sites (Yaffe and Schatz, 1984; Yaffe et al., 1985) or because the amphiphilic presequences disrupt the inner membrane. Some proteins of the intermembrane space are proteolytically processed in two steps: the first cleavage occurs in the matrix and is mediated by the matrix protease, and the second cleavage occurs in the intermembrane space. Genetic evidence suggests that there are at least two different proteases in the intermembrane space. Two independent mutants which are defective in the maturation either of cytochrome b_2 and the mitochondrially encoded cytochrome oxidase subunit II or the maturation of cytochrome $c1$ and cytochrome c peroxidase have been isolated (Pratje and Guiard, 1986). The enzyme involved in the cleavage of cytochrome $b2$ and cytochrome oxidase subunit II has been isolated and called Imp (Inner membrane protease; Schneider, A. et al., 1991). Imp is a heterodimer located in the inner membrane. Subunit 1 is a integral membrane protein of 21.4 kDa. The sequence of subunit 1 reveals some homology to the *E. coli* leader peptidase (Behrens et al., 1991). The gene for subunit 2 was recently cloned (Nunnari et al., 1993) and the heterodimer was purified (Schneider et al., 1994). Subunit 2, which carries the cleavage activity for cytochrome c_1 and cytochrome c peroxidase is termed Imp2. Mutations in Imp2 can inactivate both Imp1 and Imp2, apparently because they block assembly of the heterodimeric Imp1/Imp2 complex (Nunnari et al., 1993).

Some precursors are cleaved twice in the matrix (Sztul et al., 1987, 1989). The matrix-targeting signal is removed by the general matrix protease (MPPα/β), whereas the second cleavage is carried out by the mitochondrial intermediate peptidase (MIP; Kalousek et al., 1988, 1992). MIP specifically cleaves off an octapeptide from the once-cleaved precursor and generates the mature protein. MIP has recently been purified from rat liver (Isaya et al., 1992; Kalousek et al., 1992). It is a 75-kDa protein showing an unusual dependence on cations, suggesting that MIP belongs to a new family of peptidases. The mature N-termini of these twice cleaved precursors are suggested to be structurally incompatible with cleavage by MPP (Isaya et al., 1991).

VI. PROTEIN SORTING TO THE MITOCHONDRIAL SUBCOMPARTMENTS

Most mitochondrial proteins are transported to the matrix. As discussed above, the import pathway for such proteins is fairly well established. Less is known, however, about sorting to the other submitochondrial compartments. Here we try to summarize briefly what is known about sorting to the four submitochondrial compartments. For a more detailed discussion see: Glick and Schatz, 1991; Glick et al., 1992a; Segui-Real et al., 1992.

A. Sorting to the Matrix

Proteins destined for the matrix are usually synthesized with an N-terminal matrix-targeting signal. As discussed earlier, the precursors are kept in an import competent conformation by cytoplasmic chaperones until they bind to the outer membrane import receptors. These receptors deliver their ligands to the actual translocation machinery. Chaperones inside the matrix probably pull the precursor inside and import into the matrix is therefore ATP-dependent. Once in the matrix the presequence is removed and the proteins are folded and assembled, most of them with the help of chaperones (for a review see: Glick and Schatz, 1991).

B. Sorting to the Mitochondrial Outer Membrane

Targeting to the mitochondrial outer membrane is probably the least understood aspect of protein sorting in mitochondria. While no uniform import pathway has yet been identified, all proteins targeted to the outer membrane lack a cleavable mitochondrial presequence and do not require an electrochemical potential across the inner membrane for import (Freitag et al., 1982; Mihara et al., 1982; Gasser and Schatz, 1983). Mas70p has a bipartite N-terminal signal sequence composed of a matrix targeting signal followed by a hydrophobic stretch. Mutations in the hydrophobic stretch cause missorting to the matrix. Thus, at least some outer membrane proteins use parts of the common matrix targeting pathway (Hase et al., 1984; Pfaller and Neupert, 1987).

Porin, the 29-kDa subunit of the mitochondrial outer membrane pore VDAC (Voltage Dependant Anion Channel), also uses parts of the common matrix import pathway (Pfaller et al., 1988). On the other hand, insertion of the outer membrane import receptor MOM19 seems to be independent of the outer membrane receptor MOM72 or of preexisting MOM19 molecules, but requires MOM38 and probably the other subunits of GIP (Schneider, H. et al., 1991). The sorting pathways for Isp42p/MOM38 may be similar to that of porin since all these proteins lack a cleavable N-terminal targeting sequence and obvious transmembrane domains (Baker et al., 1990).

C. Sorting to the Mitochondrial Inner Membrane

Much of our knowledge on sorting to the mitochondrial inner membrane is derived from studies with the ADP/ATP translocator. The translocator is an integral inner membrane homodimer. It uses the common matrix import pathway, including the outer membrane receptor MOM72/MAS70 (Söllner et al., 1990; Hines and Baker, 1991), the putative channel component MOM38/Isp42p (Pfaller et al., 1988; Vestweber and Schatz, 1990), and most likely the other components of GIP as well as mhsp70 (Ostermann et al., 1990; Scherer and Krieg, 1991). It does not need hsp60, showing that the translocator is inserted into the inner membrane without passing through the matrix. Import does not require ATP in the matrix which provides further evidence that the translocator does not pass through the matrix on its way into the inner membrane (Wachter et al., 1992). How the actual insertion into the membrane occurs is not clear so far.

Other inner membrane proteins like the iron–sulfur protein of cytochrome c reductase, subunit IV of cytochrome oxidase, and subunit 9 of the F_1F_0-ATPase are synthesized with a matrix-targeting signal that transports them entirely into the matrix space, where the presequences are removed by two sequential cleavages. Thus precursor proteins destined for different locations can use the same import pathway before diverging along different routes.

Under the appropriate experimental conditions, import intermediates of these inner membrane proteins can be generated. For example, if cleavage of the presequence is inhibited, import intermediates may accumulate in the matrix. Following processing, the proteins are assembled into the inner membrane as heterooligomeric complexes. The mechanism of this assembly is completely unknown. It is tempting to propose the existence of an intermembrane space located chaperone. However, so far there is no hint for the existence of such a chaperone.

D. Sorting to the Intermembrane Space

Sorting of proteins to the intermembrane space has been the focus of intense investigation over the past few years. So far several different sorting pathways have been reported.

One of the pathways is followed by cytochrome c. The apoenzyme of cytochrome c inserts directly into the outer membrane without the help of either the outer membrane receptors MOM72/Mas70p and MOM22/Mas20p or of the general translocation machinery in the outer membrane (Stuart et al., 1990; Stuart and Neupert, 1990). Translocation is driven by interaction with cytochrome c heme lyase in the intermembrane space and the subsequent attachment of the heme group (Nicholson et al., 1988; Dumont et al., 1991). Adenylate kinase is another example for an intermembrane space enzyme that does not need the outer membrane translocation machinery for sorting into the intermembrane space (Magdolen et al., 1992). Both cytochrome c and adenylate kinase lack a cleavable presequence and do not require matrix ATP or an electrochemical potential across the inner membrane for their import.

An even more complex pathway than the one used by cytochrome c is used by cytochrome c heme lyase, a soluble protein in the intermembrane space. Cytochrome c heme lyase is imported with the help of the outer membrane translocation machinery and therefore requires neither a membrane potential across the inner membrane nor ATP hydrolysis in the matrix (Lill et al., 1992). The driving force for translocation across the outer membrane is not known. This force may be provided by folding of the polypeptide in the intermembrane space. Alternatively, the driving force could be provided by interaction with another intermembrane space protein or even with lipids on the inner side of the outer membrane or on the outer side of the inner membrane.

Another class of intermembrane space-targeted proteins has a bipartite signal sequence (for a more detailed discussion see Section II of this review). This class includes cytochrome b_2 (Guiard, 1985), cytochrome c_1 (Sadler et al., 1984), most likely cytochrome c peroxidase (Kaput et al., 1982), and mitochondrial creatine kinase (Haas and Strauss, 1990). The presequences of these precursors have an N-terminal signal that resembles a matrix-targeting signal. When this part of the presequence is fused to a passenger protein, the resulting fusion protein is transported to the matrix. The intermembrane space sorting signal is located downstream from the potential matrix-targeting signal. It is composed of a cluster of positively charged amino acids and a hydrophobic stretch (Hurt and van Loon, 1986; van Loon et al., 1986; Jensen et al., 1992; Beasley et al., 1993; Schwarz et al., 1993). Two different hypotheses for intermembrane space sorting have been suggested (for a detailed discussion see: Glick and Schatz, 1991; Glick et al., 1992b; Segui-Real et al., 1992). In the stop-transfer mechanism the sorting signal is recognized by the proteinaceous import machinery in the mitochondrial inner membrane which arrests further translocation across the inner membrane. Following this transfer stop, the precursor still bound to the inner membrane translocation machinery diffuses laterally out of the contact site, pulling the C-terminal parts of the precursor across the outer membrane (van Loon et al., 1986; Glick et al., 1992b). The conservative sorting model suggests that these precursors are first completely transported into the matrix where they interact with hsp60 that keeps them in an unfolded conformation. They are then recognized by another translocation system in the inner membrane that reexports them into the intermembrane space (Hartl et al., 1987; Hartl and Neupert, 1990). It has been proposed that reexport is closely coupled with import into the matrix (Koll et al., 1992). However, the ATP requirement for the sorting of cytochrome $c1$ (Wachter et al., 1992) and cytochrome $b2$ (Glick et al., 1993) to the intermembrane space argues strongly against the conservative sorting hypothesis. Import according conservative sorting would need ATP in the matrix, import according to stop-transfer may be independent of matrix ATP. Indeed import of cytochrome c_1 is independent of matrix ATP. Import of cytochrome b_2 requires matrix ATP, but this requirement has been traced to the unfolding of the heme-binding domain outside the mitochondria. When this domain is destabilized by a small deletion, import of cytochrome b_2 into the intermembrane space is independent of matrix ATP (Glick et al., in press). The strongest evidence in favor of conservative sorting, the apparent localization of an import intermediate of cytochrome b_2 in the matrix space (Hartl et al., 1987), seems to have been due to an invalid analysis method (Glick et al., 1992b). Furthermore, hsp60 is not required for sorting of cytochrome b_2 or cytochrome c_1 (Glick et al., 1992b; Hallberg et al., 1993) as was originally proposed (van Loon and Schatz, 1987). Although a fusion protein derivative of cytochrome b_2 was reportedly bound to hsp60 after import (Koll et al., 1992), this finding also seems to be the result of a misleading analysis method (S. Rospert, personal communication). Taken together, recent results strongly favor a stop-transfer mechanism for sorting of proteins containing bipartite presequences.

VII. ENERGETICS OF PROTEIN IMPORT INTO MITOCHONDRIA

Protein import into mitochondria involves several energy-requiring steps (for a review see: Beasley et al., 1992). Two forms of energy are used: an electrochemical potential across the mitochondrial inner membrane (Gasser et al., 1982; Schleyer et al., 1982), and ATP (Nelson and Schatz, 1979; Pfanner and Neupert, 1986; Chen and Douglas, 1987a; Eilers et al., 1987).

The requirement for a membrane potential for import was first shown by the observation that import of precursor proteins into ATP-supplemented isolated mitochondria was inhibited by agents that collapse the membrane potential (Gasser et al., 1982; Schleyer et al., 1982). Later results suggested that the electrochemical potential rather than the total proton motive force is sufficient for protein translocation (Pfanner and Neupert, 1986). The electrochemical potential is necessary only for insertion of the presequence into the inner membrane; further translocation of the rest of the polypeptide chain is potential-independent (Schleyer and Neupert, 1985; Eilers and Schatz, 1988).

The requirement for ATP was first inferred from studies with intact yeast cells (Nelson and Schatz, 1979). Recent results with *in vitro* systems that allow ATP depletion in- and outside mitochondria have now led to a more detailed understanding of the ATP requirement for import (Pfanner and Neupert, 1986; Chen and Douglas, 1987a; Chen and Tai, 1987; Eilers et al., 1987; Hwang and Schatz, 1989; Hwang et al., 1991; Wachter et al., 1992; Glick et al., 1993). Initially ATP was thought to be required only outside the mitochondria, perhaps for keeping precursor proteins import-competent (Eilers et al., 1987; Pfanner et al., 1987b). Denaturation (Pfanner et al., 1988) or destabilization of the precursor by internal deletions (Chen and Douglas, 1987b, 1988) or C-terminal truncations (Verner and Schatz, 1987) appeared to bypass the ATP requirement showing that indeed ATP outside mitochondria was required for keeping the precursor in an unfolded conformation. A major breakthrough in our understanding of ATP-requirement in mitochondrial protein import came from the observation that import of matrix-targeted DHFR-containing fusion proteins requires ATP only in the matrix (Hwang and Schatz, 1989).

It is now generally accepted that protein import into the matrix involves at least three ATP-requiring steps; one in the cytosol and one or two in the matrix. All three steps are mediated by chaperones (for review see: Beasley et al., 1992). Matrix ATP is needed for complete translocation of precursors across the inner membrane

(Wachter et al., 1992), most probably reflecting the ATP requirement of mhsp70 (Kang et al., 1990; Scherer et al., 1990; Manning-Krieg et al., 1991). The mechanism of mhsp70 action is not fully understood. It has not been directly shown that mhsp70 is necessary for import or whether the observed interaction with translocating chains occurs because they are unfolded (Manning-Krieg et al., 1991). Matrix ATP is also necessary for the import of urea-denatured fusion proteins and of the F_1-ATPase subunit β with destabilizing internal deletions (Wachter et al., in preparation).

As discussed above, the second ATP-requiring step in the matrix involves the action of hsp60. This step may not be required for all precursors; similarly, the ATP requirement in the cytosol also depends on the type of precursor. It seems that loosely folded precursors are less dependent on cytosolic ATP because they do not need assistance of cytosolic chaperones (Stuart et al., 1990; Miller and Cumsky, 1991, 1993; Lill et al., 1992). Moreover, some precursors do not require cytosolic ATP even though they contain tightly folded domains (Eilers et al., 1988; Eilers and Schatz, 1988; Endo and Schatz, 1988; Glick et al., 1993). These folded domains may be unfolded by interaction with acidic phosphoplipids on the mitochondrial surface, and by the "pulling" action of mhsp70 in the matrix. Import by a cotranslational mechanism would also not need cytosolic ATP, except for protein synthesis itself.

To summarize, import into the matrix requires matrix ATP and an electrochemical potential across the inner membrane; with some precursors, it also requires ATP in the cytosol. The same is true for import of inner membrane proteins that are first transported into the matrix. Direct import into the inner membrane does not need ATP in the matrix (Wachter et al., 1992). Import into the outer membrane requires only external ATP or is even ATP-independent. With some exceptions (Glick et al., 1993) import into the intermembrane space by the stop-transfer mechanism needs an electrochemical potential across the inner membrane, but neither cytosolic nor matrix ATP (Wachter et al., 1992; Glick et al., 1993; Wachter et al., in preparation). Studying the energy requirement for import of a precursor is therefore a valuable tool for analyzing its import pathway.

The energy requirements for the different steps of import also provided a basis for generating different translocation intermediates (Hwang et al., 1991; Jascur et al., 1992) that were useful tools for the identification of components of the mitochondrial protein translocation machinery (for a more detailed discussion see Sections V and VIII of this review; Vestweber et al., 1989; Kiebler et al., 1990; Scherer et al., 1990, 1992). These intermediates were also instrumental in demonstrating the existence of an independent translocation machinery in the mitochondrial inner membrane (Glick et al., 1991, Scherer et al., 1992; Horst et al., 1993a).

VIII. DYNAMIC INTERACTION OF THE
TWO IMPORT CHANNELS

There is general agreement that "contact sites" are the ports through which cytoplasmically-made proteins enter the mitochondrial matrix (Kellems et al., 1974; Schleyer and Neupert, 1985; Schwaiger et al., 1987; Pon et al., 1989; Glick et al., 1991). However, the molecular structure of these sites is still uncertain. Some early experimental evidence suggested that there was a single, fixed channel spanning the two membranes. This model implied that most mitochondrial proteins (except those of the outer membrane) would be initially transported into the matrix before being sorted to other submitochondrial compartments.

Recent experiments strongly argue against the "single channel" model. In one such experiment, protein import into isolated yeast mitochondria was first blocked by treating intact mitochondria with trypsin; selective disruption of the outer membrane then restored import (Ohba and Schatz, 1987a). Similar results were obtained when protein import into intact mitochondria was inhibited by adding antibodies against outer membrane proteins (Obha and Schatz, 1987b), or by jamming the import sites with a chimeric precursor protein (Hwang et al., 1989). The simplest interpretation of these results is that the inner membrane contains cryptic import sites that are unmasked by disrupting the outer membrane barrier. Indeed, purified mitochondrial inner membrane vesicles, which have the same sidedness as the inner membrane of intact mitochondria, import mitochondrial precursor proteins with the same efficiency and the same requirements as intact mitochondria (Hwang et al., 1989); import requires an electrochemical potential across the inner membrane, ATP in the matrix, and a functional mitochondrial-targeting signal on the precursor protein. Unlike import into intact mitochondria, protein import into these vesicles is sensitive to antibodies against the inner membrane and insensitive to antibodies against the outer membrane.

If these cryptic import sites in the inner membrane function during the normal transport of proteins across both mitochondrial membranes, it should be possible to experimentally dissect the import pathway of a matrix protein into two distinct steps: translocation across the outer membrane, followed by translocation across the inner membrane. This separation was first made possible by the finding that protein transport across the inner membrane requires ATP in the matrix (Hwang et al., 1990).

In related experiments, isolated yeast mitochondria were incubated with an artificial precursor protein whose C-terminus had been cross-linked to a tightly folded import-incompetent pancreatic trypsin inhibitor moiety. This chimeric precursor became stuck with its N-terminal part in the matrix and the trypsin inhibitor outside the outer membrane. When the mitochondria were then depleted of ATP and the disulfide bridges of the trypsin inhibitor were reduced (to allow transport across the outer membrane), the trypsin inhibitor moiety moved across the outer membrane into the intermembrane space; it could then be chased across the inner membrane by restoring ATP levels in the matrix (Jascur et al., 1992). This chase must have reflected the independent operation of the inner membrane translocation system in intact mitochondria. Authentic precursors can also be transported into the matrix in two discrete steps. When such precursors are imported into ATP-depleted mitochondria, they pass through the outer membrane, but get stuck across the inner one. These "ATP-depletion intermediates" are on the correct import pathway because they can be chased into the matrix simply by adding ATP, even in the absence of an electrochemical potential across the inner membrane (Hwang et al., 1991).

The separation of the two protein transport systems is even more clear-cut in experiments with cytochrome *c* heme lyase, which is imported from the cytoplasm into the intermembrane space. Import of this protein appears to be completely independent of the inner membrane translocation system because it requires neither an electrochemical potential across the inner membrane nor ATP in the matrix (Lill et al., 1992). When cytochrome *c* heme lyase is fused to an N-terminal matrix-targeting signal, the resulting hybrid protein is imported into the matrix. This process can be dissected into two stages. If the mitochondria lack a potential across the inner membrane, the fusion protein is transported only across the outer membrane into the intermembrane space. Restoration of the membrane potential then allows the protein to be transported into the matrix (Segui-Real et al., 1993).

Taken together, these experiments suggest that each of the two mitochondrial membranes contains its own protein transport system, and that these two systems can link up to form a channel spanning both membranes (Glick et al., 1991; Pfanner et al., 1992). This linkage appears to be dynamic; the two systems can disengage when transport across the inner membrane is halted by lack of ATP in the matrix (Hwang et al., 1991), or by the presence of a stop-transfer signal in the precursor (Glick et al., 1992).

For transporting a protein into the matrix, there is no obvious advantage in having two distinct, dynamically interacting import channels, since the two channels appear to separate only slowly from each other (Hwang et al., 1991; Glick et al., 1992). Import of matrix proteins is thus probably mediated by the two channels operating in tandem. However, reversible interaction of the two channels is essential for sorting imported proteins to the intermembrane space (Glick et al., 1992; Lill et al., 1992) or directly into the inner membrane (Pfanner et al., 1992). This sorting demands that precursors can "escape" from the general matrix import route once they have moved across the outer membrane; separation of the two import systems would be a convenient escape mechanism.

IX. OUTLOOK

In the last decade great progress has been made in identifying the pathways by which nuclear encoded precursor proteins are transported to their mitochondrial destinations. The energetics of protein import and the ATP requirements inside and outside mitochondria have been investigated in great detail and in many cases have been instrumental in deciphering the import pathway of certain mitochondrial proteins. A natural projection of these studies is the analysis of the energy-requiring enzymes themselves and how they mediate protein import into mitochondria.

It is very likely that most, if not all components of the outer mitochondrial import machinery have been identified. Our understanding of how precursor proteins are recognized by the mitochondrial receptors or how the various components of the outer mitochondrial import machinery interact is, however, only rudimentary. Now that the components have been identified, it should be possible to reconstitute protein import using the isolated components in order to gain further insight on the mode of action and the interaction of these components. Unlike the outer mitochondrial membrane import machinery, we are only now beginning to characterize the inner membrane translocation machinery. Three components have so far been identified and it is very likely that additional components are involved. The physical nature of the contact sites between the outer and inner translocation machineries as well as the regulation of this interaction remains far from clear. A closer look at this interaction should provide further insight into intramitochondrial sorting.

ACKNOWLEDGMENTS

We would like to thank Drs. B.S. Glick and G. Schatz for their comments and suggestions.

REFERENCES

Ades, I.Z. & Butow, R.A. (1980a). The products of mitochondria-bound cytoplasmic polysomes in yeast. J. Biol. Chem. 255, 9918–9924.

Ades, I.Z. & Butow, R.A. (1980b). The transport of proteins into yeast mitochondria. J. Biol. Chem. 255, 9925–9935.

Altmann, E. (1890). Die Elementarorganismen und ihre Beziehungen zu den Zellen. Leipzig: Veith.

Argant, C., Lusty, C.J., & Shore, G.C. (1983). Membrane and cytosolic components affecting transport of the precursor of ornithine carbamyltransferase into mitochondria. J. Biol. Chem. 258, 6667–6670.

Atencio, D.P. & Yaffe, M.P. (1992). MAS5, a yeast homolog of DnaJ involved in mitochondrial protein import. Mol. Cell. Biol. 12, 283–291.

Attardi, G. & Schatz, G. (1988). The biogenesis of mitochondria. Annu. Rev. Cell Biol. 4, 289–333.

Baker, K.P., Schaniel, A., Vestweber, D., & Schatz, G. (1990). A yeast mitochondrial outer membrane protein essential for protein import and cell viability. Nature 348, 605–609.

Baker, K.P. & Schatz, G. (1991). Mitochondrial proteins essential for viability mediate protein import into yeast mitochondria. Nature 349, 205–208.

Beasley, E.M., Wachter, C., & Schatz, G. (1992). Putting energy into mitochondrial protein import. Curr. Opinion Cell Biol. 4, 646–651.

Beasley, E.M., Müller, S., & Schatz, G. (1993). The signal that sorts yeast cytochrome-b_2 to the mitochondrial intermembrane space contains three distinct functional regions. EMBO J. 12, 2303–2311.

Becker, K., Guiard, B., Rassow, J., Söllner, T., & Pfanner, N. (1992). Targeting of a chemically pure preprotein to mitochondria does not require the addition of a cytosolic signal recognition factor. J. Biol. Chem. 267, 5637–5643.

Bedwell, D.M., Strobel, S.A., Yun, K., Jongeward, G.D., & Emr, S. D. (1989). Sequence and structural requirements of a mitochondrial protein import signal defined by saturation cassette mutagenesis. Mol. Cell. Biol. 9, 1014–1025.

Behrens, M., Michaelis, G., & Pratje, E. (1991). Mitochondrial inner membrane protease-1 of *Saccharomyces Cerevisiae* shows sequence similarity to the *Escherichia Coli* leader peptidase. Mol. Gen. Genetics 228, 167–176.

Bibus, C.R., Lemire, B.D., Suda, K., & Schatz, G. (1988). Mutations restoring import of a yeast mitochondrial protein with a nonfunctional presequence. J. Biol. Chem. 263, 13097–13102.

Bolliger, L., Deloche, O., Glick, B.S., Georgopoulos, C., Jeno, P., Kronidou, N.G., Horst, M., and Schatz, G. (1994). A mitochondrial homolog of bacterial GrpE interacts with mitochondrial hsp70 and is essential for viability. EMBO J. 13, 1998–2006.

Braun, H.P., Emmermann, M., Kruft, V. & Schmitz, U.K. (1992). The general mitochondrial processing peptidase from potato is an integral part of cytochrome-c reductase of the respiratory chain. EMBO J. 11, 3219–3227.

Caplan, A.J., Cyr, D.M., & Douglas, M.G. (1992a). YDJ1p facilitates polypeptide translocation across different intracellular membranes by a conserved mechanism. Cell 71, 1143–1155.

Caplan, A.J., Tsai, J., Casey, P.J., & Douglas, M.G. (1992b). Farnesylation of YDJ1p is required for function at elevated growth temperatures in *Saccharomyces cerevisiae*. J. Biol. Chem. 267, 18890–18895.

Chandrasekhar, G.N.; Tily, K., Woolford, C., Hendrix, R., & Georgopoulos C. (1986). Purification and properties of the groES morphogenetic protein of *Escherichia coli*. J. Biol. Chem. 261, 12414–12419.

Chen, L. & Tai, P.C. (1987). Evidence for the involvement of ATP in co-translational protein translocation. Nature 328, 164–166.

Chen, W.-J. & Douglas, M.G. (1987a). Phosphodiester bond cleavage outside mitochondria is required for the completion of protein import into the mitochondrial matrix. Cell 49, 651–658.

Chen, W.-J. & Douglas, M.G. (1987b). The role of protein structure in the mitochondrial import pathway. J. Biol. Chem. 262, 15598–15604.

Chen, W.-J. & Douglas, M.G. (1987c). The role of protein structure in the mitochondrial import pathway: Unfolding of mitochondrially bound precursors is required for membrane translocation. J. Biol. Chem. 262, 15605–15609.

Chen, W.-J. & Douglas, M.G. (1988). An F_1-ATPase β-Subunit precursor lacking an internal tetramer-forming domain is imported into mitochondria in the absence of ATP. J. Biol. Chem. 263, 4997–5000.

Cheng, M.Y., Hartl, F.-U., Martin, J., Pollock, R.A., Kalousek, F., Neupert, W., Hallberg, E.M., Hallberg, R.L., & Horwich, A.L. (1989). Mitochondrial heat-shock protein hsp60 is essential for assembly of proteins imported into yeast mitochondria. Nature 337, 620–625.

Chirico, W.J., Waters, M.G., & Blobel, G. (1988). 70K heat-shock related proteins stimulate protein translocation into microsomes. Nature 332, 805–810.

Craig, E.A., Kramer, J., & Kosic-Smithers, J. (1987). SSC1, a member of the 70-kDa heat-shock protein multigene family of *Saccharomyces cerevisiae*, is essential for growth. Proc. Natl. Acad. Sci. USA 84, 4156–4160.

Cyr, D.M., Lu, X., & Douglas, M.G. (1992). Regulation of eucaryotic Hsp70 function by a DnaJ homologue. J. Biol. Chem. 267, 20927–20931.

Dekker, P.J.T., Keil, P., Rassow, J., Maarse, A.C., Pfanner, N., & Meijer, M. (1993). Identification of MIM23, a putative component of the protein import machinery of the mitochondrial inner membrane. FEBS Lett. 330, 66–70.

Deshaies, R.J., Koch, B.D., & Schekman, R. (1988a). The role of stress proteins in membrane biogenesis. TIBS 13, 384–388.

Deshaies, R.J., Koch, B.D., Werner-Washburne, M., Craig, E.A., & Schekman, R. (1988b). A subfamily of stress proteins facilitates translocation of secretory and mitochondrial precursor polypeptides. Nature 332, 800–805.

Dumont, M.E., Cardillo, T.S., Hayes, M.K., & Sherman, F. (1991). Role of cytochrome-c heme lyase in mitochondrial import and accumulation of cytochrome-c in Saccharomyces-cerevisiae. Mol. Cell. Biol. 11, 5487–5496.

Eilers, M. & Schatz, G. (1986). Binding of a specific ligand inhibits import of a purified precursor protein into mitochondria. Nature 322, 228–232.

Eilers, M., Oppliger, W., & Schatz, G. (1987). Both ATP and an energized inner membrane are required to import a purified precursor protein into mitochondria. EMBO J. 6, 1073–1077.

Eilers, M., Hwang, S., & Schatz, G. (1988). Unfolding and refolding of a purified precursor protein during import into isolated mitochondria. EMBO J. 7, 1139–1145.

Eilers, M. & Schatz, G. (1988). Protein unfolding and the energetics of protein translocation across biological membranes. Cell 52, 481–483.

Eilers, M., Endo, T., & Schatz, G. (1989). Adriamycin, a drug interacting with acidic phospholipids, blocks import of precursor proteins by isolated yeast mitochondria. J. Biol. Chem. 264, 2945–2950.

Emtage, J.L.T. & Jensen, R.E. (1993). MAS6 encodes an essential inner membrane component of the yeast mitochondrial protein import pathway. J. Cell Biol. 122, 1003–1012.

Endo, T. & Schatz, G. (1988). Latent membrane pertubation activity of a mitochondrial precursor protein is exposed by unfolding. EMBO J. 7, 1153–1158.

Ernster, L. & Schatz, G. (1981). Mitochondria: A historical review. J. Cell Biol. 91, 227s.

Freitag, H., Janes, M., & Neupert, W. (1982). Biosynthesis of mitochondrial porin and insertion into the outer mitochondrial membrane of Neurospora crassa. Eur. J. Biochem. 126, 197–202.

Fujiki, M. & Verner, K. (1991). Coupling of protein synthesis and mitochondrial import in a homologous yeast in vitro system. J. Biol. Chem. 266, 6841–6847.

Fujiki, M. & Verner, K. (1993). Coupling of cytosolic protein synthesis and mitochondrial protein import in yeast—evidence for cotranslational import in vivo. J. Biol. Chem. 268, 1914–1920.

Gambill, D.B., Voos, W., Kang, P.J., Miao, B., Langer, T., Craig E.A., & Pfanner, N. (1993). A dual role for mitochondrial heat shock protein 70 in membrane translocation of preproteins. J. Cell Biol., 123, 109–117.

Gasser, S.M., Daum, G., & Schatz, G. (1982). Import of proteins into mitochondria: energy-dependent uptake of precursors by isolated mitochondria. J. Biol. Chem. 257, 13034–13041.

Gasser, S.M. & Schatz, G. (1983). Import of proteins into mitochondria. In vitro studies on the biogenesis of the outer membrane. J. Biol. Chem. 258, 3427–3430.

Geli, V., Yang, M., Suda, K., Lustig, A., & Schatz, G. (1990). The MAS-encoded processing protease of yeast mitochondria—overproduction and characterization of its 2 nonidentical subunits. J. Biol. Chem. 265, 19216–19222.

Georgopoulos, C. (1992). The emergence of the chaperone machines. TIBS 17, 295–300.

Glick, B. & Schatz, G. (1991). Import of proteins into mitochondria. Annu. Rev. Genetics 25, 21–44.

Glick, B.S., Beasley, E.M., & Schatz, G. (1992a). Protein sorting in mitochondria. TIBS 17, 453–459.

Glick, B.S., Brandt, A., Cunningham, K., Müller, S., Hallberg, R.L. & Schatz, G. (1992b). Cytochromes-c_1 and cytochromes-b_2 are sorted to the intermembrane space of yeast mitochondria by a stop-transfer mechanism. Cell 69, 809–822.

Glick, B.S., Wachter, C., Reid, G.A., & Schatz, G. (1993). Import of cytochrome b_2 to the mitochondrial intermembrane space: The tightly folded heme-binding domain makes import dependent upon matrix ATP. Prot. Science 2, 1901–1917.

Glover, L.A. & Lindsay, J.G. (1992). Targeting proteins to mitochondria—a current overview. Biochemical J. 284, 609–620.

Gratzer, S., Lithgow, T., Kohlwein, S.D., Haucke, V., Junne, T., Schatz, G., & Horst, M. (1994). The outer membrane protein Mas37p functions together with Mas70p as a receptor for protein import into yeast mitochondria. (in press).

Guiard, B. (1985). Structure, expression and regulation of a nuclear gene encoding a mitochondrial protein: The yeast L(+)-lactate cytochrome *c* oxidoreductase (cytochrome b_2). EMBO J. 4, 3265–3272.

Haas, R.C. & Strauss, A.W. (1990). Separate nuclear genes encode sarcomere-specific and ubiquitous human mitochondrial creatine kinase isozymes. J. Biol. Chem. 265, 6921–6927.

Hachiya, N., Alam, R., Sakasegawa, Y., Sakaguchi, M., Mihara, K., & Omura, T. (1993). A mitochondrial import factor purified from rat liver cytosol is an ATP-dependent conformational modulator for precursor proteins. EMBO J. 12, 1579–1586.

Hachiya, N., Komiya, T., Alam, R., Iwahashi, J., Sakaguchi, M., & Omura, T., & Mihara, K. (1994). MSF, a novel cytoplasmic chaperone which functions in precursor targeting to mitochondria. EMBO J. (in press).

Hallberg, E.M., Shu, Y. & Hallberg, R.L. (1993). Loss of mitochondrial hsp60 function: nonequivalent effects on matrix-targeted proteins. Mol. Cell. Biol. 13, 3050–3057.

Hartl, F.-U., Schmid, B., Wachter, E., Weiss, H., & Neupert, W. (1986). Transport into mitochondria and intramitochondrial sorting of the Fe/S protein of ubiquinol-cytochrome *c* reductase. Cell 47, 939–951.

Hartl, F.U., Ostermann, J., Guiard, B., & Neupert, W. (1987). Successive translocation into and out of the mitochondrial matrix: targeting of proteins to the intermembrane space by a bipartite signal peptide. Cell 51, 1027–1037.

Hartl, F.-U., Pfanner, N., Nicholson, D.W., & Neupert, W. (1989). Mitochondrial protein import. Bioch. Biophysica Acta 988, 1–45.

Hartl, F.-U. & Neupert, W. (1990). Protein sorting to mitochondria—evolutionary conservations of folding and assembly. Science 247, 930–938.

Hase, T., Müller, U., Riezman, H., & Schatz, G. (1984). A 70-kDa protein of the yeast mitochondrial outer membrane is targeted and anchored via its extreme amino terminus. EMBO J. 3, 3157–3164.

Hase, T., Nakai, M., & Matsubara, H. (1986). The N-terminal 21 amino acids of a 70-kDa protein of the yeast mitochondrial outer membrane direct *E. coli* beta-galaktosidase into the mitochondrial matrix space in yeast cells. FEBS Lett. 197, 199–203.

Hawlitschek, G., Schneider, H., Schmidt, B., Tropschug, M., Hartl, F.-U. & Neupert, W. (1988). Mitochondrial protein import: Identification of processing peptidase and of PEP, a processing enhancing protein. Cell 53, 795–806.

Hemmingsen, S.M., Woolford, C., van der Vies, S.M., Tilly, K., Dennis, D.T., Georgopoulos, C.P.; Hendrix, R.W., & Ellis, J. (1988). Homologues plant and bacterial protein chaperone oligomeric protein assembly. Nature 333, 330–334.

Hendrick, J.P., Hodges, P.E., & Rosenberg, L.E. (1989). Survey of amino-terminal proteolytic cleavage sites in mitochondrial precursor proteins: leader peptides cleaved by two matrix proteases share a three-amino acid motif. Proc. Natl. Acad. Sci. USA 86, 4056–4060.

Hines, V. & Baker, K.P. (1991). The protein import machinery of yeast mitochondria. Meth. Cell. Biol. 34, 377–387.

Hines, V., Brandt, A., Griffiths, G., Horstmann, H., Brutsch, H., & Schatz, G. (1990). Protein import into yeast mitochondria is accelerated by the outer membrane protein Mas70. EMBO J. 9, 3191–3200.

Hines, V. & Schatz, G. (1993). Precursor binding to yeast mitochondria—a general role for the outer membrane protein Mas70p. J. Biol. Chem. 268, 449–454.

Horst, M., Jenö, P., Kronidou, N., Bolliger, L., Oppliger, W., Scherer, P., Manning-Krieg, U., Jascur, T., & Schatz, G. (1993a). Protein import into yeast mitochondria: the inner membrane import site protein ISP45 is the *MPI1* gene product. EMBO J. 12, 3035–3041.

Horst, M., Kronidou, N.G., & Schatz, G. (1993b). Through the mitochondrial inner membrane. Curr. Biol. 3, 175–177.

Hurt, E.C., Pesold-Hurt, B., & Schatz, G. (1984a). The amino-terminal region of an imported mitochondrial precursor polypeptide can direct cytoplasmic dihydrofolate reductase into the mitochondrial matrix. EMBO J. 3, 3149–3156.

Hurt, E.C., Pesold-Hurt, B., & Schatz, G. (1984b). The cleavable prepiece of an imported mitochondrial protein is sufficient to direct cytosolic dihydrofolate reductase into the mitochondrial matrix. FEBS Lett. 178, 306–310.

Hurt, E.C., Pesold-Hurt, B., Suda, K., Oppliger, W., & Schatz, G. (1985a). The first twelve amino acids (less than half of the pre-sequence) of an imported mitochondrial protein can direct mouse dihydrofolate reductase into the yeast mitochondrial matrix. EMBO J. 4, 2061–2068.

Hurt, E.C., Müller, U., & Schatz, G. (1985b). The first twelve amino acids of a yeast mitochondrial outer membrane protein can direct a nuclear-encoded cytochrome oxidase subunit to the mitochondrial inner membrane. EMBO J. 4, 3509–3518.

Hurt, E.C. & van Loon, A.P.G.M. (1986). How proteins find mitochondria and intramitochondrial compartments. TIBS. 11, 204–207.

Hurt, E.C., Soltanifar, N., Goldschmidt-Clermont, M., Rochaix, J.D., & Schatz, G. (1986). The cleavable presequence of an imported chloroplast protein directs attached polypeptides into yeast mitochondria. EMBO J. 5, 1343–1350.

Hwang, S.T. & Schatz, G. (1989). Translocation of proteins across the mitochondrial inner membrane, but not into the outer membrane, requires nucleoside triphosphates in the matrix. Proc. Natl. Acad. Sci. USA 86, 8432–8436.

Hwang, S.T., Wachter, C., & Schatz, G. (1991). Protein import into the yeast mitochondrial matrix—a new translocation intermediate between the two mitochondrial membranes. J. Biol. Chem. 266, 21083–21089.

Isaya, G., Kalousek, F., Fenton W.A., & Rosenberg, L.E. (1991). Cleavage of precursors by the mitochondrial processing peptidase requires a compatible mature protein or an intermediate octapeptide. J. Cell Biol. 113, 65–76.

Isaya, G., Kalousek, F., & Rosenberg, L.E. (1992). Sequence analysis of rat mitochondrial intermediate peptidase—similarity to zinc metallopeptidases and to a putative yeast homologue. Proc. Natl. Acad. Sci. USA 89, 8317–8321.

Jascur, T., Goldenberg, D.P., Vestweber, D., & Schatz, G. (1992). Sequential translocation of an artificial precursor protein across the 2 mitochondrial membranes. J. Biol. Chem. 267, 13636–13641.

Jensen, R.E., Schmidt, S., & Mark, R.J. (1992). Mutations in a 19-amino-acid hydrophobic region of the yeast cytochrome c_1 presequence prevent sorting to the mitochondrial intermembrane space. Mol. Cell. Biol. 12, 4677–4686.

Kalousek, F., Hendrick, J.P., & Rosenberg, L.E. (1988). Two mitochondrial matrix proteases act sequentially in the processing of mammalian matrix enzymes. Proc. Natl. Acad. Sci. USA 85, 7536–7540.

Kalousek, F., Isaya, G., & Rosenberg, L.E. (1992). Rat liver mitochondrial intermediate peptidase (MIP)—purification and initial characterization. EMBO J. 11, 2803–2809.

Kalousek, F., Neupert, W., Omura, T., Schatz, G., & Schmitz, U.K. (1993). Uniform nomenclature for the mitochondrial peptidases cleaving precursors of mitochondrial proteins. TIBS 211, 249.

Kang, P.-J., Ostermann, J., Shilling, J., Neupert, W., Craig, E. A., & Pfanner, N. (1990). Requirement for hsp70 in the mitochondrial matrix for translocation and folding of precursor proteins. Nature 348, 137–143.

Kaput, J., Goltz, S., & Blobel, G. (1982). Nucleotide sequence of the yeast nuclear gene for cytochrome c peroxidase precursor. Functional implications of the pre-sequence for protein transport into mitochondria. J. Biol. Chem. 257, 15054–15058.

Karslake, C., Piotto, M.E., Pak, Y.K., Weiner, H., & Gorenstein, D.G. (1990). 2D NMR and structural model for a mitochondrial signal peptide bound to a micelle. Biochemistry 29, 9872–9878.

Kassenbrock, C.K., Cao, W., & Douglas, M.G. (1993). Genetic and biochemical characterization of ISP6, a small mitochondrial outer membrane protein associated with the protein translocation complex. EMBO J. 12, 3023–3034.

Kellems, R.E., Allison, V.F., & Butow, R.A. (1974). Cytoplasmic type 80 S ribosomes associated with yeast mitochondria.II Evidence for the association of cytoplasmic ribosomes with the outer mitochondrial membrane *in situ*. J. Biol. Chem. 249, 2297–3303.

Kellems, R.E. & Butow, R.A. (1972). Cytoplasmic-type 80 S ribosomes associated with yeast mitochondria. I. Evidence for ribosome binding sites on yeast mitochondria. J. Biol. Chem. 247, 8043–8050.

Kellems, R.E. & Butow, R.A. (1974). Cytoplasmic type 80 S ribosomes associated with yeast mitochondria. III Changes in the amount of bound ribosomes in response to changes in metabolic state. J. Biol. Chem. 249, 3304–3310.

Kiebler, M., Pfaller, R., Söllner, T., Griffiths, G., Horstmann, H., Pfanner, N., & Neupert, W. (1990). Identification of a mitochondrial receptor complex required for recognition and membrane insertion of precursor proteins. Nature 348, 610–616.

Kiebler, M., Keil, P., Schneider, H., van der Klei, I.J., Pfanner, N., & Neupert, W. (1993). The mitochondrial receptor complex: a central role of MOM22 in mediating preprotein transfer from receptors to the general insertion pore. Cell 74, 483–492.

Kleiber, J., Kalousek, F., Swaroop, M., & Rosenberg, L.E. (1990). The general mitochondrial matrix processing protease from rat liver: structural characterization of the catalytic subunit. Proc. Natl. Acad. Sci. USA 87, 7978–7982.

Koll, H., Guiard, B., Rassow, J., Ostermann, J., Horwich, A.L., Neupert, W. & Hartl, F.U. (1992). Antifolding activity of hsp60 couples protein import into the mitochondrial matrix with export to the intermembrane space. Cell 68, 1163–1175.

Kornfeld, S. & Mellman, I. (1989). The biogenesis of lysosomes. Annu. Rev. Cell. Biol. 5, 483–525.

Kronidou, N.G., Opllinger, W., Bollinger, L., Hannavy, K., Glick, B.S., Schatz, G., & Horst, M. (1994). Dynamic interaction between Isp45 and mitochondrial hsp70 in the protein import system of the yeast mitochondrial inner membrane. Proc. Natl. Acad. Sci USA, (in press).

Kübrich, M., Keil, P., Rassow, J., Dekker, P.J.T., Blom, J., Meijer, M., & Pfanner, N. (1994). The polytopic mitochondrial inner membrane proteins MIM17 and MIM23 operate at the same import site. FEBS Lett. 349, 222–228.

Langer, T., Lu, C., Echols, H., Flanagan, J., Hayer, M.K., & Hartl, F.-U. (1992a). Successive action of DnaK, DnaJ and GroEL along the pathway of chaperone-mediated protein folding. Nature 356, 683–689.

Langer, T., Pfeifer, G., Martin, J., Baumeister, W., & Hartl, F.-U. (1992b). Chaperonin-mediated protein folding—GroES binds to one end of the GroEL cylinder, which accommodates the protein substrate within its central cavity. EMBO J. 11, 4757–4765.

Lemire, B.D., Fankhauser, C., Baker, A., & Schatz, G. (1989). The mitochondrial targeting function of randomly generated peptide sequences correlates with predicted helical amphiphlicity. J. Biol. Chem. 264, 20206–20215.

Li, J.M. & Shore, G.C. (1992a). Protein sorting between mitochondrial outer and inner membranes— insertion of an outer membrane protein into the inner membrane. Biochimica et Biophysica Acta 1106, 233–241.

Li, J.M. & Shore, G.C. (1992b). Reversal of the orientation of an integral protein of the mitochondrial outer membrane. Science 256, 1815–1817.

Lill, R., Stuart, R.A., Drygas, M.E., Nargang, F.E., & Neupert, W. (1992). Import of cytochrome-*c* heme lyase into mitochondria—a novel pathway into the intermembrane space. EMBO J. 11, 449–456.

Lindquist, S. & Craig, E.A. (1988). The heat-shock proteins. Annu. Rev. Genet. 22, 631–677.

Lithgow, T., Junne, T., Wachter, C., & Schatz, G. (1994). Yeast mitocondria lacking the two import receptors Mas20p and Mas70p can efficienctly specifically import precursor proteins. J. Biol. Chem. 269, 152325–15330.

Lubben, T.H., Gatenby, A.A., Ahlquist, P., & Keegstra, K. (1989). Chloroplast import characteristics of chimeric proteins. Plant Mol. Biol. 12, 13–18.

Maarse, A.C., Blom, J., Grivell, L.A., & Meijer, M. (1992). MPI1, an essential gene encoding a mitochondrial membrane protein, is possibly involved in protein import into yeast mitochondria. EMBO J. 11, 3619–3628.

Maarse, A.C., Blom, J., Keil, P., Pfanner, N., & Meijer, M. (1994). Identification of the essential yeast protein MIM17, an integral mitochondrial inner membrane protein involved in protein import. FEBS Lett. 349, 215–221.

Maccecchini, M.-L., Rudin, Y., Blobel, G., & Schatz, G. (1979). Import of proteins into mitochondria: precursors forms of the extramitochondrially made F_1-ATPase subunits in yeast. Proc. Natl. Acad. Sci. USA 76, 343–347.

Maduke, M. & Roise, D. (1993). Import of a mitochondrial presequence into protein-free phospholipid vesicles. Science 260, 364–367.

Magdolen, V., Schricker, R., Strobel, G., Germaier, H., & Bandlow, W. (1992). *In vivo* import of yeast adenylate kinase into mitochondria affected by site-directed mutagenesis. FEBS Lett. 299, 267–272.

Mahlke, K., Pfanner, N., Martin, J., Horwich, A.L., Hartl, F.-U., & Neupert, W. (1990). Sorting pathways of mitochondrial inner membrane proteins. Eur. J. Biochem. 192, 551–555.

Manning-Krieg, U.C., Scherer, P.E., & Schatz, G. (1991). Sequential action of mitochondrial chaperones in protein import into the matrix. EMBO J. 10, 3273–3280.

Mayer, A., Lill, R., & Neupert, W. (1993). Translocation and insertion of precursor proteins into isolated outer membranes of mitochondria. J. Cell Biol. 121, 1233–1243.

McMullin, T.W. & Hallberg, R.L. (1987). A normal mitochondrial protein is selectively synthesized and accumulated during heat shock in *Tetrahymena thermophilia*. Mol. Cell. Biol. 7, 4414–4423.

McMullin, T.W. & Hallberg, R.L. (1988). A highly conserved mitochondrial protein is structurally related to the protein encoded by the *Escherichia coli* groEL gene. Mol. Cell. Biol. 8, 371–380.

Meyer, D.I. (1990). Receptor anti-idiotypes-mimics or gimmicks? Nature 347, 424–425.

Mihara, K., Blobel, G., & Sato, R. (1982). *In vitro* synthesis and integration into mitochondria of porin, a major protein of the outer mitochondrial membrane of *Saccharomyces cerevisiae*. Proc. Natl. Acad. Sci. USA 79, 7102–7106.

Miller, B.R. & Cumsky, M.G. (1991). An unusual mitochondrial import pathway for the yeast cytochrome *c* oxidase subunit Va. J. Cell Biol. 112, 833–841.

Miller, B.R. & Cumsky, M.G. (1993). Intramitochondrial sorting of the precursor to yeast cytochrome *c* oxidase subunit Va. J. Cell Biol. 121, 1021–1029.

Mitoma, J. & Ito, A. (1992). The mitochondrial targeting signal of rat liver monoamine oxidase-B is located at its carboxy terminus. J. Biochemistry 111, 20–24.

Miura, S., Mori, M., & Tatibana, M. (1983). Transport of ornithine carbamoyltransferase precursor into mitochondria. J. Biol. Chem. 258, 6671–6674.

Moczko, M., Dietmeier, K., Söllner, T., Segui, B., Steger, H. F., Neupert, W., & Pfanner, N. (1992). Identification of the mitochondrial receptor complex in *Saccharomyces cerevisiae*. FEBS Lett. 310, 265–268.

Munro, S. & Pelham, H.R.B. (1986). An hsp70-like protein in the ER: identity with the 78-kDa glucose-regulated protein and immunoglobin heavy chain binding protein. Cell 46, 291–300.

Murakami, H., Pain, D., & Blobel, G. (1988). 70-kD heat shock-related protein is one of at least two distinct cytosolic factors stimulating protein import into mitochondria. J. Cell Biol. 107, 2051–2057.

Murakami, H., Blobel, G., & Pain, D. (1990). Isolation and characterization of the gene for a yeast mitochondrial import receptor. Nature 347, 488–491.

Murakami, K. & Mori, M. (1990). Purified presequence binding factor (PbF) forms an import-competent complex with a purified mitochondrial precursor protein. EMBO J. 9, 3201–3208.

Murakami, K., Tokunaga, F., Iwanaga, S., & Mori, M. (1990). Presequence does not prevent folding of a purified mitochondrial precursor protein and is essential for association with a reticulocyte cytosolic factor(S). J. Biochem, 108, 207–214.

Nakai, M., Hase, T., & Matsubara, T. (1989). Precise determination of the mitochondrial import signal contained in a 70-kDa protein of yeast mitochondrial outer membrane. J. Biochemistry 105, 513–519.

Nelson, N. & Schatz, G. (1979). Energy-dependent processing of cytoplasmically made precursors to mitochondrial proteins. Proc. Natl. Acad. Sci. USA 76, 4365–4369.

Neupert, W., Hartl, F.-U., Craig, E.A., & Pfanner, N. (1990). How do polypeptides cross the mitochondrial membranes? Cell 63, 447–450.

Nguyen, M., Bell, A.W., & Shore, G.C. (1988). Protein sorting between mitochondrial membranes specified by position of the stop-transfer domain. J. Cell Biol. 106, 1499–1505.

Nguyen, M. & Shore, G.C. (1987). Import of hybrid vesicular stomatitis G protein to the mitochondrial inner membrane. J. Biol. Chem. 262, 3929–3931.

Nicholson, D.W., Hergersberg, C., & Neupert, W. (1988). Role of cytochrome c heme lyase in the import of cytochrome c into mitochondria. J.Biol. Chem. 263, 19034–19042.

Nunnari, J., Fox, T.D., & Walter, P. (1993). A mitochondrial protease with two catalytic subunits of nonoverlapping specifities. Science 262, 1997–2004.

Ohba, M. & Schatz, G. (1987a). Disruption of the outer membrane restores protein import to trypsin-treated yeast mitochondria. EMBO J. 6, 2117–2122.

Ohba, M. & Schatz, G. (1987b). Protein import into yeast mitochondria is inhibited by antibodies raised against 45-kDa proteins of the outer membrane. EMBO J. 6, 2109–2115.

Ohta, S. & Schatz, G. (1984). A purified precursor polypeptide requires a cytosolic protein factor for import into mitochondria. EMBO J. 3, 651–657.

Ono, H. & Tuboi, S. (1990a). Purification and identification of a cytosolic factor required for import of precursors of mitochondrial proteins into mitochondria. Arch. Biochem. Biophys. 280, 299–304.

Ono, H. & Tuboi, S. (1990b). Purification of the putative import-receptor for the precursor of the mitochondrial protein. J. Biochemistry 107, 840–845.

Ostermann, J., Voos, W., Kang, P.J., Craig, E.A., Neupert, W., & Pfanner, N. (1990). Precursor proteins in transit through mitochondrial contact sites interact with hsp70 in the matrix. FEBS Lett. 277, 281–284.

Ou, W.-J., Ito, A., Okazaki, H., & Omura, T. (1989). Purification and characterization of a processing protease from rat liver mitochondria. EMBO J. 8, 2605–2612.

Pain, D., Murakami, H., & Blobel, G. (1990). Identification of a receptor for protein import into mitochondria. Nature 347, 444–449.

Park, S., Liu, G., Topping, T.B., Cover, W.H., & Randall, L.L. (1988). Modulation of folding pathways of exported proteins by the leader sequence. Science 239, 1033–1035.

Pelham, H.R.B. (1986). Speculations on the functions of the major heat shock and glucose-regulated proteins. Cell 46, 959–961.

Pelham, H.R.B. (1989). Control of protein exit from the endoplasmic reticulum. Annu. Rev. Cell Biol. 5, 1–23.

Pfaller, R. & Neupert, W. (1987). High-affinity binding sites involved in the imort of porin into mitochondria. EMBO J. 6, 2635–2642.

Pfaller, R., Steger, H.F., Rassow, J., Pfanner, N., & Neupert, W. (1988). Import pathways of precursor proteins into mitochondria: multiple receptor sites are followed by a common membrane insertion site. J. Cell Biol. 107, 2483–2490.

Pfanner, N., Hartl, F-U., Guiard, D., & Neupert, W. (1987c). Mitochondrial precursor proteins are imported through a hydrophilic membrane environment. Eur. J. Biochem. 169, 289–299.

Pfanner, N., Hoeben, P., Tropschug, M., & Neupert, W. (1987a). The carboxyl-terminal two-thirds of the ADP/ATP carrier polypeptide contains sufficient information to direct translocation into mitochondria. J. Biol. Chem. 262: 14851–14854.

Pfanner, N. & Neupert, W. (1986). Transport of F1-ATPase subunit beta into mitochondria depends on both a membrane potential and nucleoside triphosphates. FEBS Lett. 209, 152–156.

Pfanner, N., Pfaller, R., Kleene, R., Ito, M., Tropschug, M., & Neupert, W. (1988). Role of ATP in mitochondrial protein import. Conformational alteration of a precursor protein can substitute for ATP requirement. J. Biol. Chem. 263, 4049–4051.

Pfanner, N., Rassow, J., van der Klei, I.J., & Neupert, W. (1992). A dynamic model of the mitochondrial protein import machinery. Cell 68, 999–1002.

Pfanner, N., Söllner, T., & Neupert, W. (1991). Mitochondrial import receptors for precursor proteins. TIBS 16, 63–67.

Pfanner, N., Tropschug, M., & Neupert, W. (1987b). Mitochondrial protein import: nucleoside triphosphates are involved in conferring import-competence to precursors. Cell 49: 815–823.

Phelps, A., Schobert, C.T., & Wohlrab, H. (1991). Cloning and characterization of the mitochondrial phosphate transporter gene from the yeast S. cerevisiae. Biochemistry 30, 248–252.

Pon, L., Moll, T., Vestweber, D., Marshallsay, B., & Schatz, G. (1989). Protein import into mitochondria: ATP-dependent protein translocation activity in a submitochondrial fraction enriched in membrane contact sites and specific proteins. J. Cell Biol. 109, 2603–2616.

Pon, L. & Schatz, G. (1991). Biogenesis of Yeast Mitochondria. In: The Molecular and Cellular Biology of the Yeast Saccharomyces: Genome Dynamics, Protein Synthesis and Energetics (Pringle, J.R., Broad, J., & Jones, E., Eds.), Vol. 1. pp. 1333–1406. Cold Spring Harbour Laboratory Press, Cold Spring Harbor, N.Y.

Prasad, T.K., Hack, E., & Hallberg, R.L. (1990). Function of the maize mitochondrial chaperonin hsp60—specific association between hsp60 and newly synthesized F1-ATPase α-subunits. Mol. Cell. Biol. 10, 3979–3986.

Randall, S.K. & Shore, G.C. (1989). Import of a mutant mitochondrial precursor fails to respond to stimulation by a cytosolic factor. FEBS Lett. 250: 561–565.

Pratje, E. & Guiard, B. (1986). One nuclear gene controls the removal of transient pre-sequences from two yeast proteins: one coded by the nuclear the other by the mitochondrial genome. EMBO J. 5, 1313–1317.

Ramage, L., Junne, T., Hahne, K., Lithgow, T., & Schatz G. (1993). Functional cooperation of mitochondrial protein import receptors in yeast. EMBO J., in press.

Randall, S.K. & Shore, G.C. (1989). Import of a mutant mitochondrial precursor fails to respond to stimulation by a cytosolic factor. FEBS Lett. 250, 561–564.

Reading, D.S., Hallberg, R.L., & Myers, A.M. (1989). Characterization of the yeast HSP60 gene coding for a mitochondrial assembly factor. Nature 337, 655–659.

Reid, G.A. & Schatz, G. (1982a). Import of proteins into mitochondria: yeast cells grown in the presence of carbonyl cyanide m-chlorophenylhydrazine accumulate massive amounts of some mitochondrial precursor polypeptides. J. Biol. Chem. 257, 13056–13061.

Reid, G.A. & Schatz, G. (1982b). Import of proteins into mitochondria: extramitochondrial pools and post-translational import of mitochondrial protein precursors in vivo. J. Biol. Chem. 257, 13062–13067.

Reid, G.A., Yonetani, T., & Schatz, G. (1982). Import of proteins into mitochondria: import and maturation of the mitochondrial intermembrane space enzymes cytochrome b_2 and cytochrome c peroxidase in intact yeast cells. J. Biol. Chem. 257, 13068–13074.

Riezman, H., Hay, R., Witte, C., Nelson, N., & Schatz, G. (1983a). Yeast mitochondrial outer membrane specifically binds cytoplasmically-synthesized precursors of mitochondrial proteins. EMBO J. 2, 1113–1118.

Riezman, H., Hase, T., van Loon, A.P.G.M., Grivell, L., Suda, K., & Schatz, G. (1983b). Import of proteins into mitochondria: a 70-kDa outer membrane protein with a large carboxy terminal deletion is still transported to the outer membrane. EMBO J. 2: 2161–2168.

Roise, D., Horvath, S.J., Tomich, J.M., Richards, J.H., & Schatz, G. (1986). A chemically synthesized pre-sequence of an imported mitochondrial protein can form an amphiphilic helix and perturb natural and artificial phospholipid bilayers. EMBO J. 5, 1327–1334.

Roodyn, D.B. & Wilkie, D. (1968). The Biogenesis of Mitochondria. Methuen, London.

Rospert, S., Glick, B.S., Jenö, P., Schatz, G., Todd, M.J., Lorimer, G.H., & Viitanen, P.V. (1993a). Identification and functional analysis of chaperonin 10, the groES homolog of yeast mitochondria. Proc. Natl. Acad. Sci. USA 90, 10967–10971.

Rospert, S., Junne, T., Glick, B.S., & Schatz, G. (1993b). Cloning and disruption of the gene encoding yeast mitochondrial chaperonin 10, the homolog of *E. coli* groES. FEBS Lett. 335, 358–360.

Rothman, J.E. & Kornberg, R.D. (1986). An unfolding story of protein translocation. Nature 322, 209.

Rothman, J.E. & Schmidt, S.L. (1986). Enzymatic recycling of clathrin from coated vesicles. Cell 46, 5–9.

Rowley, N., Prip-Buus, C., Westermann, B., Brown, C., Schwarz, E., Barrell, B., & Neupert, W. (1994). Mdj1p, a novel chaperone of the DnaJ family, is involved in mitochondrial biogenesis and protein folding. Cell 77, 249–259.

Ryan, K.R. & Jensen, R.E. (1993). Mas6p can be cross-linked to an arrested precursor and interacts with other proteins during mitochondrial protein import. J. Biol. Chem. 268, 23743–23756.

Ryan, K.R., Menold, M.M., Garrett, S., & Jensen, R.E. (1994). *SMS1*, a high copy suppressor of the yeast *mas6* mutant, encodes an essential inner membrane protein required for mitochondrial import. Mol. Biol. Cell 5, 529–538.

Sadler, I., Suda, K., Schatz, G., Kaudewitz, F., & Haid, A. (1984). Sequencing of the nuclear gene for the yeast cytochrome c_1 precursor reveals an unusual complex amino-terminal presequence. EMBO J. 3, 2137–2143.

Sager, R. (1972). Cytoplasmic Genes and Organelles. Academic Press, New York.

Sanders, S.L., Whitfield, K.M., Vogel, J.P., Rose, M.D., & Schekman, R.W. (1992). Sec61p and BiP directly facilitate polypeptide translocation into the endoplasmic reticulum. Cell 69, 353–365.

Schatz, G. (1987). Signals guiding protein to their correct locations in mitochondria. Eur. J. Biochem. 165, 1–6.

Schatz, G. (1993). The protein import machinery of mitochondria. Prot. Science 2, 141–146.

Schatz, G. & Butow, R.A. (1983). How are proteins imported into mitochondria? Cell 32, 316.

Scherer, P.E. & Krieg, U.C. (1991). Cross-linking reagents as tools for identifying components of the yeast mitochondrial protein import machinery. Meth. Cell Biol. 34, 419–426.

Scherer, P.E., Krieg, U.T., Hwang, S.T., Vestweber, D., & Schatz, G. (1990). A precursor protein partly translocated into yeast mitochondria is bound to a 70-kDa mitochondrial protein. EMBO J. 9, 4315–4322.

Scherer, P.E., Manning-Krieg, U.C., Jenö, P., Schatz, G., & Horst, M. (1992). Identification of a 45-kDa protein at the protein import site of the yeast mitochondrial inner membrane. Proc. Natl. Acad. Sci. USA 89, 11930–11934.

Schleyer, M., Schmidt, B., & Neupert, W. (1982). Requirement of a membrane potential for the posttranslational transfer of proteins into mitochondria. Eur. J. Biochem. 125, 109–116.

Schleyer, M. & Neupert, W. (1985). Transport of proteins into mitochondria: translocational intermediates spanning contact sites between outer and inner membranes. Cell 43, 339–350.

Schmid, D., Jaussi, J., & Christen, P. (1992). Precursor of mitochondrial aspartate aminotransferase synthesized in *E. coli* is complexed with heat-shock protein DnaK. Eur. J. Biochem. 208, 699–704.

Schneider, A., Behrens, M., Scherer, P., Pratje, E., Michaelis, G., & Schatz, G. (1991). Inner membrane protease I, an enzyme mediating intramitochondrial protein sorting in yeast. EMBO J. 10, 247–254.

Schneider, A., Opplinger, W., & Jenö, P. (1994). Purified inner membrane protease 1 of yeast mitochondria is a heterodimer. J. Biol. Chem. 269, 8635–8638.

Schneider, H., Söllner, T., Dietmeier, K., Eckerskorn, C., Lottspeich, F., Trulzsch, B., Neupert, W., & Pfanner, N. (1991). Targeting of the master receptor MOM19 to mitochondria. Science 254, 1659–1662.

Schulte, U., Arretz, M., Schneider, H., Tropschug, M., Wachter, E., Neupert, W., & Weiss, H. (1989). A family of mitochondrial proteins involved in bioenergetics and biogenesis. Nature 339, 147–149.

Schwaiger, M., Herzog, V., & Neupert, W. (1987). Characterization of translocation contact sites involved in the import of mitochondrial proteins. J. Cell Biol. 105, 235–246.

Schwarz, E., Seytter, T., Guiard, B., & Neupert, W. (1993). Targeting of cytochrome-b_2 into the mitochondrial intermembrane space—specific recognition of the sorting signal. EMBO J. 12, 2295–2302.

Segui Real, B., Kispal, G., Lill, R., & Neupert, W. (1993). Functional independence of the protein translocation machineries in the mitochondrial outer and inner membranes: passage of preproteins through the intermembrane space. EMBO J. 12, 2211–2218.

Segui-Real, B., Stuart, R.A., & Neupert, W. (1992). Transport of proteins into the various subcompartments of mitochondria. FEBS Lett. 313, 2–7.

Simon, S.M., Peskin, C.S., & Oster, G.F. (1992). What drives the translocation of proteins? Proc. Natl. Acad. Sci. USA 89, 3770–3774.

Slonimski, P.P. (1953). La Formation des Enzymes Respiratoires chez la Levure. Masson, Paris.

Smagula, C. & Douglas, M.D. (1988). Mitochondrial import of the ADP/ATP carrier protein in Saccharomyces cerevesiae. J. Biol. Chem. 263, 6783–6790.

Smith, B.J., & Yaffe, M.P. (1991). A mutation in the yeast heat shock factor gene causes temperature sensitive defects in both mitrochondrial import and the cell cycle. Mol. Cell. Biol. 11: 2647–2655.

Söllner, T., Griffiths, G., Pfaller, R., Pfanner, N., & Neupert, W. (1989). Mom19, an import receptor for mitochondrial precursor proteins. Cell 59, 1061–1070.

Söllner, T., Pfaller, R., Griffiths, G., Pfanner, N., & Neupert, W. (1990). A mitochondrial import receptor for the ADP/ATP carrier. Cell 62, 107–115.

Söllner, T., Rassow, J., Wiedmann, M., Schlossmann, J., Keil, P., Neupert, W., & Pfanner, N. (1992). Mapping of the protein import machinery in the mitochondrial outer membrane by cross-linking of translocation intermediates. Nature 355, 84–87.

Stuart, R.A. & Neupert, W. (1990). Apocytochrome c: An exceptional mitochondrial precursor protein using an exceptional import pathway. Biochimie 72, 115–121.

Stuart, R.A., Nicholson, D.W., & Neupert, W. (1990). Early steps in mitochondrial protein import: receptor functions can be substituted by the membrane insertion activity of apocytochrome c. Cell 60, 31–43.

Sztul, E.S., Hendrick, J.P., Kraus, J.P., Wall, D., Kalousek, F., & Rosenberg, L.E. (1987). Import of ornithine transcarbamylase precursor into mitochondria: two step processing of the leader peptide. J. Cell Biol. 105, 2631–2639.

Sztul, E.S., Chu, T.W., Strauss, A.W., & Rosenberg, L.E. (1989). Translocation of precursor proteins into the mitochondrial matrix occurs through an environment accessible to aqueous pertubants. J. Cell Sci. 94, 695–701.

Tzagaloff, A. (1982). Mitochondria. Plenum Press, New York.

van den Bosch, H., Schutgens, R.B.H., Wanders, R.J.A., & Tager, J.M. (1992). Biochemistry of peroxisomes. Annu. Rev. Biochem. 61, 157–199.

van Loon, A.P.G.M., Brandli, A.W., & Schatz, G. (1986). The presequences of two imported mitochondrial proteins contain information for intracellular and intramitochondrial sorting. Cell 44, 801–812.

van Loon, A.P.G.M. & Young, E.T. (1986). Intracellular sorting of alcohol dehydrogenase isoenzymes in yeast: a cytosolic location reflects absence of an amino-terminal targeting sequence for the mitochondrion. EMBO J. 5, 161–165.

van Loon, A.P.G.M. & Schatz, G. (1987). Transport of proteins to the mitochondrial intermembrane space: the 'sorting' domain of the cytochrome c_1 presequence is a stop-transfer sequence specific for the mitochondrial inner membrane. EMBO J. 6, 2441–2448.

van Loon, A.P.G.M., Brandli, A. W., Pesold-Hurt, B., Blank, D., & Schatz, G. (1987). Transport of proteins to the mitochondrial intermembrane space: the 'matrix-targeting' and the 'sorting' domains in the cytochrome c_1 presequence. EMBO J. 6, 2433–2439.

Vassarotti, A., Stroud, R., & Douglas, M. (1987). Independent mutations at the amino terminus of a protein act as surrogate signals for mitochondrial import. EMBO J. 6, 705–711.

Verner, K. & Schatz, G. (1987). Import of an incompletely folded precursor protein into isolated mitochondria requires an energized inner membrane, but no added ATP. EMBO J. 6, 2449–2456.

Verner K. & Schatz, G. (1988). Protein translocation across membranes. Science 241, 1307–1313.

Verner, K. & Lemire, B.D. (1989). Tight folding of a passenger protein can interfere with the targeting function of a mitochondrial presequence. EMBO J. 8, 1491–1495.

Vestweber, D. & Schatz, G. (1988). A chimeric mitochondrial precursor protein with internal disulfide bridges blocks import of authentic precursors into mitochondria and allows quantitation of import sites. J. Biol. Chem. 107, 2037–2043.

Vestweber, D., Brunner, J., Baker, A., & Schatz, G. (1989). A 42-kDa outer-membrane protein is a component of the yeast mitochondrial protein import site. Nature 341, 205–209.

Vestweber, D. & Schatz, G. (1989). DNA-protein conjugates can enter mitochondria via the protein import pathway. Nature 338, 170–172.

Vogel, J.P., Misra, L.M., & Rose, M.D. (1990). Loss of Bip/Grp78 function blocks translocation of secretory proteins in yeast. J. Cell Biol. 110, 1885–1895.

von Figura, K. & Hasilik, A. (1986). Lysosomal enzymes and their receptors. Annu. Rev. Biochem. 55, 167–193.

von Heijne, G. (1986). Mitochondrial targeting sequences may form amphiphilic helices. EMBO J. 5, 1335–1342.

Voos, W., Gambill, D.B., Guiard, B., Pfanner, N., & Craig, E.A. (1993). Presequence and mature part of preproteins strongly influence the dependence of mitochondrial protein import on heat shock protein 70 in the matrix. J. Cell Biol. 123, 119–126.

Wachter, C., Schatz, G., & Glick, B.S. (1992). Role of ATP in the intramitochondrial sorting of cytochrome-c_1 and the adenine nucleotide translocator. EMBO J. 11, 4787–4794.

Yaffe, M.P. & Schatz, G. (1984). Two nuclear mutations that block mitochondrial protein import in yeast. Proc. Natl. Acad. Sci. USA 81, 4819–4823.

Yaffe, M.P., Ohta, S., & Schatz, G. (1985). A yeast mutant temperature-sensitive for mitochondrial assembly is deficient in a mitochondrial protease activity that cleaves imported precursor polypeptides. EMBO J. 4, 2069–2074.

Yang, M., Jensen, R.E., Yaffe, M.P., Oppliger, W., & Schatz, G. (1988). Import of proteins into yeast mitochondria: the purified matrix processing protease contains two subunits which are encoded by the nuclear *MAS1* and *MAS2* genes. EMBO J. 7, 3857–3862.

Zhuang, Z. & McCauley, R. (1989). Ubiquitin is involved in the in vitro insertion of monoamine oxidase B into mitochondrial outer membranes. J. Biol. Chem. 264, 14594–14597.

Zwizinski, C. & Neupert, W. (1983). Precursor proteins are transported into mitochondria in the absence of proteolytic cleavage of the additional sequences. J. Biol. Chem. 258, 13340–13346.

Zwizinski, C., Schleyer, M., & Neupert, W. (1984). Proteinaceous receptors for the import of mitochondrial precursor proteins. J. Biol. Chem. 259, 7850–7856.

SELECTIVE SECRETION BY LYSOSOMES

Lois Isenman

I. INTRODUCTION

Most proteins in cells are subject to continuous degradation and synthesis. The constant turnover of cellular proteins reduces the probability that damaged proteins will be present. Moreover, the mechanisms that have evolved to regulate the rates of protein degradation allow the concentration of various proteins to be modulated

Membrane Protein Transport
Volume 1, pages 145–167.

Table 1. Pathways of Protein Degradation for Endogenous and Exogenous
Proteins and their Relationship to Pathways of Antigen Presentation

Endogenous Protein	Exogenous Protein
Cytosolic Pathways	
ubiquitin-dependent, ATP-dependent	
ubiquitin-independent, ATP-dependent	
ubiquitin-independent, ATP-independent	
Ca++-dependent	
Lysosomal pathways	Lysosomal or endosomal pathways
entry by microautophagy	endocytosis
entry by macroautophagy	pinocytosis
KFERQ-dependent entry	phagocytosis
Endoplasmic reticulum and	
other organelle specific degradation pathways	
Class I MHC	Class II MHC
CD8$^+$ Cells	CD4$^+$ T Cells

(Schimke and Doyle, 1970). Finally, in higher organisms, partial degradation fragments from proteins are required to serve as antigens for immune responses.

The study of protein degradation, for the most part, has focused on the production of free amino acids from degraded proteins. Only relatively recently, with increased interest in the details of immune functioning, has it become that clear that small degradation fragments, rather than intact proteins, are the antigenic determinants recognized by T cells (Schimonkevitz et al., 1983; Townsend et al., 1986). Because of the importance of understanding how antigen presentation processes and protein degradation processes interact, we chose to explore the generation of small peptide fragments during protein degradation.

Our approach has been to characterize the secretion of peptide fragments and polypeptides from intact cells and from lysosomes *in vitro* during the degradation process. We find that proteins and specific peptide fragments are released from lysosomes and released from cells. These observations characterize a pathway whereby peptides and proteins may be rescued from lysosomes and transported to the cell surface. The specificity of these events suggest that polypeptide transporters are present on both the lysosomal membrane and the plasma membrane.

II. PROTEIN DEGRADATION PATHWAYS AND ANTIGEN PRESENTATION

Many cellular proteins are degraded in the cytosol, which contains a variety of protein degradation pathways (Table 1). Although the ubiquitin-dependent degradation pathway has been specifically linked to antigen presentation (Michalek et al., 1993), the participation of other cytosolic pathways cannot be excluded. According to current theory, degradation fragments of 8 to 12 amino acids are transported, via a transporter, from the cytosol into the endoplasmic reticulum (ER) (Monaco et al., 1990; Spies and DeMars, 1991). In the ER, degradation fragments complex with major histocompatibility complex (MHC) class I molecules, and travel to the cell surface (Townsend and Bodmer, 1989). In addition, a portion of newly synthesized proteins is degraded within the ER (Lippincott-Schwartz et al., 1988), and presumably, peptide fragments from these molecules also travel to the cell surface in conjunction with class I MHC (Henderson et al., 1992). At the cell surface, CD8[+] T cells survey the peptide antigens displayed in conjunction with class I MHC and lyse those cells that demonstrate evidence of viral and other foreign protein.

In addition, cells sample the protein content of the extracellular environment by endocytosis, pinocytosis, or phagocytosis. In these processes, exogenous proteins are transported by the vacuolar system to late endosomes or lysosomes for degradation. Certain cells, (i.e. macrophages, B-lymphocytes, and dendritic cells), also express a second immune recognition molecule, MHC class II. These cells present on their surface, in conjunction with MHC class II molecules, fragments of 8–20 amino acids derived from extracellular protein to CD4[+] T cells (Germain and Margulies, 1993). CD4[+] T cells play a role in the amplification of the humoral response to foreign antigens. Both the site where MHC II and peptide meet and their mechanism of movement to the cell surface remain controversial.

Some cytosolic proteins are also degraded within lysosomes. For example, RNase A microinjected into the cytosol is degraded to free amino acids within lysosomes (Chiang et al., 1989). Transport from the cytosol into the lysosome depends on the presence of a specific amino acid motif. Moreover, a portion of most cytosolic proteins are delivered to lysosomes by non-specific processes. Internal cellular membranes round up and enclose small amounts of cytosol in vesicles that subsequently fuse with the lysosome in a process called macroautophagy (Dunn, 1990). In addition, the lysosomal membrane can round up and internalize small amounts of cytosol in a process called microautophagy (Ahlberg et al., 1982). Cytosolic proteins that are fragmented in lysosomes are presented to CD4[+] T cells (Germain and Margulies, 1993); the converse process, the presentation of exogenous proteins in conjunction with MHC I to CD8[+] T cells, has also been reported and will be discussed in detail in a latter section.

III. RELEASE OF PROTEINS AND PEPTIDES FROM PROTEIN DEGRADATION PATHWAYS

Red-cell mediated microinjection has provided an important tool for the study of protein degradation. Red-cell ghosts loaded with a radiolabeled protein of interest are fused to a large number of cells using either polyethylene glycol (McElligott

Table 2. Half-life and Percentage of Secreted Radioactivity that is
Acid-Precipitable for Microinjected Proteins and Peptides[a]

Molecule	$t_{1/2}$	Acid-Precipitable
	hr	%
[125]I-Poly(Glu:Tyr)	110 ± 10	10 ± 3
[125]I-Insulin A chain	104 ± 6	18 ± 3
[^{3}H]RNase A	80 ± 5	60 ± 4
[^{3}H]Lysozyme	59 ± 2	24 ± 4
^{125}I-IMR-90 cytosol	47 ± 10	22 ± 6
[125]I-Ovalbumin	46 ± 5	23 ± 6
[125]I-Rat liver cytosol	44 ± 4	20 ± 4
[^{3}H]RNase A S-peptide 4-13	40 ± 4	67 ± 6
[125]I-Bovine serum albumin	12 ± 2	26 ± 2

Note : [a]Radiolabeled proteins and peptides were microinjected into diploid human fibroblast. Half-lifes for
degradation to free amino acids were calculated over a 24-hr period, and the percentages of secreted
radioactivity that were acid-precipitable are given for 24 hr.

and Dice, 1984) or Sendai virus (Schlegel and Rexsteiner, 1975). The labeled
protein is delivered into the cytosol, and may undergo further distribution within
the cell. The radioactivity released into the medium during degradation can be
monitored.

Rogers and Rechsteiner (1988) individually microinjected 32 diverse ^{125}I-labeled
proteins into Hela cells. In each case, in addition to labeled acid-soluble material
(i.e., free amino acids), they found labeled acid-precipitable material (i.e., peptides
and proteins) released into the medium. To determine whether these findings could
be generalized to nontransformed cell lines, we separately microinjected nine
radiolabeled proteins or peptides, degraded by a variety of different pathways, into
the cytosol of IMR-90 cells, human diploid fibroblasts (Isenman and Dice, 1989).
Culture medium was precipitated with a mixture of phosphotungstate and hydro-
chloric acid, which precipitates all tripeptides and many dipeptides, along with
larger fragments and intact protein. Like Rogers and Rechsteiner, we found that
each molecule released acid-precipitable material into the medium. Table 2 presents
both the half-life for degradation to free amino acids and the percentage of
radiolabel in the medium that is acid-precipitable material for each of the test
substances. Because neither monoiodotyrosine (Tweto and Doyle, 1976) nor di-
methyl lysine (Kim and Paik, 1965), the respective end products of degradation of
[125]I-labeled and reductively methylated proteins, can serve as substrates for the
cell's synthetic machinery, the reutilization of the label for the *de novo* synthesis of
secretory proteins can be excluded. Taken together, these microinjection studies
suggest that acid-precipitable material is released as part of many, and perhaps all,
degradation pathways.

In subsequent studies, we chose to concentrate primarily on the production and secretion of acid-precipitable material during the degradation of RNase A. RNase A is a relatively long-lived protein that is degraded to free amino acids by lysosomes with a half-life on the order of 80 hours in the presence of serum (Dice, 1987). We chose RNase A because acid-precipitable material accounts for 60% of the label it released into the medium; this is considerably more than the other microinjected molecules we tested, with the exception of RNase A S-peptide, the N-terminal 20 amino acids of RNase A. Moreover, the degradation of RNase A to free amino acids has been a long-term interest of this laboratory (see Chapter 4).

IV. RELEASE OF PROTEINS AND PEPTIDES FROM LYSOSOMAL PATHWAYS OF PROTEIN DEGRADATION

The lysosomal membrane has generally been considered impermeable to all peptides, with the exception of free amino acids and some dipeptides. In an influential experiment in the late 1960s, Ehrenreich and Cohn (1969) used the osmotic swelling of lysosomes within cells in response to the endocytosis of high concentrations of a peptide as an indication that the lysosomal membrane was impermeable to the molecule. Of the peptides tested, they found that a high concentration of a tripeptide with a small molecular radius and a dipeptide with a larger molecular radius were able to induce lysosomal swelling. They reasoned that although other factors might influence the permeability of a peptide, increasing size would always limit this permeability. They concluded that, for the most part, in addition to free amino acids, only dipeptides would be able to freely cross the lysosomal membrane.

Although this study and a similar one (Lloyd, 1971) tested a very limited number of peptides, the results were consistent with a preexisting paradigm that asserted that proteins and peptides could not traverse most membrane barriers. Consequently most researchers accepted the idea that the lysosomal membrane was in fact impermeable to other peptides and proteins. The lysosome was thus viewed as a terminal degradation organelle for protein substrates, and all proteins delivered to it were thought to be hydrolyzed entirely to free amino acids or dipeptides.

Observations by H.L. Segal in the early 1970s (Segal et al., 1979) provided the first notable exception. Segal delivered invertase to lysosomes in hepatocytes and found a gradual increase in the cytosolic concentration of invertase over a period of days. As a possible explanation for this observation, he proposed that lysosomes are able to release proteins as well as degrade them. The model of lysosomal function he derived from these and other observations allowed for protein and peptide fluxes in both directions across the lysosomal membrane (Segal and Doyle, 1979). The work described in this chapter, along with other studies from both this and other laboratories (Buktenica et al., 1987; Chiang et al., 1989; Barrueco et al., 1992; Aniento et al., 1993), suggest that this view of lysosomal permeability may be correct.

Since the degradation of RNase A to free amino acids had been shown to occur within lysosomes (McElligot et al., 1985), it seemed likely that the labeled acid-precipitable material observed in the culture medium after the microinjection of RNase A would also originate from a lysosomal pool of the protein. This premise was strengthened by the observation that under all conditions tested, the rate of appearance of acid-soluble label in culture mededium paralleled the rate of appearance of acid-precipitable label (Isenman and Dice, 1989). For example, when the rate of release of free amino acids from RNase A was enhanced approximately twofold by serum withdrawal, the rate of release of acid-precipitable material was also enhanced twofold. Similarly, addition of ammonium chloride to cells caused parallel effects on the release of acid-soluble and acid-precipitable components in both the presence and the absence of serum. Since the rate of uptake into lysosomes was the rate-limiting step in the degradation of RNase A to free amino acids (Dice, 1987), and the release of acid-precipitable and acid-soluble material into culture medium occurred in parallel, we hypothesized that uptake of RNase A into the lysosome was also the rate-limiting step in the formation of degradation fragments from the molecule.

Other models, however, were also possible. For example, because the consensus signal for uptake into the lysosomes is close to the N-terminus, a segment of RNase A containing 60% of the radiolabeled lysines could be clipped off, with the remainder undergoing lysosomal uptake. The C-terminus fragment could then be secreted intact or in contiguous fragments across the plasma membrane. Or, if the acid-precipitable material consisted entirely of intact protein, its release could occur directly from the cytosol and the apparent linkage of the release of acid-precipitable and acid-soluble material could reflect other events.

To determine the molecular weight(s) of the released material, we applied precipitated culture medium from cells microinjected with [^{3}H]RNase A (radiolabeled on lysine) to a Sephadex G-75 gel filtration column in the presence of 6 M urea and phosphate buffered saline (PBS) (Figure 1A). The first peak was entirely unexpected; it migrated close to the void volume and represented material of a higher molecular weight than RNase A itself, which runs considerably into the column volume (relative mobility of 0.44). Because the column was run in urea to prevent aggregation of protein, the presence of high molecular weight labeled material in the culture medium indicated that high molecular covalent conjugates of either RNase A, or its degradation fragments, were generated and secreted by the cell. These high molecular conjugates appear to contain ubiquitin and will be discussed in greater detail in a later section.

The second peak of radioactivity migrated at a mobility similar to that of standard RNase A and appeared to include a number of the larger breakdown products as well as the intact molecule. The third peak centered on the inclusion volume. Because the precipitate had been washed thoroughly to remove any trapped free amino acids and dipeptides prior to its application to the column, the peak in the inclusion volume represented low molecular weight fragments from RNase A.

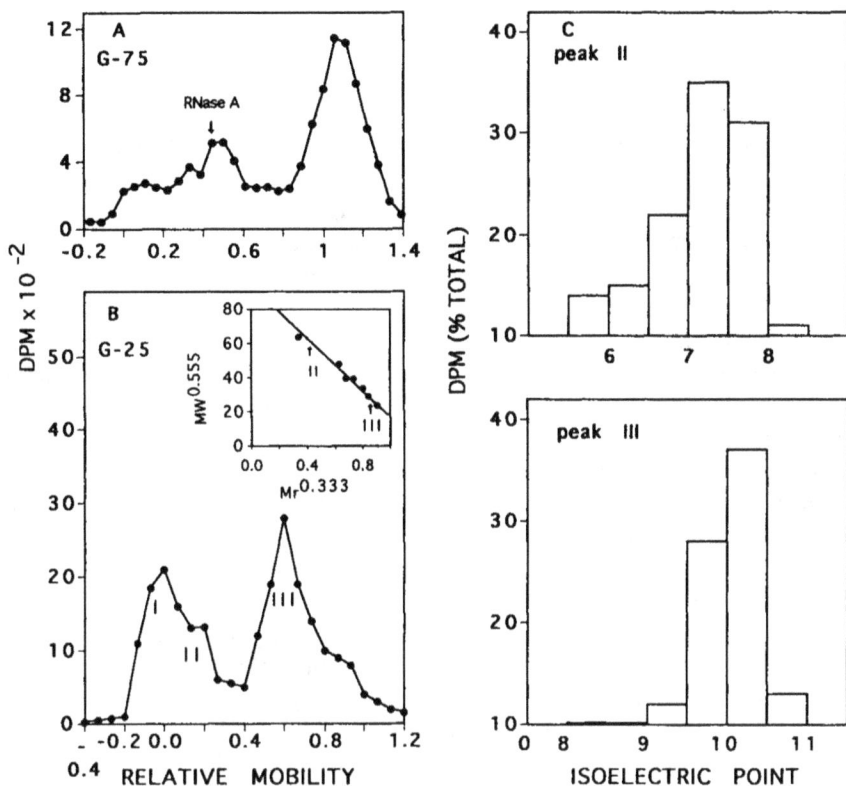

Figure 1. Size and isoelectric point of acid-precipitable material released from fibroblasts microinjected with [³H]RNase A. Final culture medium was acid-precipitated, and applied, after thorough washing to remove free amino acids, to either (**A**) a Sephadex G-75 column or (**B**) a Sephadex G-25 column, in the presence of 6 M urea containing PBS. All data from gel filtration columns are expressed in terms of relative mobility, calculated as (fraction #-void)/(inclusion-void) to facilitate comparisons between different experiments. Blue dextran (>200 kDa) and ¹⁴C-tyrosine were included in each column run to mark the void volume and inclusion volume respectively. The G-75 radioactivity profile demonstrates that high molecular weight conjugates of RNase A and low molecular weight degradation products are released from cells. The G-25 column demonstrates that the low molecular weight radioactivity resolves into two peaks (II and III). According to the G-25 column calibration (inset, **B**), the low molecular peaks correspond to fragments with approximate molecular masses of 1150 and 400 Da. (**C**) The material from each low molecular weight peak was applied to an isoelectric focusing column in the presence of 6 M urea; the radioactivity profiles, given as % of total radioactivity released into the medium, indicate relatively discrete isoelectric points. Adapted from Isenman and Dice (1989) with permission.

We considered a range of possibilities for the identity of this labeled low molecular weight secreted material. It could, on the one hand, consist of a heterogenous mixture of lysine-containing fragments. It could, on the other hand, consist of only one, or a few specific fragments, indicating that the fragment generation and/or secretion machinery exercised considerable specificity. For example, one possibility was that only fragments of a size appropriate to serve as antigens for the immune system were secreted.

To characterize the size of the low molecular weight secreted material, we applied precipitated culture medium to a Sephadex G-25 column in the presence of PBS containing 6 M urea (Figure 1B). The radioactivity profile again showed three peaks of radioactivity. The first peak, centering on the void volume, contained high molecular weight conjugates, intact RNase A, and larger breakdown products, while the low molecular weight material was distributed between the remaining two peaks. The first of the low molecular weight peaks appeared as a shoulder off the peak in the void volume and centered at a relative mobility of approximately 0.2. According to the column calibration (Figure 1B, inset), material in this peak has an average molecular weight of 1150, or 8 to 12 amino acids, and is consistent with the size of T-cell antigens. The second peak of low molecular weight material centered on a relative mobility of 0.6 and corresponded to peptides of approximately 400 Da, or 3 to 4 amino acids.

The finding that the low molecular weight radioactivity resolved itself into two discrete peaks, rather than a more equal distribution throughout the column volume, indicated that only fragments of these two sizes were generated and/or secreted by the cell. One possibility was that each peak represented a heterogenous collection of fragments with the same size. Another possibility, at the other extreme, was that each peak corresponded to the secretion of a discrete fragment. In an attempt to get better sense of the selectivity of the peptides generated from RNase A and then secreted, we determined the isoelectric point of the radioactive material in the two low molecular weight peaks from the Sephadex G-25 column.

Isoelectric focusing was performed in 6 M urea to prevent aggregation of fragments. The isoelectric point of the larger fragment was 7.0–7.5. (Figure 1C, top). The distribution of radioactivity, however, appeared considerably broadened, a finding consistent with the secretion of two or more fragments possibly differing by single charged amino acids. The isoelectric point of the smaller fragment was 10–10.5 (Figure 1C, bottom). Although the data are consistent with the secretion of a single peptide, because RNase A is a very basic protein and had been labeled on lysine whose isoelectric point is close to 10, the secretion of a variety of different small lysine containing fragments cannot be excluded. Nonetheless, the pI of the fragment appears to suggest selection against strongly acidic residues and indicates that other factors in addition to mass determine if small peptides are released into the culture medium.

In fibroblasts, material endocytosed into the vacuolar system is degraded primarily in lysosomes (Isenman and Dice, 1993). We also allowed cells to incorporate radiolabeled [^{3}H]RNase A by endocytosis in order to compare the acid-precipitable material released into the medium to that released from cells microinjected with the protein. The cells were incubated with a high concentration of labeled [^{3}H]RNase A in the culture medium for 1 hour. The medium was removed, the cells were washed thoroughly for the next 30 minutes, and the release of acid-precipitable and acid-soluble material into the medium was followed for 4 hours (Fig ure 2A). The release of acid-precipitable material was biphasic, with 12% of the total incorporated [^{3}H]RNase released in the first hour and 4% per hour released during subsequent time periods. An initial rapid release of acid-precipitable material after endocytosis has also been noted by Buktenica et al. (1987) who first presented evidence that acid-precipitable and acid-soluble material released from endocyto-sed proteins were derived from a single lysosomal pool. They attributed this initial rapid release phase to the re-release or "regurgitation" of endocytotic vesicles prior to delivery to the lysosome.

Acid-precipitable material accounted for 60% of the radiolabel released into the culture medium (Figure 2A). This value is identical to that for RNase A incorporated into cells by microinjection (see Table 2). Moreover, when precipitated culture medium from cells that had endocytosed RNase A was applied to Sephadex G-75 and G-25 columns, the radioactivity profiles were similar to those from microin-jected material. As Figure 2B demonstrates, in addition to the peak in the void volume, two peaks of low molecular weight material radiolabeled material are present and correspond to material with an average molecular mass of 1150 and 400 Da. Finally, the isoelectric points of material in the two lower molecular weight peaks are similar to those from the microinjected material (Figure 2C). The similarity of the acid-precipitable material released when RNase A is microinjected into the cytosol and when RNase A is endocytosed directly into the vacuolar system strongly supports the view that the proteins and peptides released from microinjected RNase A are derived from a lysosomal pool of the enzyme.

Either a one-step or a two-step transport model could account for the observed release of peptides and proteins from lysosomes during the degradation process. In the one-step model, lysosomes would simply fuse with the plasma membrane and deliver their contents to the exterior of the cell. This model would limit lysosomes to generating only those fragments that appear in secretion. Because lysosomal enzymes in their processed intralysosomal form are not released from human fibroblasts (Hasilik and Neufeld, 1980), a one-step fusion model cannot easily explain the appearance of degradation fragments in culture medium.

Instead, the two membrane barriers between the lysosome and the culture medium, the lysosomal and the plasma membrane, appear to be crossed in sequence. Either a membrane transport or a vesicular trafficking model is consistent with the observations. A membrane transport model would require the presence of transporters on each membrane whose specificities include the released fragments. Other fragments in addition to those found in secretion

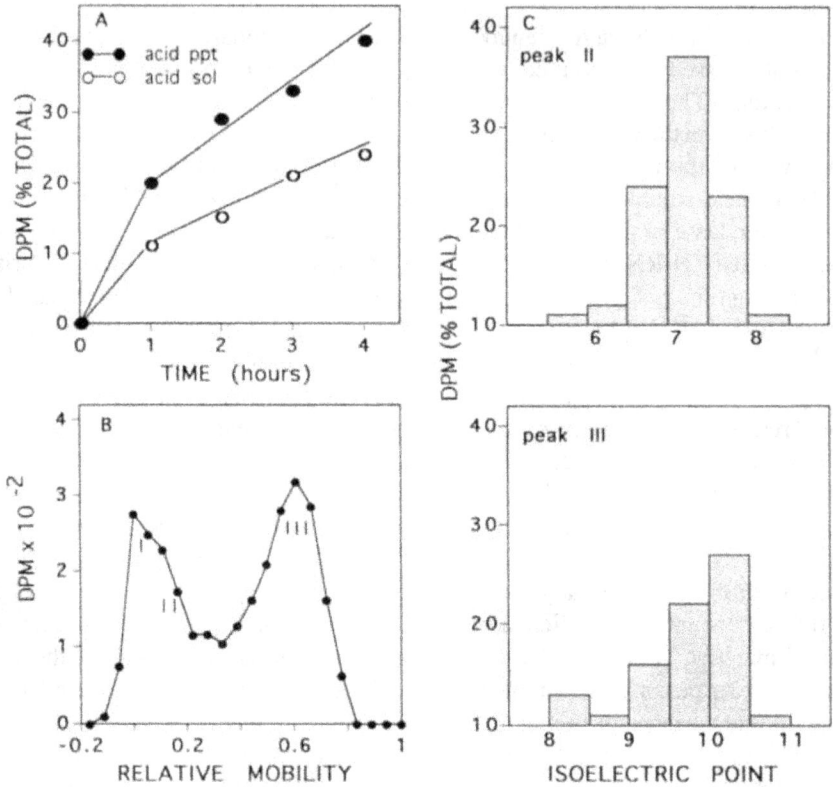

Figure 2. Release of acid-precipitable material after endocytosis of [3H]RNase A by fibroblasts. (**A**) Release of acid-soluble and acid-precipitable material into culture medium after 1 hr of endocytosis and wash, given as a percentage of total incorporated radioactivity. (**B**) A Sephadex G-25 radioactivity profile of precipitated culture medium demonstrates that after endocytosis, the low molecular weight material released by cells also resolves into two peaks corresponding to fragments of approximately 1150 and 400 Da. (**C**) The distribution of radioactivity following isoelectric focusing of material from each peak in 6 M urea is similar to that seen for the microinjected material. Adapted from Isenman and Dice (1989) with permission.

could be formed in substantial quantity within lysosomes, and might even be released from them, since ultimate selectivity would reside with the cytoplasmic transporters. In contrast, with vesicular trafficking, selectivity would be determined entirely during vesicle loading. With either model, it is likely that the degradation fragments, either free or enclosed within a vesicle, would reside, even if very briefly, within the compartment intermediate to the two membranes, the cytosol.

V. RELEASE OF PEPTIDES FROM LYSOMES *IN VITRO*

We hoped to confirm the initial step of the process—transport from the lysosome into an intermediate compartment—by demonstrating that lysosomes *in vitro* secrete the same fragments that are released into culture medium by intact cells. We first attempted to show that the fragments are present in lysosomes *in vitro* after incubation. Cells were allowed to endocytose radiolabeled RNase A overnight, the label was chased into the lysosomes during a 50-minute wash period, and a 10,000 x g pellet containing the lysosomes was generated (Isenman and Dice, 1993). The resuspended pellet was incubated for two hours at 24 °C and aliquots were applied to either a Sephadex G-75 or a Sephadex G-25 column in the presence of 6 M urea containing PBS.

The G-75 radiolabel profile from isolated lysosomes demonstrated the same three peaks that appeared in the profile from precipitated culture medium, although, as might be expected, the second peak containing intact RNase A was now the largest peak. The G-25 radiolabeled profile, however, was a surprise. Only one peak of low molecular weight radiolabeled material was apparent. Moreover, rather than centering at a relative mobility of 0.6, the relative mobility of the small fragment seen in culture medium, this peak centered at a relative mobility of 0.5, corresponding to fragments of approximately 550 Da (Figure 3, top). The apparent failure of the *in vitro* lysosomes to generate the 400-Da peptides could not have resulted from the altered endocytosis conditions, since overnight endocytosis, as opposed to 1 hour of endocytosis, (for unknown reasons) greatly enhanced the relative amount of the smaller fragment released into the culture medium (Figure 3, bottom; compare to Figure 2B). Nor was the presence of the 550-Da fragments in lysosomes an artifact of *in vitro* degradation; we found that the major peak of radiolabeled material remaining in cells following endocytosis and 5 hours of chase also consisted of larger fragments (Figure 3, bottom).

We considered several other explanations. One possibility was that only small amounts of the 400-Da peptides are generated in lysosomes *in vitro* under these conditions and could not be detected in the presence of so much more of the slightly larger peptides. Another was that lysosomes *in vivo* release 550-Da fragments into the cytosol and they are processed there into the 400-Da peptides before being released into the culture medium. A third possibility was that intact cells secrete a portion of the 550-Da peptides into the culture medium, where they undergo further rapid processing by another secreted moiety into the 400-Da peptides.

The first two appeared more likely. To test for these, it was necessary to analyze the radiolabeled material secreted from the *in vitro* lysosomes independent of the material that remained with them. To this end, we respun the lysosomes after the 2-hour incubation period and analyzed the supernatant and the resuspended pellet separately. In order to adjust the data for lysosomal breakage, we also monitored the presence of a lysosomal enzyme in both fractions. Supernatant and pellet fractions were initially analyzed on a Sephadex G-75 column in the presence of 6

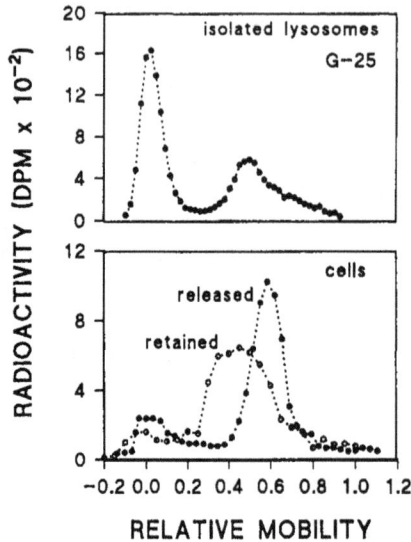

Figure 3. The low molecular weight radioactivity from lysosomes *in vitro* compared to the low molecular weight radioactivity that is released and retained by intact cells. (A) A 10,000 x g fraction derived from cells that had endocytosed [³H]RNase A 16 hr was incubated for 2 hr at 24 °C and applied to a Sephadex G-25 column in 6 M urea containing PBS. The only low molecular weight peak present corresponds to fragments with an approximate molecular mass of 550 Da. (B) Sephadex G-25 radioactivity profiles of precipitated culture medium (closed circles) and cells harvested in 1% Triton X-100 (*open circles*). Cells were incubated with [³H]RNase A for 16 hr and chased for 5 hr. Under these endocytosis conditions, large amounts of the 400-Da fragment are secreted into the culture medium during a 5 hr chase, but most of the radiolabel remaining in cells also appears in slightly larger fragments. Adapted from Isenman and Dice (1993) with permission.

M urea containing PBS (Figure 4). The radiolabel profile for the supernatant indicated that 24% of the total radioactivity associated with the high molecular weight material (relative mobility 0.0–0.2) and 46% of the total radioactivity associated with the low molecular weight material (relative mobility 0.8–1.0) were found in the supernatant. Unexpectedly, in contrast to what we had observed from intact cells, little or no intact RNase A and/or its larger breakdown products were released into the supernatant. The fact that the most abundant molecules within the lysosomes were not found in the incubation medium argues that the integrity of the membrane remained intact during *in vitro* incubation, and suggests that the release of higher and lower molecular weight material result from a physiological rather than an artifactual process.

The reason that RNase A is not released from *in vitro* lysosomes remains unclear. With intact cells, it is possible that the RNase A found in chase medium after

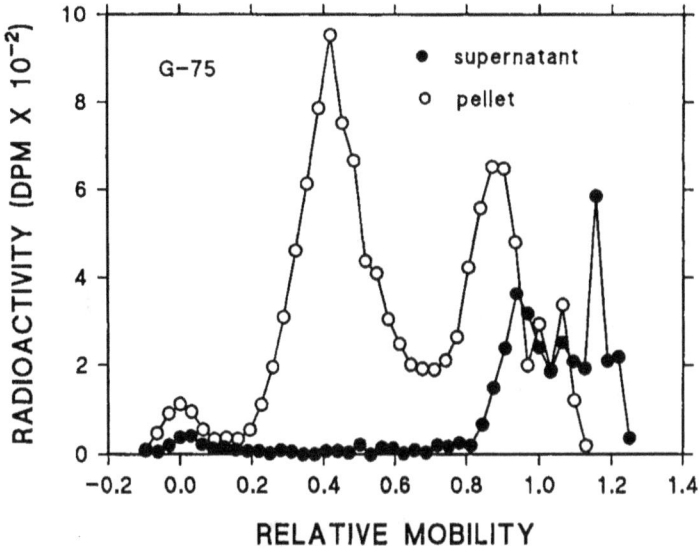

Figure 4. Products of [³H]RNase A that are released and products that are retained by lysosomes *in vitro*. After incubation, *in vitro* lysosomes were repelleted, and the pellet (*open circles*) and the supernatant (*closed circles*) analyzed separately on a Sephadex G-75 column in the presence of 6 M urea containing PBS. The radioactivity profiles demonstrate that little RNase A or its larger breakdown products are released, although substantial amounts of both high and low molecular weight material are found in the supernatant. Reprinted from Isenman and Dice (1993) with permission.

endocytosis results exclusively from re-release of endocytotic vesicles *prior* to their delivery to the degradation compartment. Likewise, release of RNase A after microinjection into cytosol of cells could reflect secretion via a plasma membrane transporter. However, since control experiments with conditioned medium indicate that RNase A is not fragmented in the culture medium, these explanations cannot easily account for release of larger degradation fragments into chase medium. Alternatively, the release of RNase A and its higher molecular weight breakdown fragments from the lysosome could require factors not present under these *in vitro* incubation conditions.

In order to resolve the low molecular weight material released from *in vitro* lysosomes, supernatant and pellet fractions were analyzed on Sephadex G-25 columns in the presence of 6 M urea containing PBS. At 0 °C, little radioactivity was evident in the incubation medium at a relative mobility of 0.6. However, about 10% of the total radioactivity that migrates at a relative mobility of 0.5 was released into the medium during the 2-hour incubation (Figure 5, top). At 24 °C, the peak of released radioactivity was shifted from a relative mobility of 0.5 to a relative

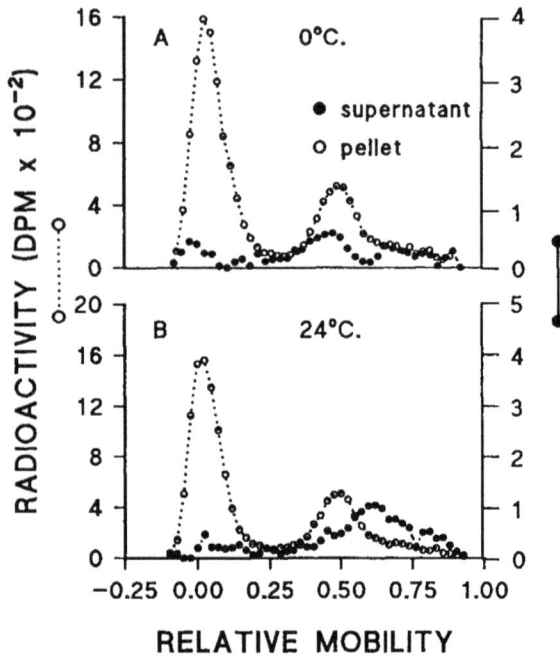

Figure 5. Sephadex G-25 radioactivity profiles for resuspended pellets and supernatants fractions from lysosomes *in vitro* showing that 400 Da peptides are generated and released into the incubation medium when *in vitro* lysosomes are incubated at 24 °C (**B**), but not at 0 °C (**A**). Reprinted from Isenman and Dice (1993) with permission.

mobility of 0.6 (Figure 5, bottom), the relative mobility of the small fragments secreted by intact cells. At 24 °C but not at 0 °C, the 400-Da peptides are apparently produced by isolated lysosomes and readily released from them.

We used reverse-phase high performance liquid chromatography to confirm that the 400-Da fragments released from lysosomes *in vitro* are similar to those released into the culture medium by intact cells. In both cases one major peak of radioactivity was apparent. The fact that lysosomes *in vitro* both released and retained degradation fragments similar to those released and retained by intact cells argues very strongly that the release of peptides from lysosomes *in vitro* reflects a physiological process.

The relative permeability of the lysosomal transport mechanism to the 400- and 550-Da fragments at 24 °C can be estimated from the amount of each fragment that is released into the media by *in vitro* lysosomes relative to the amount that is retained. The ratio of secreted:retained material varies greatly from experiment to experiment at each relative mobility. When the values are compared within each experiment, however, much of the variability disappears, and the lysosomes appear to be at least fourfold to ninefold more permeable to the 400-Da peptides than to

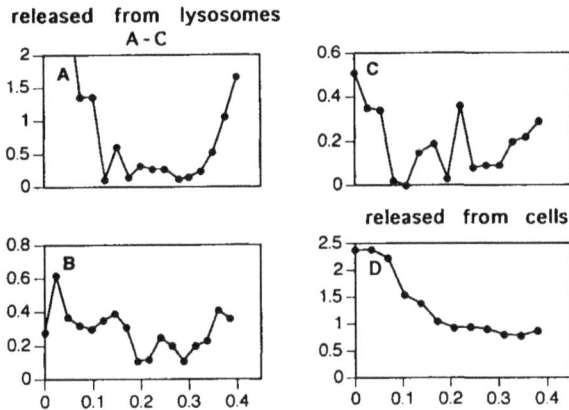

Figure 6. (A–C) Partial Sephadex G-25 radioactivity profiles showing the lysosomal secretion of a range of fragments derived from [³H]RNase A of an appropriate size to serve as antigens. (D) A partial Sephadex G-25 radioactivity profile of precipitated chase medium from intact cells also suggests several peaks in this range. Adapted from Isenman and Dice (1993) with permission.

the 550-Da peptides. This analysis gives a minimum estimate of the permeability difference of the lysosomal membrane to the two fragments because much of the radioactivity appearing in lysosomes at a relative mobility of 0.6 can be attributed to peaks appearing at a relative mobility of 0.5 and 0.7 (Figure 5, bottom).

The initial portion of the G-25 columns from *in vitro* lysosomes consistently demonstrates the presence of several peaks or shoulders of radiolabeled material running between a relative mobility of 0.05 and 0.3 (Figure 6). This material corresponds to peptides with approximate molecular masses of 800 to 1900 Da and covers the size range of peptides that can serve as antigens for the cell based immune system. Although these peptides are not present in large quantities, less than 100 peptides bound to class II MHC per cell are required to elicit a maximum helper T-cell response (Demotz et al., 1990).

Multiple peaks are also suggested in a high resolution profile of the initial portion of a G-25 profile of culture medium from intact cells (Figure 6D). These peaks, however, are much less distinct than the peaks in supernatant from lysosomes *in vitro*. This difference may only be apparent. Lysosomes *in vitro* secrete little intact RNase A or its larger breakdown fragments into the incubation medium. Since this material runs in the void volume, it tends to mask nearby peaks at low relative mobilities. Cytosolic processing of peptides released from lysosomes *in vivo* may also account for this discrepancy.

To explore the permeability of the lysosomal membrane to a series of similar molecules, we synthesized AYK, MMK, and RCK, three tripeptides from RNase A with the same charge as the small natural degradation fragment. We incubated cells

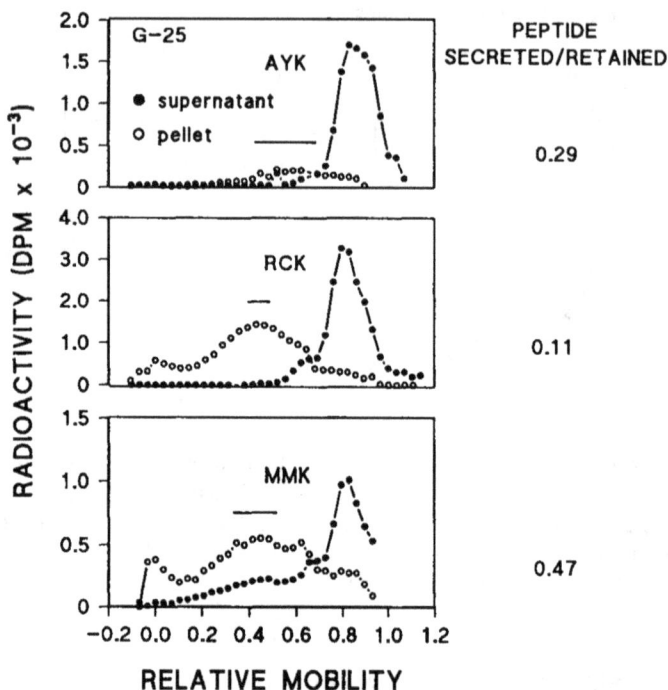

Figure 7. Selective release of synthetic tripeptide probes by lysosomes. Lysosomal fractions generated from cells that had endocytosed AYK, RCK or MMK were incubated for 30 min. The pellets (*open circles*) and supernatants (*closed circles*) were applied to a Sephadex G-25 column equilibrated with PBS containing urea. The intact peptides migrate between a relative mobility of 0.2 and 0.7. The bars above the graphs indicate the positions of the peaks. The effective permeability of the lysosomal membrane to the peptides covered a 4- to 5-fold range. Reprinted from Isenman and Dice (1993) with permission.

in the presence of high concentrations of the reductively methylated probes and prepared lysosomal fractions. Because we anticipated that release might be very rapid, these fractions were incubated *in vitro* for only 30 minutes before we separated the incubation medium and the pellet.

Unexpectedly, each peptide was handled differently by the lysosomes as determined by the Sephadex G-25 radioactivity profiles (Figure 7). AYK was predominantly degraded, RCK was predominantly retained, while a substantial amount of intact MMK was released into the incubation medium. We compared the effective permeability of the lysosomal membrane to the three probes by determining the amount of secreted to retained peptide at the end of the incubation period. This measure, which is independent of the degree of degradation to free amino acids, indicates that MMK was transported four to five times more readily, and AYK two to three times more readily than RCK.

Both the formation and release of partial degradation fragments by lysosomes appear to be selective processes. Apparently, only a small number of lysine containing fragments are generated in substantial enough quantity to be detected by our assay system. One possible explanation is that only these fragments are excised from the protein, while other portions are degraded directly to amino acids. Alternatively, a large variety of partial degradation fragments could be generated, but only these few are spared further lysosomal degradation. Some fragments might be spared because of an inherent resistance of their amino acid sequences to lysosomal hydrolases. For example, the 550-Da degradation fragments from RNase A, because of their low transport as well as low degradation rate, can be identified as candidates for this type of natural resistance to degradation. Some partial degradation fragments may be protected by binding to other molecules. For example, MHC, or an accessory molecule with similar binding specificity, could protect antigenic fragments from further degradation while they remain in the degradation compartment. An additional possibility is that some fragments are spared lysosomal degradation because they undergo rapid transport out of the degradation compartment. The 400-Da peptides released from RNase A and MMK may owe their survival to such a mechanism.

Some fragments that are spared lysosomal degradation, such as the 550-Da fragments from RNase A, apparently are largely retained by the lysosome, while others, such as the 400-Da peptides, are mostly secreted. Moreover, the mechanism(s) responsible for the transport of the small peptides are capable of fine distinctions, since one peptide probe from RNase A of the same size and charge as the 400-Da degradation fragment was primarily retained within the lysosome, while a substantial amount of another probe of the same size and charge was released. We know much less about the selectivity of transport of the 8 to 18 amino acid fragments; we are in the process of characterizing these important molecules further and determining their distribution between the lysosome and the incubation medium.

VI. SECRETION OF UBIQUITIN CONJUGATES OF ENDOCYTOSED PROTEINS

The presence of high molecular weight conjugates of [³H]RNase A in lysosomes (see Figure 4) and culture medium (see Figures 1A, 3B, and 4) suggested a possible involvement of the ubiquitin conjugating system. Ubiquitin is an 8-kDa protein that is activated and attached via a carrier protein to a lysine group on a variety of protein acceptors. Ubiquitin also binds to specific lysine residues of other ubiquitin molecules, and the presence of a multiubiquitin chain on a protein appears to mark the protein for degradation (Finley and Chau, 1991; Hershko and Ciechanover,

1992; Jentsch, 1992). The manner in which ubiquitin promotes this degradation is not known; in most cases, however, degradation appears to be carried out by the multicatalytic proteosome (Finley and Chau, 1991; Hershko and Ciechanover, 1992; Jentsch, 1992). The ubiquitin-conjugating enzymes described initially were cytosolic, and the ubiquitin degradation system is generally viewed as a pathway that functions in the degradation of cytosolic and nuclear proteins (Finley and Chau, 1991; Hershko and Ciechanover, 1992; Jentsch, 1992). Ubiquitin conjugates have been observed by immunocytochemistry in lysosomes (Schwartz et al., 1988; Laszlo et al., 1990), but because conjugating enzymes have not been detected (Schwartz et al., 1992), most researchers assume this material has a cytosolic origin. It is interesting to note, however, that the ubiquitin activating enzyme has been localized to the cytoplasmic face of lysosomes (Schwartz et al., 1992).

Oxidized RNase A is known to be a much better substrate for the ubiquitin system than the native molecule (Dunten and Cohen, 1989). We hypothesized that the ubiquitin system had a role in the genesis of the high molecular weight conjugates we had observed, since much more of this material was formed and secreted when cells were allowed to endocytose oxidized RNase A. Moreover, analysis of the high molecular weight material indicated it consists of molecules with a wide range of molecular weights, as is also characteristic of ubiquitin-conjugated proteins. Incubation of the high molecular weight radioactive material in culture medium with antibodies to ubiquitin chains indeed confirmed that at least 62% of the high molecular weight radioactivity, and possibly more, could be attributed to ubiquitin conjugates.

When cells were allowed to endocytose radiolabeled ubiquitin, high molecular weight radiolabeled conjugates were also produced and secreted. Although a large portion of the high molecular weight material generated following endocytosis of either [³H]ubiquitin or [³H]RNase A was degraded, 80% of what remained was secreted. Moreover, most of this secretion occurred in the first hour after the wash period. Therefore, at least one component of the initial rapid release of acid-precipitable material following endocytosis, first described by Buktenica et al. (Buktenica et al., 1987), is composed of ubiquitin conjugates. Conjugates of a specific general size range do appear to predominate in secretion, but more work is required to characterize the origin and nature of this specificity.

Several mechanisms could account for the ubiquination of endocytosed proteins. Intact proteins or degradation fragments released from lysosomes could undergo ubiquitination in the cytosol. However, to explain the presence of conjugates in lysosomes and their release from them, some conjugates would then have to be transferred back from the cytosol into lysosomes. Another possibility is that autophagic processes import the required ubiquination machinery from the cytosol, although very different pH optima are required of cytoplasmic and lysosomal enzymes. A third possibility is that ubiquination can occur within lysosomes and is carried out by enzymes that have not yet been described. Such a scheme not only would account for the present data but could also explain the hitherto puzzling phenomenon that some cell surface receptors are ubiquinated on their extracellular domain (Siegelman and Weissman, 1988). It is important to point out that these different explanations are not necessarily mutually exclusive. Endocytosed proteins and/or their degradation fragments could serve as potential substrates for ubiquination first in the lysosome and then in the cytosol.

VII. TRANSPORT MECHANISMS

The task of constructing a vesicular trafficking model to account for the data in its entirety appears formidable. Vesicles large enough to accept very large conjugates would be necessary, and all vesicles, both large and small, would require binding mechanisms capable of selecting only certain molecules. Larger vesicles, moreover, would need mechanisms to limit the amount of fluid phase they trap. Finally, vesicles would require a means to actively load certain molecules that are found predominantly in secretion, such as the 400-Da natural degradation fragment from RNase A.

We propose instead that a series of selective molecular transporters are present in the lysosomal membrane and determine the selectivity of the degradation products released by the lysosomes. Several additional lines of evidence support the idea that the lysosomal membrane has been adapted to facilitate membrane protein transport. Chiang et al. (1989) report that RNase A and other KFERQ-containing proteins are transported from the cytosol across the lysosomal membrane when the 73-kD heat-shock cognate protein is present. And, Barrueco et. al. (1992), while studying the turnover of the cancer drug methotrexate, discovered a lysosomal uptake system for polyglutamic acid derivatives of the drug. Further characterization indicated that it was a bidirectional carrier, possibly with broad specificity for peptides containing a C-terminal glutamic acid.

Many of the protein membrane transporters described to date are constrained, by either recognition or binding requirements, to transport only polypeptides containing a specific signal sequence or consensus motif. Other transporters appear to be completely dedicated to the transport of one molecule. In contrast, the transporters responsible for releasing lysosomal contents, like the transporter for glutamate-containing peptides described above, are likely to have very broad-specificity substrate recognition and binding mechanisms. This appears to be necessary because the substrate pool for transport out of the lysosome includes the whole range of possible exogenous proteins and their degradation products, as well as all endogenous proteins and their degradation products.

A well-understood precedent exists for this type of broad-specificity peptide binding to an effector molecule. MHC molecules bind peptides with a specificity broad enough to include a sampling of the whole array of possible peptides. The specificity of this binding is determined by pockets in the hydrophobic binding groove on the floor of the MHC molecule. These pockets bind with high affinity the side chains of specific amino acids, called anchoring residues, in peptide molecules whose other features do not prevent them from orienting correctly (Matsumura et al., 1992). It is likely that the transporter mechanisms responsible for releasing polypeptides from the lysosome (as well as the transporter responsible for ER import) would be required to use principles as simple as these in selecting transport substrates.

VIII. LYSOSOMAL PROTEIN DEGRADATION AND ANTIGEN PRESENTATION

Although fibroblasts do not express MHC class II molecules, the lysosomal transport processes described here may play a role in antigen presentation in cells that do. There is disagreement as to where in the class II pathway antigens bind to MHC. MHC localization has implicated endosomes (Guagliardi et al., 1989) and novel lysosomal compartments (Harding and Geuze, 1993; Peters et al., 1991, Amigorena et al., 1994; Tulp et al., 1994; West et al., 1994). Although antigens targeted directly to the lysosome are presented very efficiently (Harding et al., 1991), a mechanism for the transport of MHC–peptide complexes from the lysosome to the cell surface has yet to be described. We have shown that processes exist to transport both large and small peptides from the lysosome to the cell surface. It is possible that peptides complexed to MHC (or an accessory molecule), as well as peptides alone, could use these processes to arrive at the cell surface. Alternatively, peptides that leave the lysosome could be taken up by either a pre-lysosomal or endosomal compartment containing the class II MHC (Broadsky and Guagliardi, 1991).

Our data also bear on another aspect of antigen presentation. Microinjected RNase A serves as an example of a class of cytosolic proteins that are degraded in lysosomes. Unless crossover from class II to the class I pathway can occur, cells containing viral proteins of this type could escape detection by cytotoxic T cells. Moreover, as Jin et al. (1988) point out, such a crossover pathway is necessary to explain how vaccination with exogenous protein can prime an organism for a cytotoxic T-cell response. This type of crossover has now been described in a number of instances. For example, Harding et al. (1992) describe the presentation of protein from a parasite that resides in the lysosome by the class I pathway, and Rock et al. (1993) describe the presentation of exogenous ovalbumin by a specific subset of macrophages. The release of polypeptide from lysosomes described above provides a potential mechanism to account for this phenomenon. Peptides released into the cytosol by the lysosome could be transported into the ER by the peptide transporter on the ER membrane and join the class I pathway. The release of ubiquitinated proteins and/or peptides from lysosomes is particularly interesting in this regard, since cytosolic ubiquitin degradation processes have been implicated in the generation of class I antigens.

An objection to such a scheme might be that it would lead to the needless killing of many healthy cells that happen to take up viral protein from the extracellular environment. Though this may happen, killing even many healthy cells would appear less of a risk to the organism than allowing a few infected cells to remain unchecked. More to the point perhaps, under most conditions, endogenous proteins continually synthesized by the cell, and taken up and degraded by lysosomes, would supply the bulk of crossover antigens. It will be particularly interesting to determine whether fibroblasts that have endocytosed RNase A can be lysed by appropriately primmed CD8[+] T cells, or whether the crossover phenomenon is restricted to certain immune effector cells.

ACKNOWLEDGMENTS

I wish express deep appreciation to my collaborator, Paulo Dice. I also wish to thank Debra Osnowitz, Rachel Skvirsky, and Paulo Dice for critical review of the manuscript.

REFERENCES

Ahlberg, A., Marzella, L., & Glauman, H. (1982). Uptake and degradation of proteins by isolated rat liver lysosomes: suggestion of a microautophagic pathway of proteolysis. Laboratory Investigation 6, 523–531.

Amigorena, S., Drake, J.R., Webster, P., & Mellman, I. (1994). Transient accumulation of new class II MHC molecules in a novel endocytic compartment in B lymphocytes. Nature (London) 369, 113–119.

Aniento, F., Roche, E., Cuervo, A.M., & Knecht, E. (1993). Uptake and degradation of glyceraldehyde-3-phosphate dehydrogenase by rat liver lysosomes. J. Biol. Chem. 268, 10463–10470.

Barrueco, J.R., O'Leary, D.F., & Sirotnak, F.M. (1992). Facilitated transport of methotrexate polyglutamates into lysosomes derived from S180 cells. J. Biol. Chem. 28, 19986–19991.

Brodsky, F.M. & Guagliardi, L.E. (1991). The cell biology of antigen processing and presentation. Annu. Rev. Immunol. 9, 707–744.

Buktenica, S., Olenick, S.J., Salgia, R., & Frankfater, A. (1987). Degradation and regurgitation of extracellular proteins by cultured mouse peritoneal macrophages and baby hamster kidney fibroblasts. J. Biol. Chem. 262, 9469–9476.

Chiang, H-L., Terlecky, S.R., Plant, C.P., & Dice, J.F. (1989). A role for a heat shock 70-kD protein in lysosomal degradation of intracellular proteins. Science 246, 382–385.

Demotz, S., Grey, H.M., & Sette, A. (1990). The minimal number of class II MHC-antigen complexes needed for T Cell Activation. Science 249, 1028–1030.

Dice, J.F., (1987). Molecular determinants of protein half-lives in eukaryotic cells. Faseb J. 1, 349–357.

Dunn, W.A. (1990). Studies on the mechanisms of autophagy: formation of the autophagic vacuole. J. Cell. Biol. 110, 35–45.

Dunten, R.L. & Cohen, R.E. (1989). Recognition of modified forms of ribonuclease A by the ubiquitin system. J. Biol. Chem. 264, 16739–16747.

Ehrenreich, B.A. & Cohn, Z.A. (1969). The fate of peptides pinocytosed by macrophages *in vitro*. J. Exp. Med. 129, 227–245.

Finley, D. & Chau, V. (1991). Ubiquination. Annu. Rev. Cell Biol. 7, 25–69.

Germain, R.N. & Margulies, D.H. (1993). The biochemistry and cell biology of antigen processing and presentation. Annu. Rev. Immunol. 11, 403–450.

Guagliardi, L.E., Koppleman, B., Blum, J.S., Marks, M.S., Cresswell, P., & Brodsky, F.M. (1989). Co-localization of molecules involved in antigen processing and presentation in an early endocytic compartment. Nature (London) 343, 133–139.

Harding, C.V., Collins, D.S., Slot, J.W., Geuze, H.J., & Unanue, E.R. (1991). Liposome-encapsulated antigens are processed in lysosomes, recycled, and presented to T cells. Cell 64, 394–401.

Harding, C.V. & Geuze, H.J. (1993). Immunogenic peptides bind to class II MHC molecules in an early lysosomal compartment. J. Immunol. 151, 3988–3998.

Hasilik, A. & Neufeld, E.F. (1980). Biosynthesis of lysosomal enzymes in fibroblasts. J. Biol. Chem. 255, 4946–4950.

Henderson, R.A., Michael, H., Sakaguchi, K., Shabanowitx, J., Appella, E., Hunt, D.F., & Engelhard, H. (1992). HLA-A2.1-associated peptides from a mutant cell line: a second pathway of antigen presentation. Science 255, 1214–1215.

Hendil, K.B. (1981). Autophagy of metabolically inert substances injected into fibroblasts in culture. Exp. Cell Res. 135, 57–166.

Hershko, A. & Ciechanover, A. (1992). The ubiquitin system for protein degradation. Annu. Rev. Biochem. 61, 761–807.

Isenman, L.D. & Dice, J.F. (1989). Secretion of intact proteins and peptide fragments by lysosomal pathways of protein degradation. J. Biol. Chem. 264, 21591–21596.

Isenman, L.D. & Dice, J.F. (1993). Selective release of peptides by lysosomes. J. Biol. Chem. 268, 23856–23859.

Jentsch, S. (1992). The ubiquitin-conjugating system. Annu. Rev. Genet. 26, 179–207.

Jin, Y., Shih, J.W.-K., & Berkower, I. (1988). Human T cell response to the surface antigen of hepatitis B virus. J. Exp. Med. 168, 293–306.

Kim, S.J. & Paik, W.K. (1965). Studies on the origin of e-n-methyl-l-lysine in protein. J. Biol. Chem. 240, 4629–4634.

Laszlo, L., Doherty, F.J., Osborn, N.U., & Mayer, R.J. (1990). Ubiquinated protein conjugates are specifically enriched in the lysosomal system of fibroblasts. FEBS Lett. 261, 365–368.

Lippincott-Schwartz, J., Bonafacinio, J.S., Yuan, L.C., & Klausner, R.D. (1988). Degradation from the endoplasmic reticulum: Disposing of newly synthesized proteins. Cell 54, 209–220.

Lloyd, J.B. (1971). A study of permeability of lysosomes to amino acids and small peptides. Biochem. J. 121, 245–248.

Matsumura, M., Fremont, D.H., Peterson, P.A., & Wilson, I.A. (1992). Emerging principles for the recognition of peptide antigens by MHC class I molecules. Science 257, 927–934.

McElligot, M.A., Miao, P., & Dice, F. (1985). Lysosomal degradation of ribonuclease s-protein microinjected into the cytosol of human fibroblasts. J. Biol. Chem. 260, 11986–11993.

McElligott, M.A. & Dice, J.F. (1984). Microinjection of cultured cells using red-cell-mediated fusion and osmotic lysis of pinosomes: a review of methods and applications. Bioscience Reports 4, 451–466.

Michalek, M.T., Grant, E.P., Gramm, C., Golberg, A.L., & Rock, K.L. (1993). A role for the ubiquitin-dependent proteolytic pathway in MHC class I-restricted antigen presentation. Nature 363, 552–554.

Monaco, J.J., Cho, S., & Attaya, M. (1990). Transporter protein genes: possible implications for antigen processing. Science 250, 1723–1726.

Peters, P.J., Neefjwa, J.J., Oorschot, V., Plough, H.L., & Geuze, H.J. (1991). Segregation of MHC class II molecules from MHC class L molecules in the Golgi complex for transport to lysosomal compartments. Nature (London) 349, 669–676.

Pfeifer, J.D., Wick, M.J., Roberts, R.L., Findlay, K., Normark, S.J., & Harding, C.V. (1993). Phagocytotic processing of bacterial antigens for class I MHC presentation to T-cells. Nature (London) 361, 359–362.

Rock, K.L., Rothstein, L., Gamble, S., & Fleischacker, C. (1993). Characterization of antigen-presenting cells that present exogenous antigens in association with class I MHC molecules. J. Immunol. 150, 438–446.

Rogers, S.W. & Rechsteiner, M. (1988). Degradation of structurally characterized proteins injected into HeLa cells. J. Biol. Chem. 263, 19833–19842.

Schimke, R.T. & Doyle. (1970). Control of enzyme levels in animal tissues. Annu. Rev. Biochem. 39, 929–976.

Schlegel, R.A. & Rexsteiner, M.C. (1975). Microinjection of thymidine kinase and bovine serum albumin into mammalian cells by fusion with red blood cells. Cell 5, 371–379.

Schwartz, A.L., Ciechanover, A., Brandt, R.A., & Geuze, H. (1988). Immunoelectron microscopic localization of ubiquitin in hepatoma cells. EMBO J. 2961–2966.

Schwartz, A.L., Trausch, J.S., Ciechanover, A., Slot, J.W., & Geuze, H. (1992). Immunoelectron microscopic localization of the ubiquitin-activating enzyme E1 in Hep G2 cells. Proc. Natl. Acad. Sci. USA. 89, 5542–5546.

Segal, H.L. & Doyle, D. (1979). In: Protein Turnover and Lysosomal Function (Segal, H.L. & Doyle, D., Eds.), pp. 1–6 Academic Press, New York.

Segal, H.L., Brown, J.A., Dunaway, G.A., Winkler, J.R., Madnick, H.M., & Rothstein, D.M. (1979). In: Protein Turnover and Lysosomal Function (Segal, H.L. & Doyle, D., Eds.), pp. 9–27 Academic Press, New York.

Shimonkevitz, R., Kappler, J.W., Marrack, P., & Grey, H.M. (1983). Antigen recognition by H-2-restricted T cells. J. Exp. Med. 158, 303–316.

Siegelman, M. & Weissman, I. (1988). In: Ubiquitin (Rechsteiner, M., Ed.) pp. 239–269, Plenum, New York.

Spies, T. & DeMars, R. (1991). Restored expression of major histocompatibility class I molecules by gene transfer of a putative transporter. Nature (London) 351, 323–324.

Townsend, A. & Bodmer, H. (1989). Antigen recognition by class I-restricted T lymphocytes. Annul. Rev. Immunol. 7, 601–624.

Townsend, A.R., Rothbard, J., Gotch, F.M., Bahadur, G., Wraith, D., & McMichael, A.J. (1986). The epitopes of influenza nucleoprotein recognized by cytotoxic T lymphocytes can be defined with short synthetic peptides. Cell 44, 959–968.

Tulp, A., Verwoerd, D., Dobberstein, B., Ploegh, H.L., & Pieters, J. (1994). Isolation and characterization of the intracellular MHC class II compartment. Nature (London) 369, 120–126.

Tweto, J. & Doyle, D. (1976). Turnover of the plasma membrane proteins of hepatoma tissue culture cells. J. Biol. Chem. 251, 872–882.

West, M.A., Lucocq, J.M., & Watts, C. Antigen processing and class II MHC peptide-loading compartments in human B-lymphoblastoid cells. Nature (London) 369, 147–151.

COLICIN TRANSPORT

Claude J. Lazdunski

Membrane Protein Transport
Volume 1, pages 169–199.
Copyright © 1995 by JAI Press Inc.
All rights of reproduction in any form reserved.
ISBN: 1-55938-907-9

I. INTRODUCTION

A tremendous amount of effort has been devoted over the last 15 years to improving our understanding of the multiple mechanisms underlying protein insertion into and across membranes.

This is not a trivial pursuit, since these mechanisms underlie the biogenesis of the multiple organelles of the cell that are necessary for cell life and our own to continue. Progress was rapid at the beginning with the discovery of signaling systems (Blobel, 1980; Neupert and Schatz, 1981) and pathways that showed great variety as well as an apparently high specificity (Neupert and Schatz, 1981; Silhavy et al., 1983). More recent research has shown that although protein translocation does not involve a single universal mechanism, some common features are to be found. The translocation process consists of three basic steps: the association of the protein with receptors on the appropriate membrane, translocation through that membrane, and the folding and in some cases covalent modification on the opposite membrane surface. Our understanding of the energetics of translocation is slowly progressing.

Colicins are toxic proteins produced by and active against *E. coli* and other closely related bacteria. They are produced in large amounts and are generally secreted across the cell envelope into the extracellular medium. They can adsorb onto specific receptors located at the external surface of the outer membrane of sensitive cells and are then translocated to their specific targets within these cells. Colicins were discovered a long time ago and were studied intensively up to the 1960s. After a period of relative neglect, they have now emerged as powerful model systems for investigating the underlying mechanisms and the energetics of protein insertion into and across membranes.

The cell envelope of *E. coli* consists of two membranes: the cytoplasmic, or inner membrane, and the outer membrane. The outer layer of the outer membrane consists of lipopolysaccharide (LPS), while its inner layer contains phospholipids. This membrane is therefore asymmetrical. Between the outer and inner membranes lies the periplasmic space, which contains the bacterial cytoskeleton, peptidoglycan, conferring its shape and osmotic resistance on the cell envelope. This peptidoglycan is covalently attached to the outer membrane via the so-called major lipoprotein (for a review, see: Lugtenberg and van Alphen, 1983).

The main proteins present in the outer membrane have been called porins since they form pores through which small hydrophilic solutes with molecular weights of up to 650 (in most cases) can pass (Nikaido and Vaara, 1985). The two main porins are named OmpF and OmpC, but another pore protein, PhoE, is induced when cells are grown under phosphate limitation (Tommassen and Lugtenberg, 1980). In addition to the porins, the outer membrane contains proteins (LamB, BtuB, Tsx) which facilitate the uptake of specific solutes (maltodextrins, vitamin B_{12}, nucleosides).

OmpA is another abundant outer membrane protein, which together with the major lipoprotein appears to be involved in maintaining the structural integrity of the outer membrane and the rod shape of the cell.

In the following sections, experiments designed to elucidate colicins' release and entry mechanisms will be described. We shall focus mainly on colicin A, which is the model system on which we have been working at our laboratory.

II. COLICIN RELEASE ACROSS THE CELL ENVELOPE

A. Colicin Release Mediated by Lysis Proteins

Most of the plasmids which encode colicins also encode a small protein, named lysis protein (or bacteriocin release protein), which is required for colicins to be released by the cells of origin into the extracellular medium (for a review, see: Pugsley, 1984a; de Graaf and Oudega, 1986; Lazdunski et al., 1988). The genes of these small, 28–33 amino acid proteins form an operon with the colicin structural gene and are coregulated under SOS control (Hakkart et al., 1981; Cole et al., 1985; Toba et al., 1986; Lloubès et al., 1988). The lysis proteins are produced in lower amounts than the colicins, however, due to the presence of a transcription terminator after the colicin gene (Hakkart et al., 1981; Ebina and Nakazawa, 1983; Cole et al., 1985; Luirink et al., 1986).

The primary structures of many colicin lysis proteins have been determined from the nucleotide sequence of the corresponding colicin operons (de Graaf and Oudega, 1986; Lazdunski et al., 1988). They all contain a signal peptide with a cysteine residue at the cleavage site and in all the cases examined so far, they have been found to undergo modifications after the addition of lipid, probably via the same pathway as the major lipoprotein of *E. coli* (Cavard et al., 1987; Regue and Wu, 1990). The acylated form of the colicin A lysis protein is specifically cleaved by the Deg protease into two acetylated fragments with molecular masses of 6 and 4.5 kDa, thus decreasing the available amounts of the mature form of the lysis protein that is able to promote cell permeabilization (Cavard et al., 1989a).

The CloDf13 lysis protein has been detected in both inner and outer membranes of *E. coli* (Hakkart et al., 1981; Oudega et al., 1984), while the colicin N lysis protein has been observed only in the outer membrane (Pugsley, 1988a). In addition, in

Citrobacter freundii the mature form of the colicin A lysis protein (Cal) was found to be partly released into the extracellular medium (Cavard et al., 1985).

The induction of lysis proteins has been found to activate the normally dormant phospholipase A located in the outer membrane (Pugsley and Schwartz, 1984; Luirink et al., 1986; Cavard et al., 1987), thereby increasing the permeability of this membrane. It now emerges that lysis proteins indirectly activate the phospholipase A (Howard et al., 1989). The mechanism whereby lysis proteins contribute to the transfer of colicins across both the inner and outer membranes is not yet fully understood, however. It has nevertheless been demonstrated that the release process is nonspecific with respect to the colicin itself, which was found not to contain any topogenic export signal in the case of colicin A (Baty et al., 1987).

The colicin A lysis protein (Cal) is a 33 amino acid lipoprotein which is responsible for "quasi-lysis", involving the release of colicin A and many other proteins from induced colicinogenic cells (Cavard et al., 1985). It is a 51 amino acid precursor which undergoes a remarkably slow processing, during which the unmodified precursor, modified precursor, mature Cal, and signal sequence are visible for at least 30 min after pulse-labeling (Cavard et al., 1987).

The amino acid sequences of almost all the lysis proteins are very homologous, and even identical in the case of some of the E colicins (Lazdunski et al., 1988). There clearly exists a conserved 18 amino acid region while the C-terminal region is not conserved at least in colicin A and colicin N lysis proteins (Figure 1). This pattern of conservation has suggested that some structural and functional factors may be correlated. With a view to answering this question, we used *in vitro* mutagenesis to alter selected amino acids or regions of the colicin A lysis protein (Cal). We constructed an initial series of mutants into which the highly conserved amino acids—Arg-7, Gly-11, and Val-14—were substituted using a cassette mutagenesis technique (Howard et al., 1989).

A second series of mutants was designed to test the functional importance of the C-terminal region of Cal (Figure 1). Two mutant plasmids coded for truncated Cal proteins which contained only the very homologous first 16 or 18 (S16 and S18) amino acids of the normal 33 amino acid lysis proteins. Another construct FS2 was made to produce a Cal protein in which the C-terminal region (amino acids 17 to 33) following the conserved amino-terminal portion of Cal was completely altered, since it differed from the wild-type sequence both in length and in its sequence (Howard et al., 1989).

The three major phenotypic effects of colicin operon induction that have been found to be strictly dependent on lysis gene expression are a large but incomplete drop in the optical density of the colicinogenic culture (quasi-lysis), the concurrent release of large quantities of colicin and other soluble cell proteins, and a pronounced degradation of the cellular lipids, due to the activation of phospholipase A in the outer membrane.

−1 ↓ +1   7   11   14
Ala Cys Gln Val Asn Asn Val <u>Arg</u> Asp Thr Gly <u>Gly</u> Gly Ser <u>Val</u> Ser Pro Ser Ser Ile Val Thr

Gly Val Ser Met Gly Ser Asp Gly Val Gly Asn Pro
                                    33

Figure 1. Primary structure of the Cal point mutants. The primary structure of the wild-type (WT) Cal protein is given, and the positions where substitutions were introduced, have been underlined. The numbering of the amino acids is that of the mature form of Cal.

Cells containing the mutant plasmids (encoding the mutant Cal proteins) were therefore induced with mitomycin C (an inducer of SOS responses) and examined from the point of view of these effects which can be used as a functional index to the various mutant proteins involved.

The results with the point mutants in the conserved region indicated that although the amino-terminal amino acid sequence was the most highly conserved in the colicin lysis proteins, the sequence specificity in this region is very loose. Not only could each of the conserved amino acids tested be replaced by another amino acid, but they could also be replaced by amino acids with entirely different chemical properties. Arg 7 could be replaced for example by Val, an amino acid with a large hydrophobic side chain, while Val 14 could be replaced by the positively charged Arg, without any detrimental functional effects on the protein.

Among the substitutions tested, only the replacement of Arg 7 or Gly 11 by Glu had significant effects on the lytic activity of Cal. Pulse-chase analysis of these mutants indicated that the functional impairment was not due to inhibition of the normal modification and processing of these mutants (Howard et al., 1989). These results suggest that the charge of the amino-terminal region of Cal may play some role in its function. Although there is already an aspartate in this region, adding a further negative charge appears to have adverse effects. In contrast, the adding of a positive charge in this region in the valine mutant, and the presence of many positive and negative charges in the carboxy-terminal region of FS2, did not impair the function of Cal. It should also be mentioned that the lysis protein of colicin N contains one additional positive charge in the amino-terminal region as compared with the other lysis proteins, and five additional positive charges in its carboxy-terminal region (Pugsley, 1988).

One possible role of the leader sequence of lysis proteins in the lytic process has been suggested on the basis of the loose homology existing with portions of phage lysis proteins (Lau et al., 1987). This idea was supported on the one hand by the finding that the signal sequences of Cal (Cavard et al., 1985) and the Col E2 lysis protein (Pugsley and Cole, 1987) are apparently very stable, and on the other by the fact that when Cal or the Col E2 lysis protein were modified so that they could not be processed (thus preventing the release of the stable signal sequence), the lytic events subsequent to their induction were either sharply attenuated or delayed

(Cavard et al., 1987; Pugsley and Cole, 1987). Further data (Howard et al., 1989) argued against this hypothesis, however, since the Cal mutants Ala-2 and Gly-4, while their lytic effects were impaired, were not affected in their processing, and they therefore accumulated the signal sequence in just the same way as the wild-type. The question remains as to why the signal sequences of lysis proteins are so stable. It may be that for structural reasons the lysis protein signal sequences are not susceptible to cleavage by signal peptide peptidase, and another possibility is that they are for some reason not able to contact the peptidase in the envelope. A study on the properties of various lipoprotein signal sequences has indicated that the Cal signal sequence is more hydrophobic than those of other lipoproteins (Klein et al., 1988), a factor which might contribute decisively to its stability.

The fact that truncated proteins consisting of 16 and 18 amino acids (S16 and S18) were neither modified nor processed (Howard et al., 1989) is very interesting. Like the mutant Cal proteins that we previously constructed, which were altered at the processing and cleavage sites (Cavard et al., 1987), the S16 and S18 proteins were completely inactive. The ability to be modified and processed was entirely restored by the addition of an 11 amino acid carboxy-terminal sequence in a frameshift mutant (Howard et al., 1989). This addition also restored its activity to the mutant protein. This result again supports the hypothesis that the modification is the key requirement for the activity of the lysis proteins. Moreover, this suggests that the carboxy-terminal sequence, which is not conserved in the lysis proteins, functions only in the assembly of the nascent lysis protein polypeptide chains.

We do not yet know why the truncated Cal proteins are unable to be post-translationally processed while the full length protein can be. It is possible that the explanation lies in a coupling between the synthesis and export of the lysis protein. About 35 amino acids are masked within the large subunit of the ribosome during translation (Bernabeu and Lake, 1982). When the polypeptide chain is truncated, as in S16 and S18, with the signal sequence of 18 amino acids, the nascent chains (totaling 34 or 36 amino acids) may not emerge from the ribosome and thus may not be accessible to the cellular machinery involved in export and modification. Consequently, these nascent chains may be released into the cytoplasm. On the other hand, the addition of a "spacer sequence" of 12 amino acids to the frameshift mutant may allow the nascent chain to emerge from the ribosome, and to be recognized and inserted into the export and modification pathway.

All the results obtained so far indicate that apart from the structural features required for correct assembly, the modification of Cal which occurs upon the addition of lipid to the amino terminal cysteine seems to be the most important prerequisite for quasi-lysis and the release of colicin A. In this context, it is worth noting that a family of lipopeptides extracted from the culture media of various microorganisms has been shown to exhibit antibiotic properties. For example, iturin A and bacillomycin L and D produced by *B. subtilis*, and peptidolipin NA produced by Nocardia asteroides all have antifungal activity. These lipopeptides have been reported to dramatically increase the electrical conductance across planar lipid membranes (Maget-Dana et al., 1985a,b). In view of the large range of conductance values, it has been suggested that local modifications in the structure of the bilayer may be induced by interactions with lipopeptide micelles (Maget-Dana et al., 1985b). We have recently observed that the induction of Cal causes depolarization of the inner membrane in *E. coli* cells (unpublished result). Disorganization of the lipid structure of the envelope may be the primary effect of Cal, the activation of phospholipase A being a consequence of this perturbation. If this is the case it is not surprising that the modification of Cal by lipid, rendering it amphiphilic, is a functional prerequisite.

A critical concentration of Cal within the cells appears to be required to promote quasi-lysis. This has been demonstrated by constructing plasmids yielding a high level expression of Cal. Two different types of plasmids of this kind have been constructed. In those of the first type, pAT1, the T1 terminator located downstream of the colicin A gene (*caa*) which normally arrests about 80% of the transcripts initiated at the promoter of the *caa-cal* operon (Lloubès et al., 1988) was deleted. In those of the second type (designated pCK4), the *cal* gene was inserted downstream of the *tac* promoter of an expression vector (Cavard et al., 1989c). These two plasmids were transformed into PldA$^+$ and PldA$^-$ strains producing an active or inactive phospholipase A (PldA). Cal overproducing clones were induced in wild-type and *pldA* mutant cells. The decrease in absorbance after induction was similar with all the various cultures, while the timing of this decrease differed. With the overproducing clones pAT1 and pCK4, the decrease in absorbance started after 70 minutes of induction, i.e. almost one hour earlier than with the normal Cal producer (pColA9 cells). In each case, the quasi-lysis of *pldA* cells was reduced, which confirms the role of phospholipase activation in the process. We also observed that the quali-lysis of lipoprotein-deficient cells was similar to that of lipoprotein-containing cells whatever the plasmid carried by the bacteria (Cavard et al., 1989c). This suggests that the attachment of peptidoglycan to the outer membrane is not important for Cal to work.

Cal overproduction caused not only early quasi-lysis but also early release of cellular proteins

The fact that Cal must reach a critical concentration within the cells before the occurrence of quasi-lysis and colicin release again supports the idea that Cal directly affects the inner membrane. Local modifications in the structure of the bilayer may be induced by the interactions with lipopeptide micelles, as previously suggested in the case of Iturin A and bacillomycin L (Maget-Dana et al., 1985b).

Phospholipase A may be located in contact sites between inner and outer membranes, as reported by Bayer and Bayer (1985). The activation of this phospholipase after the local modification of the bilayer may have a synergic effect involving lysophospholipids produced which have a detergent-like effect. This hypothesis is supported by the fact that the addition of Triton X-100 to *pldA* cells can supplement the lack of formation of lysophospholipids (Pugsley and Schwartz, 1984; Cavard et al., 1989b). The combined effects of Cal and lysophospholipids may result in local membrane permeabilization without any transient accumulation of colicin in the periplasmic space, as demonstrated by electron microscope studies (Cavard et al., 1984).

More recent studies have provided evidence that the colicin A lysis protein has phospholipase A-independent effects on the integrity of the *E. coli* envelope (Howard et al., 1991). In particular, it has been investigated whether phospholipase (PldA) activation is the main effect of Cal or whether it reflects more generalized damage to the envelope resulting from the presence of large quantities of this small acylated protein. *E. coli tolQ* cells, which were found to be leaky with periplasmic proteins, were transduced to *pldA* and then transformed with the recombinant colicin A plasmid pKA. Both the *pldA* and *pldA*$^+$ strains released large quantities of colicin A after induction, which indicated that in these cells phospholipase A activation is not required for colicin release. This release was still dependent on a functional Cal protein, however. The assembly and processing of Cal *in situ* in the cell envelope was studied by combining pulse-chase labeling with isopycnic sucrose density gradient centrifugation of the cell membranes. Precursor Cal and lipid-modified precursor Cal were detected in the inner membrane at early chase times, and gave rise to mature Cal which accumulated in both the inner and outer membrane after further chase. The signal peptide was also visible on these gradients, and its distribution too was restricted to the inner membrane. Gradient centrifugation of the envelopes of cells which were overproducing Cal resulted in very poor separation of the membranes.

As mentioned above, the "lysis proteins" also called bacteriocin release proteins (BRP)s (De Graaf and Oudega, 1986), or "kil proteins", have stable signal sequences which are not proteolitically degraded after maturation (Cavard et al., 1987; Pugsley and Cole, 1987; Luirinketal et al., 1989).

The BRP mediates the secretion of cloacin DF13 (De Graaf and Oudega, 1986). It is slowly processed to yield the mature BRP and its stable signal peptide which is also involved in cloacin DF13 secretion. The role of the stable BRP signal peptide was analyzed by constructing two plasmids (Luirink et al., 1991). First, the stable BRP signal peptide was fused to the murein lipoprotein, and second, a stop codon was introduced after the BRP signal sequence. Exchanging the unstable murein lipoprotein signal peptide for the stable BRP signal peptide resulted in an accumulation of precursors of the hybrid murein lipoprotein. This indicated that the BRP signal peptide, as part of this hybrid precursor, is responsible for the slow processing. The stable BRP signal peptide itself was not able to direct the transfer of cloacin DF13 into the periplasmic space or into the culture medium. Overexpression of the BRP signal peptide was lethal and caused lysis. Subcellular fractionation experiments revealed that the BRP signal sequence is located exclusively in the cytoplasmic membrane, whereas the mature BRP, targeted by either the stable BRP signal peptide or the unstable Lpp signal peptide, is located in both the cytoplasmic and outer membrane.

These results are in agreement with the hypothesis that the stable signal peptide and the mature BRP together are required for the passage of cloacin DF13 across the cell envelope. The signal peptide alone causes lethality but is unable to induce the translocation of the bacteriocin across the cytoplasmic membrane (Van der Wal et al., 1992).

To conclude, the mechanism of colicin release into the extracellular medium mediated by lysis proteins has by now been mostly elucidated. It does not involve any topogenic export signal in the polypeptide chain of colicins or bacteriocins in general. Lysis proteins promote a nonspecific increase in the permeability of the cell envelope due to their direct effects on the inner membrane and their indirect effects on the outer membrane, mediated by phospholipase A activation.

B. Colicin Release Mediated by the Three-Component (ABC) System

Although most colicins are released across the cell envelope via the pathway described above, the shorter ones, particularly colicin V, have a different secretion pathway.

Four plasmid genes from pColV-K30, spanning 4.5 kilobases of DNA, have been found to be required for ColV synthesis, export, and immunity (Gilson et al., 1987). These genes are arranged in two converging operons, the transcription of which is induced under iron limitation under the control of the regulatory protein, Fur (Chehade and Braun, 1988; Gilson, 1990). The immunity gene, *cvi*, and the structural gene *cvaC*, constitute one of these operons.

The other operon contains *cvaA* and *cvaB*, the products of which are necessary for the export of the antibiotic. The product of the unlinked chromosomal gene *tolC* is also required for the export of ColV. The three protein components of the dedicated ColV export system showed a significant degree of amino acid sequence similarity with proteins performing similar functions in the extracellular secretion of α-hemolysins (Wagner et al., 1983; Felmlee et al., 1985; Wandersman and Delepelaire, 1990) *Erwinia* and *Pseudomonas* proteases (Delepaire and Wandersman, 1990; Guzzo et al., 1991), and *Bordetella* cyclolysin (Glaser et al., 1988). These protein complexes are not only similar but also functionally homologous, since mutations in *cvaA* or *cvaB* or both can be complemented, at the cost of a drop in the efficiency, by the dedicated exporters of α-hemolysin and the *Erwinia* protease (Fath et al., 1992).

The CvaB protein presumably spans the membrane six times and its C-terminal cytoplasmic domain seems to contain an ATP-binding domain (Gilson et al., 1990). The amino acid sequence of this ATP-binding domain is highly conserved in many of the proteins involved in export processes, including: the MDR family of drug exporters in mammals (Endicott and Ling, 1989); STE6, the α-factor exporter of *Saccharomyces cerevisiae* (McGrath and Varshavsky, 1989); and CFTR, the chloride channel which is impaired in patients with cystic fibrosis (Riordan et al., 1989;

Higgins, 1992). A similar domain has been described in a protein thought to export the lantibiotic subtilin (Klein et al., 1992).

The export signal of ColV, which is recognized by the CvaA/CvaB exporter, is located within the N-terminal 39 residues of the product of *cvaC* (Gilson et al., 1990). These 39 residues do not show any of the features typical of an export signal sequence. When these CvaC residues were fused to an alkaline phosphatase moiety devoid of its signal sequence, a CvaA/CvaB-dependent mode of hybrid transloca-tion was observed. An N-terminal processing step occurs in ColV synthesis and export which results in the cleavage of 15 amino acids. We have observed that OmpT protease can efficiently cleave colicin A derivatives constructed by genetic engineering upon their release from producing cells (Cavard and Lazdunski, 1990). This protease may therefore well be responsible for colicin V cleavage.

Two different colicin secretion strategies have therefore evolved: one of them does not involve any topogenic signals in the colicin polypeptide chain, and the other one does.

III. IMPORT OF COLICINS IN SENSITIVE CELLS

Two different pathways are involved in the transport of nutrients across the permeability barrier of the outer membrane. The passive diffusion pathway is that taken by small hydrophilic molecules ($M_r < 700$) that can diffuse across porins (Nikaido and Vaara, 1985). Bulky nutrients such as iron-bearing siderophores or vitamin B_{12}, which exceed the diffusion limit of the outer membrane, are imported through high-affinity energy-dependent receptors which make it possible for them to cross the outer membrane. The TonB protein provides the energy for the latter pathway.

Colicins have parasitized both the passive diffusion pathway and the energy-dependent pathway. On the basis of genetic data, the proteins involved in the translocation step have been identified as follows.

Group B colicins (B, D, G, H, Ia, Ib, M, Q, V) in addition to their specific receptors, require TonB, ExbB, and ExbD (and probably TolQ) (for a review, see Postle, 1990; Braun et al., 1991), while group A colicins (A, E1, E2, E3, K, L, N, S4) require TolA, B, Q, R, and TolC in the case of colicin E1 (for a review, see: Davies and Reeves, 1975a,b; Webster, 1991). There is no relationship between the lethal activity of the colicin and the translocation pathway used.

Consistent with the three steps in the mode of colicin action (receptor binding, translocation, and lethal activity), their polypeptide chains comprise three linearly organized domains, each corresponding to one of these steps. The N-terminal domain is involved in translocation, the central domain is involved in binding to the receptor, and the C-terminal domain carries the lethal activity (Baty et al., 1988, 1990).

Recombinant DNA techniques were used at our laboratory in the case of colicin A to construct chimeric colicins lacking specific regions, and to define the boundaries and the functions of the various domains indicated above (Baty et al., 1988).

A. Components of the Translocation Machineries

Various systems have evolved for dealing with the uptake of macromolecules such as colicins or the genomes of certain bacteriophages. In general, the transport of these molecules to their target depends on interactions with high-affinity receptors in the outer membrane. Most of the outer membrane porins involved in nutrient uptake are used as colicin receptors (Datta et al., 1977; Braun and Hantke, 1977; Kadner et al., 1979; Konisky, 1982). For instance, the polypeptide that serves as the receptor for colicin E1, E2, and E3 takes part in the uptake of vitamin B_{12}, whereas the colicin K receptor (Tsx) serves as a specific diffusion pathway for nucleosides (Hantke, 1976).

Several colicin receptors are involved in iron uptake, serving as siderophore-binding proteins. For example, FhuA (formerly called TonA) is the receptor for colicin M and ferrichrome (Hantke and Braun, 1975a; Wayne and Neilands, 1975; Braun et al., 1980), whereas enterochelin and colicins B and D utilize FepA for adsorption (Hantke and Braun, 1975b; Pugsley and Reeves, 1977; Hollifield and Neilands, 1978). The colicin Ia, Ib receptor (Cir) is also involved in iron accumulation (Konisky, 1982; Nau and Konisky, 1985). IutA is the receptor for aerobactin and cloacin D13 (Krone et al., 1985).

Bacteria carrying specific mutations in the genes coding for these outer membrane receptors do not bind their respective colicins and have been termed colicin resistant bacteria. Mutations have been isolated in other loci which led to the colicins adsorbing to their receptors while being insensitive or refractory to their effects. Many of these mutant bacteria have been termed colicin tolerant (*tol*). The mutations causing this tolerance are presumably located in genes which encode functions necessary for translocating the colicin to its target after it has bound to its outer membrane receptor.

The colicins have been assigned to one of two groups, A or B, based on whether or not each colicin is active towards various mutant bacteria (Davies and Reeves, 1975a,b). For example. the group B colicins (B, D, G, H, Ia, Ib, M, Q, V) are inactive against bacteria containing *tonB* mutations. The TonB protein, as mentioned above, is part of a family of high-affinity systems transporting vitamin B_{12}, iron–siderophore complexes, and many group B colicins, whereas, the group A colicins (A, E1, E2, E3, K, L, N, S4) are inactive against strains with *tolA* gene lesions. TolA is part of another system, termed the Tol import system, composed of several proteins which are involved in the import of group A colicins as well as the DNA of the filamentous bacteriophage. Generally speaking, bacteria containing mutations in the Tol import system are tolerant to only one or more of the group A colicins, while those in the TonB system are tolerant to any colicin in group B.

B. The TonB-Dependent Pathway

The *tonB* mutants are resistant to phage T1 since the TonB protein plays a role in the adsorption rather than in the uptake of phage DNA (Hancock and Braun, 1976). Energy is generated in the cytoplasm and the cytoplasmic membrane, and charged molecules such as ATP and phosphoenol pyruvate are not exported. By providing cytoplasmic membrane energy to the active transport of iron-bearing siderophores and vitamin B_{12} across the outer membrane, the TonB protein gives gram-negative bacteria the benefits of an outer membrane, and helps them to overcome the permeability barrier to iron and vitamin B_{12} (Braun et al., 1991).

The uptake of colicin M also involves TonB requirements. This colicin binds to the same FhuA receptor as phage T1. It binds and stays bound to the receptor in *tonB* mutant cells and in unenergized cells (Braun, 1989). Since *tonB* mutants became colicin M-sensitive upon the application of osmotic shocks which temporarily permeabilized the outer membrane, colicin M transport across the outer membrane seems likely to depend on the TonB protein.

The *tonB* gene was cloned and sequenced (Postle and Good, 1983; Hannavy et al., 1990). This gene encodes a 26-kDa protein with a high prolyl residue content (17%). A particularly proline-rich region is located in the amino-terminal third of TonB (a 33-residue peptide segment) which appears to adopt an extended constrained conformation presumably spanning the periplasmic space. This region has recently been found to interact specifically with the FhuA protein (Brewer et al., 1990), the outer membrane ferrichrome–iron receptor and colicin M receptor. As TonB is anchored to the cytoplasmic membrane (Postle, 1990), these data together with those from studies mentioned above suggested a model whereby TonB serves to convey conformational information over long distances, from the cytoplasmic membrane to the outer membrane (Hannavy et al., 1990).

The interactions between the TonB protein and various receptors seem likely to involve common receptor structure which is recognized by the TonB protein. One such structure comprising a pentapeptide close to the N-terminus, which has been termed the "TonB box" structure, has in fact been found to occur in all the receptors with a TonB-dependent activity (Braun et al., 1991).

The TonB box was again found to occur in colicins which are taken up by a TonB-dependent mechanism (Ross et al., 1989). This might mean that the colicin release mechanism is the same here as in the case of the other substrates, but the further transport across the outer membrane requires an additional TonB-dependent step. Apart from the TonB box, there exists in fact little, if any, similarity between the amino acid sequences of these receptors.

C. Auxiliary Proteins in the TonB-Dependent Pathway

Unlinked genes undergo mutations which seem to affect the action of TonB. The most fully documented of these mutated genes is the *exbB* gene. The term exb was originally used to mean mutations which rendered cells insensitive to colicin B through a compound (enterochelin/enterobactin) which was excreted into the culture medium and competed for FepA with colicin B. The *exbB* mutants have a "leaky" TonB phenotype with low but detectable levels of vitamin B_{12} and siderophore-mediated iron transport (Hantke and Zimmerman, 1981). The *exbB* mutants are insensitive to B-group colicins, but less so than *tonB* mutants. The functional role of ExbB protein seems to consist of stabilizing TonB protein (Fischer et al., 1989).

ExbB is a 26-kDa protein which has been located in the cytoplasmic membrane. Its amino acid sequence is 26% identical to that of TolQ, a protein involved in the uptake of group A colicins. The *tolQ-exbB* strains are completely insensitive to group B colicins, φ 80, and T1 (Braun, 1989). The stabilization of TonB observed in the presence of ExbB/TolQ proteins can be taken to show that mechanical interactions occur between the TonB protein and ExbB/TolQ.

There are two spatially distinct, functional interactions between ExbB and TonB. First, these two proteins seem to interact through their membrane-spanning helices (Karlsson et al., 1993a). Second, ExbB interacts with TonB in the cytoplasm during synthesis and/or localization of TonB, suggesting a chaperone-like role for ExbB (Karlsson et al., 1993b). It has been recently demonstrated that ExbB is a normal component of the energy coupling system for the transport of cobalamin across the outer membrane (Bradbeer, 1993).

Other genes whose products probably modify or enhance TonB activity are *exbC* and *exbD*. Except for the finding that *exbC* mutants hyperexcrete enterochelin, many questions still remain to be answered about these genes and their locus. The *exbD* gene was discovered while performing sequence analysis on an *exbB* clone (Eick-Helmerich and Braun, 1989). It is located downstream of the *exbB* gene and encodes a 15.5-kDa protein localized in the cytoplasmic membrane. ExbD shares 25% amino acid identity with TolR. The structural similarities between ExbBD proteins and TolQR proteins suggest the existence of a common ancestor. Unlike ExbB, ExbD does not stabilize TonB (Fisher et al., 1989).

The hypothesis that a functional connection may exist between Exb and Tol proteins gained support with the finding that the TonB-dependent sensitivity to colicins, which is considerably reduced but not fully abolished when mutations were induced in the *exbBD* locus, was completely lost when an additional mutation occurred in the tolQ gene (Braun, 1989). The double *exbBD tolQ* mutants were completely insensitive to colicins B, D, M, and to phages T1 and φ 80. This insensitivity was found to occur both in colicins with a TonB-dependent and in those with a TonB-independent uptake mechanism.

There is an evolutionary relationship between the two uptake systems for biopolymers in *E. coli*. Cross-complementation between the TonB-ExbB-ExbD and the TolA-TolQ-TolR proteins has been evidenced (Braun and Herrmann, 1993) which, in addition to sequence homology, suggests the evolution of the two import system from a single one.

D. The TonB-Dependent Colicin Uptake Mechanism

There exists both genetic and biochemical evidence that TonB directly interacts with the outer membrane receptors and group B colicins which both contain the so-called "TonB box" (Brewer et al., 1990).

The TonB-dependent colicin uptake mechanism is still poorly understood so far. It is likely that the first step, the binding of colicins to the receptors, may as with the group A colicins (Benedetti et al., 1991b) lead to a conformational change with partial unfolding, making the TonB box available for interactions with TonB in the second step. The mechanism underlying the translocation of the polypeptide chain across the outer membrane has not yet been elucidated. Recent evidence indicates however that FepA (ferric enterobactin receptor) is probably a gated porin and that TonB acts as a gate keeper, thus suggesting that the FepA channel itself may be used (Rutz et al., 1992). Similar results have been recently described with FhuA (the protein through which ferrichrome and the structurally similar antibiotic albomycin are taken up) (Killmann et al., 1993). The transition of the translocation complex from the binding conformation to the translocating conformation may be induced by energization of the cell (potential across the inner membrane). The conformational change in the receptor may be triggered by a conformational change in TonB transmitted via the X-Pro region from the inner to the outer membrane (Hannavy et al., 1990).

The role of auxiliary proteins such as ExbBD and TolQR has not yet been established. It seems likely that these proteins may form a complex with TonB and modify its functional half-life. Nothing is known beyond this, however.

E. The Tol-Dependent Pathway

The above translocation pathway is also that used by filamentous phages M13, fd, and f1. Tol proteins (TolA, B, Q, R, and C) participate in the uptake of phage DNA and colicins, but they are not always all indispensable for the entry of each colicin: TolA and TolQ are necessary for the entry of colicins A, E1, E2, E3, K, and N, whereas TolB and TolR are necessary for the entry of colicins A, E2, E3, and K only, and TolC is only required for that of colicin E1 (Nagel del Zwaig and Luria, 1967; Davies and Reeves, 1975a,b; Sun and Webster, 1986, 1987).

The *tolQ, R, A,* and *B* genes constitute a cluster localized at 16.8 min on the chromosomal map of *E. coli* (Sun and Webster, 1986). This cluster has been completely sequenced (Sun and Webster, 1987; Levengood and Webster, 1989). Each of the genes can be transcribed separately but the possible existence of an operon cannot be ruled out (Sun and Webster, 1987). The gene products are only weakly expressed but they were characterized by cloning the gene under the control of a strong promoter and using the "maxicell" and "minicell" systems (Sun and

Webster, 1987; Levengood and Webster, 1989). Tol Q, R, A, and B are proteins having 230, 142, 421, and 431 amino acid residues, respectively.

Sun and Webster (1986) have reported that TolQ remained associated with the membranes after fractionation and that this protein had three putative transmembrane hydrophobic segments. The protein was fused to the 30 N-terminal residues of colicin A, forming an epitope, and was overproduced under the control of the colicin A promoter (Bourdineaud et al., 1989). It was found to be preferentially localized in adhesion zones between inner and outer membranes. TolR seems to span the inner membrane only once through a single hydrophobic region (Sun and Webster. 1987). As previously mentioned, TolQ and TolR are homologous to ExbB and ExbD.

After fractionation, TolA is still associated with the inner membrane where it is probably anchored through a N-terminal 21 amino acid hydrophobic region. The rest of the protein is periplasmic and can be degraded when trypsin gains access to this compartment. TolA also has a region of 223 residues containing very high alanine, lysine, glutamate, and aspartate levels (45, 20, and 16%, respectively). The latter consists of 10 repeating units: ED(K)1-2 (A)2-4, which strongly stabilize a α-helical structure (Levengood and Webster, 1989). This region of the protein may therefore form an uninterrupted α-helix with a length of approximately 34 nm which makes it possible for the periplasmic space to be spanned by the protein (Webster, 1990).

The case of TolB is more complex. In "minicells", two different products of *tolB* with molecular masses of 43 and 47.5 kDa were characterized but no degradation kinetics was observed in the 47.5-kDa into 43-kDa transformation. However, recent evidence indicates that the short form is produced after processing of a signal peptide and is located in the periplasm where it interacts with the TolQ-R-A complex (Isnard et al., unpublished result).

The TolC protein plays an important part in the translocation of colicin E1. Its gene, *tolC*, is located at 66 min, far from the locus of other *tol* genes. It is a minor outer membrane protein produced in the precursor form. The mature hydrophilic form contains 467 amino acid residues (Hackett and Reeves, 1983). On the basis of sequencing data, the central region of the protein (i.e. 90 amino acid residues) has recently been corrected (Niki et al., 1990). Besides its role in colicin E1 translocation, TolC is also involved in the secretion of hemolysin, a toxin with no signal sequence but which is nevertheless released into the extracellular medium through a two-component secretion process (Wandersman and Delepaire, 1990). TolC is involved in the transcriptional regulation of micF and consequently strongly affects OmpF synthesis (Misra and Reeves, 1987). It has been recently reported that TolC functions as an outer membrane channel (Benz et al., 1993).

IV. COMPONENTS AND DYNAMICS OF TRANSPORT THROUGH THE TOL SYSTEM

A. Transport of Colicin A through the Tol System

The import of colicin A, a pore-forming colicin, has been extensively investigated over recent years. The receptor is composed of two proteins, BtuB (vit. B_{12} receptor) and OmpF (major porin) (Cavard and Lazdunski, 1981). These two proteins do not both play the same role, however; BtuB is used only as a receptor, whereas OmpF is used both as a receptor and for translocation purposes across the outer membrane (Benedetti et al., 1989).

The kinetics of the K^+ efflux induced by colicin A in *E. coli* sensitive cells have been investigated using a K^+-selective electrode. The dependence of K^+ efflux upon the multiplicity, pH, temperature, and membrane potential ($\Delta\psi$) was determined. The translocation of colicin A from the outer membrane receptor to the inner membrane and its insertion into the inner membrane required a fluid membrane, but once inserted, the channel properties showed little dependence upon the state of the lipids. At a given multiplicity, the lag time before the onset of K^+ efflux was found to reflect the time required for translocation and/or insertion of colicin into the cytoplasmic membrane. Opening of the channel occurred only above a threshold value of $\Delta\psi$ of 85 mV at pH 6.8. The conditions were defined for channel closing and reopening *in vivo*. We used these conditions to separately test the $\Delta\psi$ requirements for translocation and channel opening: translocation and/or insertion did not appear to require $\Delta\psi$ (Bourdineaud et al., 1990). It seems likely that the import of group A colicins across the outer membrane, unlike what occurs in the case of group B colicins, does not require energy.

B. The N-Terminal Domain of Colicins is Responsible for Uptake Specificity

Group A colicins with a known sequence carry an N-terminal region rich in glycine residues (Pugsley, 1987). A similar situation is encountered with the filamentous phage protein G3p responsible for the binding of these phages to their receptor (F pilus) (Boeke et al., 1982). G3p is involved in the Tol-dependent translocation of phage DNA across the cell envelope, and the N-terminal glycine-rich sequences appear to play a leading role in this process (Stengele et al., 1990).

About 15 different hybrid colicins have been constructed by recombining various domains of group A and group B colicins through genetic engineering. These hybrid colicins were purified and their properties were studied (Frenette et al., 1991; Benedetti et al., 1991a). The results clearly demonstrated that information specifying the uptake pathway beyond the receptor (Tol-dependent or TonB-dependent pathway) was contained in the N-terminal domains of colicins. A given colicin could be shifted from the "passive diffusion" pathway that depends upon porins and Tol proteins to the energy-requiring pathway that depends upon a high-affinity receptor (in the case of siderophore or vit. B_{12}) and TonB (Benedetti et al., 1993).

C. Colicin A Unfolds during its Translocation in *E. coli* Cells and Spans the Whole Cell Envelope when its Pore has Formed

As mentioned above, adding of colicin A to *E. coli* cells results in an efflux of cytoplasmic potassium. This efflux occurs after a lag time which corresponds to the time needed for the translocation of the toxin through the envelope. Denaturing colicin A with urea before adding it to the cells did not affect the properties of the pore but decreased the lag time. After renaturation, the lag time was similar to that of the native colicin. This suggests that the unfolding of colicin A accelerates its translocation (Benedetti et al., 1992). Adding trypsin, which has access neither to the periplasmic space nor to the cytoplasmic membrane, resulted in an immediate arrest of the potassium efflux induced by colicins A and B. The possibility that trypsin may act on a bacterial component required for colicin reception and/or translocation was ruled out. It is thus likely that the arrest of the efflux may correspond to a closing of the pores and that the colicin polypeptide chain remains at the translocation sites when the pore has formed in the inner membrane (Benedetti et al., 1992).

D. Colicins A and E1 Interact with a Component of their Translocation System

Since the N-terminal domain of colicins specifies their interactions with components of the import machinery, it seemed likely that direct interactions between this domain and one or several of these components may occur. To test this hypothesis, a system was designed for obtaining specific labeling and overexpression of Tol protein. The T7 RNA polymerase system was used (Tabor and Richardson, 1985). Plasmids were constructed which led to an overproduction of the Tol proteins involved in the import of group A colicins. *In vitro* binding of overexpressed Tol proteins to either Tol-dependent (group A) or TonB-dependent (group B) colicins was analyzed. The Tol-dependent colicins A and E1 were able to interact with TolA but the TonB-dependent colicin B was not. The C-terminal region of TolA, which is necessary for colicin uptake, was also found to be necessary for colicin A and E1 binding to occur. Furthermore, only the isolated N-terminal domain of colicin A, which is involved in the translocation step, was found to bind to TolA (Benedetti et al., 1991b). These results demonstrate the existence of a correlation between the ability of group A colicins to translocate and their *in vitro* binding to TolA protein, suggesting that these interactions might be part of the colicin import process.

E. A Hypothetical Model for the Translocation of Colicin A

A hypothetical model for the import of colicin A (a Tol-dependent colicin), based upon the results reported, can be proposed. In the first step, the central domain (R) of colicin A may bind to BtuB, the outer membrane receptor and OmpF. This binding would cause partial unfolding, thereby leading the N-terminal domain of colicin A (T domain) to interact with the OmpF region involved in translocation (Fourel et al., 1990). The translocation of the polypeptide chain might then be initiated through the binding of the T domain to the C-terminal region of TolA (Benedetti et al., 1991b). After transport across the outer membrane, the pore-forming domain might undergo the first step in the insertion process into the inner membrane. This step is the electrostatic interaction between the ring of positive charges located on the face of the protein defined by the loop connecting the two hydrophobic helices, and negatively charged phospholipid headgroups (Parker et al., 1990).

We do not yet know the molecular mechanisms involved in the transport across the outer membrane. Whether the OmpF pore itself is used or additional proteins are required is not clear but it is very likely that this transport may occur in an aqueous environment. The TolA and TolQ proteins seem to play a central role in the translocation of group A colicins since they are required in each case, whereas other proteins (TolB, TolR and TolC) are not required for some colicins.

The TolC protein, in the case of colicin E1, may constitute a surrogate for OmpF in the translocation process across the outer membrane: TolC has been found to contribute to the secretion of hemolysin (Wandersman and Delepelaire, 1990) and to act as an outer membrane channel for peptides (Benz et al., 1993). The Tol proteins may form an aqueous pore complex connecting the inner and outer membranes through their interactions with OmpF. Transport through this aqueous pore may require only minimum activation energy, in agreement with our results (Bourdineaud et al., 1990). This "passive diffusion" pathway is in fact the porin diffusion pathway which is taken by small hydrophilic molecules of nutrients and has been parasitized by group A colicins.

Several lines of evidence indicate that colicin A remains at the translocation sites while the C-terminal domain has already been inserted and has formed a pore. First, adding trypsin, which has access neither to the periplasmic space nor to the cytoplasmic membrane, resulted in an immediate arrest of the potassium efflux induced by colicins A and B (Benedetti et al., 1992). Secondly, competition experiments between colicins A and N indicated that after the penetration of colicin A into bacteria immune to colicin A, colicin N could no longer reach its target via the same translocation pathway (H. Benedetti, Ph.D. thesis, 1991a).

F. Possible Existence of Targeting Sequences and Translocation Contact Sites

The two nutrient transport pathways across the outer membrane have been parasitized by colicins. The passive diffusion pathway taken by low $M_r(< 700M_r)$ nutrients is the Tol-dependent pathway used by group A colicins. The energy-requiring pathway is the TonB-dependent pathway taken by nutrients (siderophores and vit. B_{12}) that exceed the diffusion limit of the outer membrane. This pathway has been parasitized by group B colicins. Whether a colicin is channeled through one pathway or another is entirely specified by the nature of its N-terminal domain. This domain contains specific sequences (glycine-rich sequence and TonB-box) resulting in interactions with either the TolA or TonB components of the transloca-tion machinery. It has been established in genetic studies that a physical interaction occurs between outer membrane receptors and TonB (Heller et al., 1988; Schöffler and Braun, 1989). Since TonB is anchored to the inner membrane (Postle and Skare, 1988), these results suggest the existence of translocation contact sites between the inner and outer membranes. It has furthermore been demonstrated that TolQ is also preferentially located at contact sites (Bourdineaud et al., 1989). Recent evidence suggests that in the presence of colicin A, other Tol proteins also have the same location (Guihard, Boulanger, Benedetti, Lloubès, Besnàrd and Letellier, manu-script in preparation).

The components of colicins translocation machinery have therefore now been identified and located in the cell envelope. The targeting signals have also been identified. The field is moving steadily ahead with the emphasis visibly shifting from the signals to the molecular mechanisms using *in vitro* reconstituted systems.

V. INSERTION OF THE C-TERMINAL DOMAINS OF COLICINS INTO MEMBRANES AND PORE-FORMATION

The sequences of various channel-forming colicins (A, B, E1, I_a, I_b, N) have been determined. Their C-terminal domains are homologous (Lazdunski et al., 1988). These 18–20-kDa domains can easily be isolated using limited proteolytic digestion procedures (Cleveland et al., 1983; Martinez et al., 1983). One paradoxical property is that they are able to exist in water-soluble as well as membrane-protein form. This paradox was resolved by determining the structure of the pore-forming domain of colicin A at 2.5 Å resolution (Parker et al., 1989; Tucker et al., 1989). This fragment (204 residues) consists of 10 α-helices organized in a three-layer struc-ture. Eight of these helices are amphipathic, and two of them are completely buried within the protein fold and form a hydrophobic hairpin loop similar to that proposed in the case of the signal sequences involved in translocation. A similar structure deduced from NMR data has been proposed in connection with colicin E1 (Wor-mald et al., 1990).

The colicin system provides useful material for studying the molecular mechanism underlying protein insertion into membranes. It is necessary that the hydrophobic regions of the C-terminal domain that are shielded from the solvent in the water-soluble form should become accessible when the molecule makes the transition to the membrane-bound state. The molecular rearrangement and unfolding that exposes hydrophobic and amphipathic domains to the membrane must be critical events in this transition. Both colicin A and colicin E1 have been extensively studied with respect to this point using model membrane systems (Pattus et al., 1983a; Frenette et al., 1989; Massotte et al., 1989; Merrill et al., 1990; Goormagthigh et al., 1991).

The mechanism of colicin A insertion into the lipid bilayer of the cytoplasmic membrane can probably be divided into two steps: binding to the membrane surface and insertion of the protein into the hydrophobic core of the bilayer. The first step is probably electrostatic in nature, whereas the second may be driven by the partioning of hydrophobic residues into the lipid environment of the bilayer. Colicin A and its thermolytic fragment do not bind to neutral bilayers but display a great affinity for negatively charged lipids (Pattus et al., 1983a; Frenette et al., 1989; Massote et al., 1989). As mentioned above, the colicin A polypeptide chain (592 amino acid residues) contains three domains which are linearly organized and participate in the sequential steps involved in colicin entry and action. We have compared the penetration ability into phospholipid monolayers of colicin A with that of protein derivatives containing various combinations of its domains. The NH_2-terminal domain (171 amino acid residues) required for translocation across the outer membrane was found to have little affinity for dilauroyl phosphatidyl glycerol (DLPG) monolayers at all the pHs tested. The central domain has a pH-dependent affinity, although it is lower than that of the entire colicin A. The COOH-terminal domain contains a high affinity lipid binding site, but an electrostatic interaction is required in addition as the first step in the penetration process into negatively charged DLPG films.

Analysis of the distribution of the surface side chains of the pore-forming peptide structure shows that they are randomly distributed with respect to various potential interaction parameters such as the charge, hydrophobicity, and aromaticity. The exception to this finding occurs at the surface defined by the loop that connects the two hydrophobic helices 8 and 9 in the structure (Parker et al., 1989). This loop is surrounded by a ring of eight positively charged side chains. The fact that the interface defined by this loop has a mean positive charge explains the affinity of colicin A and its thermolytic fragment for negatively charged lipids.

The occurrence of a translocation competent state *in vitro* at acidic pH, facilitating the insertion of channel-forming domain into membranes, has been widely documented with both colicin A and colicin E1 (Cramer et al., 1990; Lazdunski et al., 1992). An important feature in this process is the retention of the secondary structure in the shift from the water-soluble form to the membrane form which has been found to occur using both Fourier-transform infrared (FTIR) and circular dichroism

(CD) spectroscopy (Pattus et al., 1985; Merrill et al., 1990; Goormagthigh et al., 1991; Lakey et al., 1991a; Rath et al., 1991).

A role of an acidic pH in inducing an import-competent state or more mobile conformation was further suggested in the case of the C-terminal colicin E1 domain by (1) the increased accessibility of channel peptide to protease at low pH, (2) the increased partition into non-ionic detergent, and (3) the increased polarity and solvent exposure of Cys-505, as indicated by a red shift in the emission spectrum and by the acrylamide quenching of the IAEDANS probe attached to this cysteine.

Similar results were observed with the colicin A C-terminal domain. The kinetics of insertion were followed by the use of brominated lipid quenchers of tryptophan fluorescence (Gonzalez Manas et al., 1992). When vesicles of brominated dioleyaphosphatidylglycerol were added to a solution of colicin A, an exponential decrease in the fluorescence occurred. This was because membrane insertion brings the tryptophans of the ColA C-terminal fragment close to the bromine atoms attached to the membrane lipids.

The insertion kinetics and the extent of insertion into lipid monolayers were both found to depend on the lipid–protein ratio as well as on the pH and the proportion of negatively charged lipids (Frenette et al., 1989). A fast initial binding step occurs followed by a slow insertion process.

The existence of an intermediate state on the pathway of the conformational reorganization and the mechanism destabilizing the native state to provide access to the intermediate state, have been recently investigated (van der Goot et al., 1991). The pH at the surface of a negatively charged surface is lower than in the bulk solution, owing to the electrical surface potential. This surface pH appears to bring about a transition to the molten globule state at which partitioning of the helices into the membrane can more readily occur. The molten globular state, by preserving the secondary structure (as opposed to complete unfolding), provides the best pathway for minimizing the energetic cost of making a compact protein competent for membrane insertion.

The first model for the insertion was proposed by Parker et al. (1990): in the first phase of membrane insertion, after an electrostatic interaction between a ring of positively charged side chains (positioned on a well-defined surface of the protein) and phospholipid head groups, the two hydrophobic helices are assumed by the model to partition into the hydrophobic core of the membrane, leaving the two charged outer layers of the protein embedded in the surface of the bilayer rather like the opening of an umbrella. For this reason, the model was denoted the "umbrella model" (Parker et al., 1990).

To test the umbrella model, which assumes the pore to be in the closed state, an experimental approach was used involving the fluorescent labeling of single cystein residues introduced by site directed mutagenesis (Lakey et al., 1991b). The spectral properties of the probe 1,5-N′-(iodoacetyl)-N′-(5 sulfonaphthyl) ethylenediamine (IAEDANS) provided information about the validity of the model as regards the inferred protein-membrane contacts. The three tryptophan residues of the pore forming domain are clustered in one of the outer layers and form a spatially defined source of donor energy. Using energy transfer, it was therefore possible to measure the distance between the tryptophan cluster and the single IAEDANS probe on each mutant. The results showed that two of the amphipathic helices of the C-terminal domain open out on the surface of the lipid bilayer during the initial phase of membrane insertion (Lakey et al., 1991b). These results were compatible with the umbrella model in which the hydrophobic interior of the molecule becomes exposed, and the three folding units of the high resolution structure (helices 1 and 2; helices 3, 4, 5, 6 and 7; and the hydrophobic helices 8 and 9) are preserved, resulting in a membrane-bound conformation.

Recent distance measurements carried out with additional cysteine mutants, raised some doubts however as to the orientation of putative transmembrane helices 8 and 9. The expected distance increase between these helices and the tryptophans did not occur when the colicin insertion into lipids took place. This indicates that only the outer layer consisting of helices 1 and 2 opens out upon insertion, while helices 8 and 9 run parallel to the membrane surface (Lakey et al., 1993). The existence of a distinct hydrophobic helical hairpin does not therefore explain the spontaneous membrane insertion activity of colicin A. This hairpin provides neither the energy nor the structural stabilization necessary for this process. The surface-pH-induced molten globule state (van der Goot et al., 1991) in fact facilitates the insertion without any expense of energy.

The conclusions drawn from fluorescence spectroscopy studies were fully con-firmed using a second experimental approach (Duché et al., 1993). Four double-cysteine mutants were constructed to connect with disulfide bonds—the helices 1 and 9, 5 and 6, 7 and 8, and 9 and 10—in order to prevent either membrane insertion or pore formation, in line with the umbrella model described above. None of the single-cysteine mutants inhibited the colicin A pore-forming activity. The oxidized double-cysteine mutants did not affect either *E. coli*-sensitive cells or planar lipid bilayer conductance. Upon adding a reducing agent, the pore-forming activity of the inactive double-cysteine mutants was restored. Moreover, the double-cysteine mutants—H5-H6, H7-H8, and H9-H10—were able to insert into the lipids in the absence of reducing agent, unlike the double-cysteine mutant H1-H9 connecting the helices 1 and 9. These results demonstrate that the colicin A pore-formation is a two-step process involving potential-independent membrane insertion and poten-tial-dependent channel opening. They confirmed the model whereby, at the first step, colicin A unfolds extending the helices 1 and 2 away from the helical hairpin which remains closely packed against the helices 3 and 10 and does not insert perpendicularly to the membrane surface. These results also suggest that the channel opening is a sensitive step which requires a high level of helix mobility.

The third experimental approach used to test the mode of insertion of the pore-forming domain involved the use of NMR techniques (Geli et al., 1992). ^{2}H and ^{31}P NMR techniques were used to study the effects on acyl chain order and lipid organization of this domain upon insertion in model membrane systems derived from the *E. coli* fatty acid auxotrophic strain K1059, which was grown in the presence of ^{2}H-labeled oleic acid. Adding of the protein to dispersions of the *E. coli* total lipid extract, in a 1:70 molar ratio of peptide to lipids, resulted in a large pH-dependent decrease in the quadrupolar splitting of the ^{2}H NMR spectra. The decrease in the quadrupolar splitting obtained at the various pH values was correlated with the pH dependence of the insertion of the protein into monolayer films using the same *E. coli* lipid extracts. The p*K* governing the perturbing effects on the order of the fatty acyl chains was around 5, which is in agreement with the values of the pH-dependent conformational changes in the pore-forming domain of colicin A required for membrane insertion reported by van der Goot et al. (1991). ^{31}P NMR measurements show that, upon addition of the protein, the bilayer organization remained insensitive in dispersions of lipid extract. Surprisingly, ^{31}P NMR measurements as a function of the temperature indicate that the pore-forming domain of colicin A actually stabilizes the bilayer lipid structure at pH 4. Both the strong effects of the protein on the acyl chain order and its bilayer-stabilizing activity indicate that the protein was located at the surface.

Using three independent techniques—fluorescence spectroscopy, constraint through disulfide bridges, and NMR—it was therefore demonstrated that the initial insertion in the absence of $\Delta\psi$ does not involve a transmembrane insertion of the hydrophobic helical hairpin. Helices 1 and 2 open out onto the membrane surface, while this hairpin remains closely packed against the rest of the structure and mostly embedded in the membrane (Lakey et al., 1991a).

In fact, it has been shown through various techniques that after binding to the membrane, the colicin A pore-forming domain undergoes the same changes that take place after exposure either to low pH (van der Goot et al., 1991) or high temperature (Muga et al., 1993). This membrane-induced destabilization of protein structure may facilitate its insertion and/or translocation since in addition the insertion itself has an effect on the lipid bilayer (Geli et al., 1992).

Because the transmembrane potential moves a large proportion of the colicin polypeptide chain from the surface into the membrane phase and some of it even to the opposite side (Slatin et al., 1986; Merrill and Cramer, 1990), these intramembrane segments must form the ion channel. Very little is known at present about the structure of this channel, apart from the fact that a single colicin probably makes the transmembrane channel (Bruggeman and Kayalar, 1986; Slatin, 1988; Bourdineaud et al., 1990), that the C-terminal 136 residues of colicin A alone form a wild-type channel (Baty et al., 1990), and that 88 residues of colicin E1 have a channel-forming ability (Liu et al., 1986). Whatever the channel structure may be, however, the results obtained on colicin A mutants containing disulfide bridges indicate that a major reorganization and highly mobile helices are prerequisites for channel opening (Duché et al., 1993).

VI. CONCLUSIONS AND OUTLOOK

An unfolded state is known to be required for protein to be imported into organelles and exported from bacteria (Verner and Schatz, 1988; Lazdunski and Benedetti, 1990), but the structural details of this unfolded state are not yet known in any system. The colicin system provides a valuable model in approaching this problem. It has already provided evidence that a major means of inducing protein insertion and/or protein translocation consists of bringing the protein into a molten globule state. This reduces the energetic cost of the conformational changes required for translocation across and/or insertion into membranes.

The so-called "translocation competent state" is usually maintained in prokaryotic and eukaryotic organisms by interactions with chaperones. A similar role may be assigned to colicin receptors and Tol proteins. In particular, interactions between the colicin A and E1 N-terminal domain and the TolA C-terminal domain may facilitate further unfolding and translocation of other colicin A domains.

The molecular events associated with the import translocation across the outer membrane are as yet poorly understood. Whether the OmpF porin channel is taken by colicin A or colicin N (Fourel et al., 1990; Benedetti et al., 1991) has not yet been established. There exists compelling evidence that translocation occurs in eukaryotic cells through a protein-conducting channel comprising a complex of ER-associated proteins (Simon and Blobel, 1991) and that bidirectional movements of nascent polypeptides across microsomal membrane are possible (Ooi and Weiss, 1992). It is worth noting that different pathways are used to release colicins into the extracellular medium and to introduce them into sensitive cells.

As regards the translational dynamics of colicin A, tools are now available for further investigating mechanisms underlying import into sensitive cells. Mutants with disulfide bridges preserve the integrity of the protein and are also able to block the protein at the various steps (receptor binding, translocation, potential-independent insertion, and potential-dependent channel opening) of the activity. Since disulfide bridges can be cleaved using a reducing agent *in vivo* after colicin binding to sensitive cells, these mutants have now been used to study the respective effects of the various bridges introducing different constraints on translocation, insertion, and channel opening (D. Duché, personal communication). The cystein mutants and disulfide bond mutants blocked at various steps should also provide useful means of cross-linking the proteins (receptor, porin, and Tol proteins) involved in the translocation mechanism.

ACKNOWLEDGMENTS

I wish to thank Mrs. Payan for her careful work on the preparation of the manuscript. Most of the research described in this article has been the fruit of a long-time collaboration between various groups, including those headed by D. Baty, F. Pattus, D. Tsernoglou, L. Letellier, A.

Killian, J.M. Ruysschaert. Many co-workers have been involved over recent years. I am especially grateful to D. Duché, D. Baty, V. Géli, H. Benedetti, R. Lloubès, D. Cavard. and J.M. Pagès for communicating unpublished results and for helpful discussions. These studies were supported by funds from the CNRS, the Foundation pour la Recherche Médicale, the European Economic Community under the "Science" program (contract no SC-10334 C), and the GDR no G-0964 of the CNRS.

REFERENCES

Baty, D., Lloubès, R., Geli, V., Lazdunski, C., & Howard, S.P. (1987). Extracellular release of colicin A is non-specific. EMBO J. 6, 2463–2468.

Baty, D., Frenette, M., Lloubès, R., Geli, V., Howard, S.P., Pattus, F., & Lazdunski, C. (1988). Functional domains of colicin A. Mol. Microbiol. 2, 807–811.

Baty, D., Lakey, J., Pattus, F., & Lazdunski, C. (1990). A 136-amino acid residue COOH-terminal fragment of colicin A is endowed with ionophoric activity. Eur. J. Biochem. 189, 409–413.

Bayer, M.H. & Bayer, M.E. (1985). Phosphoglycerides and phospholipase C in membrane fractions of *Escherichia coli* J. Bacteriol. 162, 50–54.

Benedetti, H., Frenette, M., Baty, D., Lloubès, R., Geli, V., & Lazdunski, C. (1989). Comparison of uptake systems for the entry of various BtuB group colicins into *Escherichia coli*. J. Gen. Microbiol. 135, 3413–3420.

Benedetti, H. (1991a). Importation des Colicines à Travers l'Enveloppe d'*Escherichia coli*. Ph.D. Thesis, Université d'Aix-Marseille I.

Benedetti, H. (1991b). Protein import into *Escherichia coli*: colicins A and E1 interact with a component of their translocation system. EMBO J. 10, 1989–1995.

Benedetti, H., Lloubès, R., Lazdunski, C., & Letellier, L. (1992). Colicin A unfolds during its translocation in *Escherichia coli* cells and spans the whole cell envelope when its pore has formed. EMBO J. 11, 441–447.

Benedetti, H., Letellier, L., Lloubès, R., Geli, V., Baty, D. & Lazdunski, C. (1992). Study of the import mechanisms of colicins through protein engineering and K+ efflux kinetics. In: Bacteriocins, Microcins, and Lantibiotics (James, R., Lazdunski, C., & Pattus, F., Eds.) Heidelberg, Springer-Verlag, 215–223.

Benz, R., Maier, E., & Gentschev, I. (1993). TolC of *Escherichia coli* functions as an outer membrane channel. Zbl. Bakt. 278, 187–196.

Bernabeu, C. & Lake, J.A. (1982). Nascent polypeptide chains emerge from the exit domain of the large ribosomal subunit. Proc. Natl. Acad. Sci. USA 79, 3111–3115.

Blobel, G. (1980). Intracellular protein topogenesis. Proc. Natl. Acad. Sci. USA 77, 1496–1500.

Boeke, J., Model, P., & Zinder, N. (1982). Effects of bacteriophage f1 gene III protein on the host cell membrane. Mol. Gen. Genet. 186, 185–192.

Bourdineaud, J.P., Howard, S.P., & Lazdunski, C. (1989). Localization and assembly into the *Escherichia coli* envelope of a protein required for entry of colicin A. J. Bacteriol. 171, 2458–2465.

Bourdineaud, J.P., Boulanger, P., Lazdunski, C., & Letellier, L. (1990). *In vivo* properties of colicin A: channel activity is voltage dependent but translocation may be voltage independent. Proc. Natl. Acad. Sci. USA 87, 1037–1041.

Bradbeer, C. (1993). The proton motive force drives the outer membrane transport of cobalamin in *Escherichia coli*. J. Bacteriol. 175, 3146–3150.

Braun, V. & Hantke, K. (1977). In: Microbial Interactions (Reissing, J.L., Ed.), pp. 101–137. Chapman and Hal, London.

Braun, V. (1989). The structurally related *exbB* and *tolQ* genes are interchangeable in conferring *tonB*-dependent colicin, bacteriophage, and albomycin sensitivity. J. Bacteriol. 171, 6387–6390.

Braun, V., Frenz, S., Hantke, K., & Schaller, K. (1980). Penetration of colicin M into cells of *Escherichia coli*. J. Bacteriol. 142, 162–168.

Braun, V., Günter, K., & Hantke, K. (1991). Transport of iron across the outer membrane. Bio. Metals 4, 14–22.

Braun, V. & Herrmann, C. (1993). Evolutionary relationship of uptake systems for biopolymers in *Escherichia coli*: cross-complementation between the TonB-ExbB-ExbD and the TolA-TolQ-TolR proteins. Mol. Microbiol. 8, 261–268.

Brewer, S., Tolley, M., Trayer, I., Barr, G., Dorman, C., Hannavy, K., Higgins, C., Evan, J., Levine, B., & Wormald, M. (1990). Structure and function of X-Pro dipeptide repeats in the TonB proteins of *Salmonella typhimurium* and *Escherichia coli*. J. Mol. Biol. 216, 883–895.

Bruggeman, E.P. & Kayalar, C. (1986). Determination of the molecularity of the colicin E1 channel by stopped-flow ion flux kinetics. Proc. Natl. Acad. Sci. USA 83, 4273–4276.

Cavard, D. & Lazdunski, C. (1981). Involvement of BtuB and OmpF proteins in binding and uptake of colicin A. FEMS Microbiol. Lett. 12, 311–316.

Cavard, D., Bernadac, A., Pagès, J.M., & Lazdunski, C. (1984). Colicins are not transiently accumulated in the periplasmic space before release from colicinogenic cells. Biol. Cell. 51, 79–86.

Cavard D., Lloubès R., Morlon J., Chartier M., & Lazdunski, C. (1985). Lysis protein encoded by plasmid ColA-CA31. Gene sequence and export. Mol. Gen. Genet. 199, 95–100.

Cavard, D., Baty, D., Howard, S.P., Verheij, H.M., & Lazdunski, C. (1987). Lipoprotein nature of the colicin A lysis protein: effect of amino acid substitution at the site of modification and processing. J. Bacteriol. 169, 2187–2194.

Cavard, D., Lazdunski, C., & Howard, S.P. (1989a). The acylated precursor form of the colicin A lysis protein is a natural substrate of the Deg protease. J. Bacteriol. 171, 6316–6322.

Cavard, D., Howard, S.P., & Lazdunski, A. (1989b). Functioning of the colicin A lysis protein is affected by Triton X-100 divalent cations and EDTA. J. Gen. Microbiol. 135, 1715–1726.

Cavard, D., Howard, S.P., Lloubès, R., & Lazdunski, C. (1989c). High-level expression of the colicin A lysis protein. Mol. Gen. Genet. 217, 511–519.

Cavard, D. & Lazdunski, C. (1989d). Colicin cleavage by OmpT protease during both entry into and release from *Escherichia coli* cells. J. Bacteriol. 172, 648–652.

Chehade, H. & Braun, V. (1988). Iron-regulated synthesis and uptake of colicin V. FEMS Microbiol. Lett. 52, 177–182.

Cleveland, M., Slatin, S., Finkelstein, A., & Levinthal, S. (1983). Structure-function relationships for a voltage-dependent ion channel: properties of COOH-terminal fragments of colicin E1. Proc. Natl. Acad. Sci. USA 80, 3706–3710.

Cole, S.T., Saint-Joanes, B., & Pugsley, A.P. (1985). Molecular characterization of the colicin E2 operon and identification of its products. Mol. Gen. Genet. 198, 465–472.

Cramer, W.A., Cohen, F.S., Merrill, A.R., & Song, H.Y. (1990). Structure and dynamics of the colicin E1 channel. Mol. Microbiol. 4, 519–526.

Datta, D.B., Arden, B., & Henning, U. (1977). Major proteins of the *Escherichia coli* outer cell envelope membrane as bacteriophage receptors. J. Bacteriol. 131, 821–829.

Davies, J.K. & Reeves, P. (1975a). Genetics of resistance to colicins in *Escherichia coli* K12: cross-resistance among colicins of group B. J. Bacteriol. 123, 96–101.

Davies, J.K. & Reeves, P. (1975b). Genetics of resistance to colicins in *Escherichia coli* K12 cross-resistance among colicins of group A. J. Bacteriol. 123, 102–117.

De Graaf, F.K. & Oudega, B. (1986). Production and release of cloacin DF13 and related colicins. Curr. Top. Microbiol. Immunol. 125, 183–205.

Delepelaire, P. & Wandersman, C. (1990). Protein secretion in Gram-negative bacteria. J. Biol. Chem. 265, 17118–17125.

Duché, D., Parker, M.W., Gonzalez-Manas, J.M., Pattus, F., & Baty, D. (1994). Uncoupled steps of the colicin A pore formation demonstrated by disulfide bond engineering. J. Biol. Chem. 269, 6332–6339.

Ebina, Y. & Nakazawa, A. (1983). Cyclic AMP-dependent initiation and ρ-dependent termination of colicin E1 gene transcription. J. Biol. Chem. 258, 7072–7078.

Eick-Helmerich, K. & Braun, V. (1989). Import of biopolymers into *Escherichia coli*: nucleotide sequences of the *exbB* and *exbD* genes are homologous to those of the *tolQ* and *tolR* genes, respectively. J. Bacteriol. 171, 5117–5123.

Endicott, J.A. & Ling, V. (1989). The biochemistry of P-glycoprotein-mediated multidrug resistance. Annu. Rev. Biochem. 58, 137–171.

Fath, M., Skvirsky, R., & Kolter, R. (1992). Functional complementation between bacterial MDR-like export systems: colicin V, α-hemolysin, and *Erwinia* protease. J. Bacteriol. 173, 7549–7556.

Fath, M., Skvirsky, R., Gilson, L., Mahanty, H.K., & Kolter, R. (1992). The secretion of colicin V. In: Bacteriocins, Microcins, and Lantibiotics (James, R., Lazdunski, C., & Pattus F., Eds.). Heidelberg, Springer-Verlag, 331–348.

Femlee, T., Pellet, S., & Welch, R.A. (1985). Nucleotide sequence of an *Escherichia coli* chromosomal hemolysin. J. Bacteriol. 163, 94–105.

Fischer, E., Günter, K., & Braun, V. (1989). Involvement of ExbB and TonB in transport across the outer membrane of *E. coli*: phenotype complementation of *exbB* mutants by overexpressed *tonB* and physical stabilization of TonB by ExbB. J. Bacteriol. 171, 5127–5134.

Fourel, D., Hikita, C., Bolla, J.M., Mizushima, S., & Pagès, J.M. (1990). Characterization of OmpF domains involved in *Escherichia coli* K-12 sensitivity to colicins A and N. J. Bacteriol. 172, 3675–3680.

Frenette, M., Knibiehler, M., Baty, D., Geli, V., Pattus, F., Verger, R., & Lazdunski, C. (1989). Interactions of colicin A domains with phospholipid monolayers and liposomes: relevance to the mechanism of action. Biochemistry 28, 2509–2514.

Frenette, M., Benedetti, H., Bernadac, A., Baty, D., & Lazdunski, C. (1991). Construction, expression and release of hybrid colicins. J. Mol. Biol. 217, 421–428.

Geli, V., Koorengerel, M.C., Demel, R.A., Lazdunski, C., & Killian, A.J. (1992). Acidic interaction of the colicin A pore-forming domain with model membranes of *Escherichia coli* lipid results in a large perturbation of acyl chain order and stabilization of the bilayer. Biochemistry 31, 11089–11094.

Gilson, L. (1990). Signal-Sequence-Independent Export of Colicin V. Ph.D. Thesis, Harvard University.

Gilson, L., Mahanty, H.K., & Kolter, R. (1987). Four plasmid genes are required for colicin V synthesis, export, and immunity. J. Bacteriol. 169, 2466–2470.

Gilson, L., Mahanty, H.K., & Kolter, R. (1990). Genetic analysis of an MDR-like export system: the secretion of colicin V. EMBO J. 9, 3875–3884.

Glaser, P., Sakamoto, H., Bellalon, J., Ullmann, A., & Danchin, A. (1988). Secretion of cyclolysin, the calmodulin-sensitive adenylate cyclase-haemolysin bifunctional protein of *Bordella pertussis*. EMBO J. 7, 3997–4004.

Gonzales Manas, J.M., Lakey, J.H., & Pattus, F. (1992). Brominated phospholipids as a tool to study the membrane insertion of colicin A. Biochemistry 31, 7294–7300.

Goormagtghigh, E., Vigneron, L., Knibiehler, M., Lazdunski, C., & Ruysschaert, J.M. (1991). Secondary structure of the membrane-bound form of the pore forming domain of colicin A. An attenuated total-reflection polarized FTIR spectroscopy study. Eur. J. Biochem. 202, 1299–1305.

Guzzo, J., Duong, F., Wandersman, C., Murgier, M., & Lazdunski, A. (1991). The secretion genes of *Pseudomonas aeruginosa* alkaline protease are functionally related to those of *Erwinia chrysanthemi* proteases and *Escherichia coli* α-hemolysin. Mol. Microbiol. 5, 447–453.

Hackett, J. & Reeves, P. (1983). Primary structure of the *tolC* gene that codes for an outer membrane protein of *Escherichia coli* K-12. Nucleic Acids Res. 11, 6487–6495.

Hakkart, M.J., Veltkamp, E., & Nijkamp, H.J. (1981). Protein H encoded by plasmid Clo DF13 involved in lysis of the bacterial host. I. Localization of the gene and identification and subcellular localization of the gene H product. Mol. Gen. Genet. 183, 318–325.

Hancock, R.E. & Braun, V. (1976). Nature of the energy requirement for the irreversible adsorption of bacteriophages T1 and φ80 to *Escherichia coli*. J. Bacteriol. 125, 309–415.

Hannavy, K., Barr, G., Dorman, C., Adamson, J., Mazengera, L., Gallagher, M., Evans, J., Levine, B., Trayer, I., & Higgins, C. (1990). TonB protein of *Salmonella typhimurium*: A model for signal transduction between membranes. J. Mol. Biol. 216, 897–910.

Hantke, K. (1976). Phage T6-colicin K receptor and nucleotide transport in *Escherichia coli*. FEBS. Lett. 70, 109–112.

Hantke, K. & Braun, V. (1975a). Membrane receptor dependent iron transport in *Escherichia coli*. FEBS Lett. 49, 301–305.

Hantke, K. & Braun, V. (1975b). A function common to iron-enterochelin transport and action of colicins B, I, V in *Escherichia coli*. FEBS Lett. 59, 277-281.

Hantke, K. & Zimmerman, L. (1981). The importance of *exbB* gene for vitamin B_{12} and ferric iron transport. FEMS Microbiol. Lett. 12, 31–35.

Heller, K.J., Kadner, R.J., & Günther, K. (1988). Suppression of the *btuB451*: mutations in the *tonB* gene suggests a direct interaction between TonB and TonB-dependent receptor proteins in the outer membrane of *Escherichia coli*. Gene 64, 147–153

Higgins, C.F. (1992). ABC transporters: from microorganisms to man. Ann. Rev. Cell Biol. 8, 67–113.

Hollifield, W.C. & Neilands, J.B. (1978). Ferric enterobactin transport system in *Escherichia coli* K12. Extraction, assay and specificity of outer membrane receptor. Biochemistry 17, 1922–1929.

Howard, S.P., Cavard, D., & Lazdunski, C. (1989). Amino acid sequence and length requirements for assembly and function of the colicin A lysis protein. J. Bacteriol. 171, 410–418.

Howard, S.P., Cavard, D., & Lazdunski, C. (1991). Phospholipase A-independent damage caused by the colicin A lysis protein during its assembly into the inner and outer membranes of *Escherichia coli*. J. Gen. Microbiol. 137, 81–89.

Kadner, R.J., Bassford, P.J., & Pugsley, A.P. (1979). Colicin receptors and mechanism of colicin uptake. Zentralbl. Bakteriol. Parasitenkd. Infektionskr. Hy. Abt. 244, 90–104.

Karlsson, M., Hannavy, K., & Higgins, C. (1993a). A sequence-specific function of the N-terminal signal-like sequence of the TonB protein. Mol. Microbiol. 8, 379–388.

Karlsson, M., Hannavy, K., & Higgins, C.F. (1993b). ExbB acts as a chaperone-like protein to stabilize TonB in the cytoplasm. Mol. Microbiol. 8, 389–396.

Killmann, H., Benz, R., & Braun, V. (1993). Conversion of the FhuA transport protein into a diffusion channel through the outer membrane of *Escherichia coli*. EMBO J. 12, 3007–3016.

Klein, P.R., Somorjai, R.I., & Lau, P. (1988). Distinctive properties of signal sequences from bacterial lipoproteins. Protein. Eng. 2, 15–20.

Klein, C., Kaletta, C., Schnell, N., & Entian, K.D. (1992). Analysis of genes involved in the biosynthesis of the lantibiotic subtilin. Appl. Environ. Microbiol. 58, 132–142.

Konisky, J. (1982). Colicins and other bacteriocins with established modes of action. Ann. Rev. Microbiol. 36, 125–144.

Krone, W.J., Stegehuis, F., Koningstein, G., Doorn, C.V., Roosendaal, B., de Graaf, F.K., & Oudega, B. (1985). Characterization of the pColV-k30 encoded cloacin DF13/aerobactin outer membrane receptor protein of *Escherichia coli*: isolation and purification of the protein and analysis of its nucleotide sequence and primary structure. FEMS Microbiol. Lett. 26, 153–161.

Lakey, J., Massote, D., Heitz, F., Dasseux, J.L., Faucon, J.F., Parker, F., & Pattus, F. (1991a). Membrane insertion of the pore-forming domain of colicin A, a spectroscopic study. Eur. J. Biochem. 196, 599–607.

Lakey, J., Baty, D., & Pattus, F. (1991b). Fluorescence energy transfer distance measurements using site-directed single cysteine mutants: the membrane insertion of colicin A. J. Mol. Biol. 219, 639–653.

Lakey, J., Duché, D., Gonzalez-Manas, J.M., Baty, D., & Pattus, F. (1993). Insertion of a distinct helical hairpin is not required for the spontaneous membrane insertion of colicin A. J. Mol. Biol. 230, 1055–1067.

Lau, P.C., Hefford, M.A., & Klein, P. (1987). Structural relatedness of lysis proteins from colicinogenic plasmids and icosahedral coliphages. Mol. Biol. Evol. 4, 544–556.

Lazdunski, C., Baty, D., Geli, V., Cavard, D., Morlon, J., Lloubès, R., Howard, S.P., Knibiehler, M., Chartier, M., Varenne, S., Frenette, M., Dasseux, J.L., & Pattus, F. (1988). The membrane channel-forming colicin A: synthesis, secretion, structure, action and immunity. Biochim. Biophys. Acta 947, 445–464.

Lazdunski, C. & Benedetti, H. (1990). Insertion and translocation of proteins into and through membranes. FEBS Lett. 268, 408–414.

Lazdunski, C., Baty, D., Geli, V., Lloubès, R., Benedetti, H., Letellier, L., Duché, D., & Pattus, F. (1992). In: Membranes Proteins: Structures, Interactions and Models (Pullman, A., et al., Eds.), pp. 413–425. Kluwer Academic Publishers, Dordrecht, Holland.

Levengood, S. & Webster, R. (1989). Nucleotide sequences of the *tolA* and *tolB* genes and localization of their products, components of a multistep translocation system in *Escherichia coli*. J. Bacteriol. 171, 6600–6609.

Liu, Q.R., Crozel, V., Levinthal, F., Slatin, S., Finkelstein, A., & Levinthal, C. (1986). A very short peptide makes a voltage-dependent ion channel: the critical length of the channel domain of colicin E1. Proteins 1, 218–219.

Lloubès, R., Baty, D., & Lazdunski, C. (1988). Transcriptional terminators in the *caa-cal* operon and *cai* gene. Nucleic Acids Res. 16, 3739–3749.

Lugtenberg, B. & van Alphen, L. (1983). Molecular architecture and functioning of the outer membrane of *Escherichia coli* and other gram-negative bacteria. Biochim. Biophys. Acta 737, 51–115.

Luirink, J., van der Sande, C., Tommassen, J., Veltkamp, E., de Graaf, F.K., & Oudega, B. (1986). Effects of divalent cations and of phospholipase A activity on excretion of cloacin DF13 and lysis of host cells. J. Gen. Microbiol. 132, 825–834.

Luirink, J., Duim, B., De Gier, J., & Oudega, B. (1991). Functioning of the stable signal peptide of the pCloDF13 encoded bacteriocin release protein. Mol. Microbiol. 5, 393–399.

Macgrath, J.P & Varshavsky, A. (1989). The yeast STE6 gene encodes a homologue of the mammalian multidrug resistance P-glycoprotein. Nature 340, 400–404.

Maget-Dana, R., Heitz, F., Ptak, M., Peypoux, F., & Guinand, M. (1985a). Bacteria lipopeptides induce ion-conducting pores in planar bilayers. Biochem. Biophys. Res. Commun. 129, 965–971.

Maget-Dana, R., Ptak, M., Peypoux, F., & Michel, G. (1985b). Pore-forming properties of iturin A, a lipopeptide antibiotic. Biochim. Biophys. Acta 815, 405–409.

Martinez, M.C., Lazdunski, C., & Pattus, F. (1983). Isolation, molecular and functional properties of the C-terminal domain of colicin A. EMBO J. 2, 1501–1507.

Massotte, D., Dasseux, J.L., Sauve, P., Cyrklaff, M., Leonard, K., & Pattus, F. (1989). Interaction of the pore-forming domain of colicin A with phospholipid vesicles. Biochemistry 28, 7713–7719.

Merrill, A.R., Cohen, F.S., & Cramer, W.A. (1990). On the nature of the structural change of colicin E1 channel peptide necessary for its translocation competent state. Biochemistry 29, 5829–5836.

Merrill, A.R. & Cramer, W.A. (1990). Identification of the voltage-responsive segments of the potential gated colicin E1 ion channel. Biochemistry 29, 8529–8534.

Misra, R. & Reeves, P. (1987). Role of *micF* in the *tolC* mediated regulation of OmpF a major outer membrane protein of *Escherichia coli* K-12. J. Bacteriol. 169, 4722–4730.

Muga, A., Gonzalez-Manas, J.M., Lakey, J.H., Pattus, F., & Surewicz, W.Z. (1993). pH-dependent stability and membrane interaction of the pore-forming domain of colicin A. J. Biol. Chem. 268, 1553–1557.

Nagel del Zwaig, R. & Luria, J.E. (1967). Genetics and physiology of colicin-tolerant mutants of *Escherichia coli*. J. Bacteriol. 94, 1112–1123.

Nau, C.D. & Konisky, J. (1989). Evolutionary relationship between the TonB-dependent outer membrane transport proteins: nucleotide and amino acid sequences of the *E. coli* colicin I receptor gene. J. Bacteriol. 171, 1041–1047.

Neupert, W. & Schatz, G. (1981). How proteins are transported into mitochondria. TIBS 6, 1–4.

Nikaido, H. & Vaara, M. (1985). Molecular basis of bacterial outer membrane permeability. Microbiol. Rev. 49, 1–32.

Niki, H., Imamura, R., Ogura T., & Hiraga, S. (1990). Nucleotide sequence of the *tolC* gene of *Escherichia coli*. Nucleic Acids Res. 18, 5547.

Ooi, C.E. & Weiss, J. (1992). Bidirectional movement of a nascent polypeptide across microsomal membranes reveals requirements for vectorial translocation of proteins. Cell 71, 87–96.

Oudega, B., Ykema, A., Stegehnis, F., & de Graaf, F.K. (1984). Detection and subcellular localization of mature protein H involved in excretion of cloacin DF13. FEMS Microbiol. Lett. 22, 101–109.

Parker, M.W., Pattus, F., Tucker, A.D., & Tsernoglou, D. (1989). Structure of the membrane-pore-forming fragment of colicin A. Nature 337, 93–96.

Parker, M., Tucker, A., Tsernoglou, D., & Pattus, F. (1990). Insights into membrane insertion based on studies of colicin. TIBS 15, 126–129.

Pattus, F., Martinez, M.C., Dargent, B., Cavard, D., Verger, R., & Lazdunski, C. (1983a). Interaction of colicin A with phospholipid monolayers and liposomes. Biochemistry 22, 5698–5703.

Pattus, F., Cavard, D., Verger, R., Lazdunski, C., Rosenbuch, J., & Schindler, H. (1983b). Formation of voltage-dependent pores in planar bilayers by colicin A. In: Physical Chemistry of Transmembrane Ions Notions (Spach, G., Ed.), pp. 407–413. Elsevier Biochemical Press, Amsterdam.

Pattus, F., Heitz, F., Martinez, C., Provencher, S.W., & Lazdunski, C. (1985). Secondary structure of the pore-forming colicin A and its C-terminal fragment. Eur. J. Biochem. 152, 681–689.

Postle, K. (1990). TonB and the gram-negative dilemma. Mol. Microbiol. 4, 2019–2025.

Postle, K. & Good, R. (1983). DNA sequence of the *Escherichia coli* tonB gene. Proc. Natl. Acad. Sci. USA 80, 5235–5339.

Postle, K. & Skare, J. (1988). *Escherichia coli* TonB protein is exported from the cytoplasm without proteolytic cleavage of its amino terminus. J. Biol. Chem. 263, 11000–11007.

Pugsley, A. & Reeves, P. (1977). The role of colicin receptors in the uptake of ferrienterochelin by *Escherichia coli* K12. Biochem. Biophys. Res. Commun. 74, 903–911.

Pugsley, A.P. (1984a). The ins and outs of colicins. Part I: Production and translocation across membranes. Microbiol. Sci. 1, 168–175.

Pugsley, A.P. (1984b). The ins and outs of colicins. Part II: Lethal action, immunity and ecological implications. Microbiol. Sci. 1, 203–205.

Pugsley, A.P. & Schwartz, M. (1984). Colicin E2 release: lysis leakage or secretion? Possible role of a phospholipase. EMBO J. 3, 2393–2397.

Pugsley, A.P. (1987). Nucleotide sequencing of the structural gene for colicin N reveals homology between the catalytic C-terminal domains of colicins A and N. Mol. Microbiol. 1, 317–325.

Pugsley, A.P. (1988). The immunity and lysis genes of Col N plasmid pCHAP4. Mol. Gen. Genet. 211, 335–341.

Pugsley, A.P. & Cole, S.T. (1987). An unmodified form of the ColE2 lysis protein, an envelope lipoprotein, retains reduced ability to promote colicin E2 release and lysis of producing cells. J. Gen. Microbiol. 133, 2411–2420.

Rath, P., Bousché, O., Merrill, A.R., Cramer, W.A., & Rothschild, K.J. (1991). FTIR evidence for a predominantly α-helical structure of the membrane-bound channel forming C-terminal peptide of colicin E1. Biophys. J. 59, 516–522.

Regue, M. & Wu, H.C. (1988). Synthesis and export of lipoproteins in bacteria. In: Protein Transfer and Organelle Biogenesis (Das, R.C. & Robbins, P.W., Eds.). Academic Press, London.

Riordan, J.R., Rommens, J.M., Kerem, B.S., Alon, N., Rozmahel, R., Grzelczak, Z., Zielenski, Lok, S. Plavsic, N., Chou, J.L., Drumm, M.L., Ianmuzzi, M.C., Collins, F.S., & Tsui, L.C. (1989). Identification of the cystic fibrosis gene: cloning and characterization of complementary DNA. Science 245, 1066–1072.

Ross, U., Harkness, R., & Braun, V. (1989). Assembly of colicin genes from a few DNA fragments. Nucleotide sequence of colicin D. Mol. Microbiol. 3, 891–902.

Rutz, J.M., Liu, J., Lyons, J.A., Goranson, J., Amstrong, S.K., McIntosh, M.A., Feix, J.B., & Klebba, P.E. (1992). Formation of a gated channel by a ligand-specific transport protein in the bacterial outer membrane. Science 258, 471–475.

Schöffler, A. & Braun, V. (1989). Transport across the outer membrane of *Escherichia coli* via the Fhu A receptor is regulated by the TonB protein of the cytoplasmic membrane. Mol. Gen. Genet. 217, 378–383.

Silhavy, T.J., Benson, S.A., & Emr, S.D. (1983). Mechanism of protein localization. Microbiol. Rev. 47, 313–344.

Simon, S.M. & Blobel, G. (1991). A protein conducting channel in the endoplasmic reticulum. Cell 65, 371–380.

Slatin, S., Raymond, L., & Finkelstein, A. (1986). Gating of a voltage-dependent channel (colicin E1) in planar lipid bilayers. The role of protein translocation. J. Membr. Biol. 92, 247–254.

Slatin, S. (1988). Channels formed by colicin E1 in planar lipid bilayers are monomers. Biophys. J. 53, 155a.

Stengele, I., Bross, P., Garces, X., Giray, J., & Rasched, I. (1990). Dissection of functional domains in phage fd adsorption protein: discrimination between attachment and penetration sites. J. Mol. Biol. 212, 143–149.

Sun, T.P. & Webster, R.E. (1986). *fii* (*tolQ*) a bacterial locus required for filamentous phage infection and its relation to colicin-tolerant *tol*A *tol*B. J. Bacteriol. 165, 107–115.

Sun, T.P. & Webster, R.E. (1987). Nucleotide sequence of a gene cluster involved in the entry of the E colicins and the single stranded DNA of infecting filamentous phage into *Escherichia coli*. J. Bacteriol. 169, 2667–2674.

Tabor, S. & Richardson, C. (1985). A bacteriophage T7 DNA polymerase/promoter system for controlled exclusive expression of specific genes. Proc. Natl. Acad. Sci. USA 70, 3160–3164.

Toba, M., Masaki, H., & Ohta, T. (1986). Primary structures of ColE2-P9 and ColE3-CA38 lysis genes. J. Biochem. 99, 591–596.

Tommassen, J. & Lugtenberg, B. (1980). Outer membrane protein of *Escherichia coli* is co-regulated with alkaline phosphatase. J. Bacteriol. 143, 151–157.

Tucker, A.D., Baty, D., Parker, M.W., Pattus, F., Lazdunski, C., & Ternoglou, D. (1989). Crystallographic phases through genetic engineering: experiences with colicin A. Prot. Eng. 2, 399–405.

Van der Goot, F., Gonzalez-Manas, J.M., Lakey, J., & Pattus, F. (1991). A membrane insertion intermediate of the pore-forming domain of colicin A. Nature 364, 408–410.

Van der Wal, F.J., Oudega, B., Kater, N.M., Ten Hagen-Jongman, C., de Graaf, F.K., & Luirink, J. (1992). The stable BRP signal peptide causes lethality but is unable to provoke the translocation of cloacin DF13 across the cytoplasmic membrane of *E. coli*. Mol. Microbiol. 6, 2309–2318.

Verner, K. & Schatz, G. (1988). Protein-translocation across membranes. Science 241, 1307–1313.

Wagner, W., Vogen, M., & Goebel, W. (1983). Transport of hemolysin across the outer membrane of *Escherichia coli* requires two functions. J. Bacteriol. 154, 200–210.

Wandersman, C. & Delepelaire, P. (1990). TolC, an *E. coli* outer membrane protein required for hemolysin secretion. Proc. Natl. Acad. Sci. USA 87, 4776–4780.

Wayne, R. & Neilands, J.B. (1975). Evidence for common binding sites for ferrichrome compounds and bacteriophage Ø 80 in the cell envelope of *Escherichia coli*. J. Bacteriol. 121, 459–503.

Webster, R. (1991). The *tol* gene products and the import of macromolecules into *Escherichia coli*. Mol. Microbiol. 5, 1005–1011.

Wormald, M.R., Merrill, A.R., Cramer, W.A., & Williams, R.J.P. (1990). Solution NMR studies on colicin E1 C-terminal thermolytic peptide. Structural comparison with colicin A and the effects of pH changes. Eur. J. Biochem. 191, 155–161.

THE MECHANISM OF DIPHTHERIA TOXIN TRANSLOCATION ACROSS MEMBRANES

Erwin London

Membrane Protein Transport
Volume 1, pages 201–227.
Copyright © 1995 by JAI Press Inc.
All rights of reproduction in any form reserved.
ISBN: 1-55938-907-9

I. INTRODUCTION

Many bacterial protein toxins must cross biological membranes in order to function. Such toxins are invaders, and do not appear to use the ordinary machinery for protein translocation in order to gain entry into the cytoplasm. This requires that much or all of the information for their translocation across membranes be locked within their own primary sequence. This is in contrast to translocation of ordinary cellular proteins, which involves a complex sequence of events in which many proteins play a role. Comparatively, protein toxins represent much simpler experimental systems, amenable to detailed biochemical and biophysical analysis.

One might question whether it will be possible to extrapolate of the behavior of protein toxins to ordinary cellular proteins. It is the author's belief that all mechanisms of protein translocation are likely to share a number of features. As described below, there are already very strong parallels between translocation of toxins and ordinary cellular proteins, and it has already been possible to use the principles learned from toxin translocation successfully predict the behavior of an *E. coli* protein involved in ordinary protein translocation. In addition, it is fascinating that there is a protein toxin structure that allows us to predict a structure that may be found in many translocator proteins.

In this chapter, I have concentrated on the protein diphtheria toxin because it is the toxin for which we have the most information concerning membrane translocation. A detailed review of the literature on diphtheria toxin translocation has appeared (London, 1992a). This article will concentrate on more recent studies.

II. DIPHTHERIA TOXIN STRUCTURE AND FUNCTION

Diphtheria toxin (M_r 58,348) is a protein secreted by *Corynebacterium diphtheriae*. It is synthesized as a single polypeptide, but at a site which loops out into solution it can be readily cleaved by proteolysis into A and B chains. The A and B chains remain joined by a disulfide bond (Figure 1).

The structure of the toxin in its native conformation at neutral pH has been solved by X-ray crystallography (Choe et al., 1992). The crystal structure shows the protein has three domains of roughly equal size (Figure 2). At the N-terminal side of the protein is the A chain (also called the C (for catalytic) domain). This domain forms the N-terminal third of the protein. Structurally, the A chain appears to be an ordinary globular protein, containing both α-helices and β-sheets, and having a distinct active site cleft. The A chain inhibits protein synthesis and kills cells by catalyzing the ADP-ribosylation of diphthamide, a modified His residue found on protein synthesis elongation factor 2 (Collier, 1982).

One group has suggested that the toxin can induce internucleosomal DNA breakdown (Chang et al., 1989a), and that the A chain has a nuclease activity which may be responsible for this process (Chang et al., 1989b). This proposal is very controversial because a contaminating nuclease has been reported in some preparations of the toxin (Wilson et al., 1990). Furthermore, no DNase-like features have been reported in the crystal structure of the toxin, and the breakdown of chromosomal DNA has been reported to be simply a consequence of the inhibition of protein synthesis (Morimoto and Bonavida, 1992; Kochi and Collier, 1993). On the other hand, it has been claimed in a recent study that the nuclease activity associated with the toxin is not due to a contaminant (Lessnick et al., 1992).

The B chain is composed of the other two domains of the toxin. The N-terminal portion of the B chain forms the T ("translocation") domain. It is an α-helical region (Choe et al., 1992) containing several long hydrophobic sequences. These sequences are the most hydrophobic within the toxin, and tend to be localized within helices which are largely buried in the native state. It is believed the exposure of these hydrophobic helices is the central event in membrane-insertion (see below). The C-terminal portion of the B chain forms the R ("receptor-binding") domain, a β-sheet dominated structure responsible for toxin–receptor interaction (Choe et al., 1992). The receptor-binding sequence appears to be restricted to the final 54 residues of the toxin (Rolf and Eidels, 1993).

The possibility that the B chain has additional biological activities has not been ruled out. It appears to contribute to the ability of the toxin to interact with anionic ligands via a loosely defined portion of the R domain which has been named the P-site (Collier, 1982). The P-site seems to include part of the C-terminal region of the R domain (Proia et al., 1980). The ability of the B chain to interact with polyphosphoinositides is particularly intriguing in terms of possible implications for toxin effects on signal transduction (Kagan, 1991).

The interaction of diphtheria toxin with ligands is sometimes complex. For example, the toxin binds the dinucleotide ApUp with extremely high affinity (Collins et al., 1984). ApUp binds to the A chain at the same site as NAD^+, but there is strong evidence suggesting that its especially tight binding is due to additional interactions with the B chain, probably at the P-site (Collier, 1982). Some other

GlyAlaAspAspValValAspSerSerLys SerPheValMetGluAsnPheSerSerTyr HisGlyThrLysProGlyTyrValAspSer IleGlnL
    CH1   CB1   CB2   CH2

GlyThrGlnGlyAsnTyrAspAspAspTrp LysGlyPheTyrSerThrAspAsnLysTyr AspAlaAlaGlyTyrSerValAspAsnGlu AsnPro
                      CB3   .   CH3

ValLysValThrTyrProGlyLeuThrLys ValLeuAlaLeuLysValAspAsnAlaGlu ThrIleLysLysGluLeuGlyLeuSerLeu ThrGlu
  CB4   CB5   CH4

GluGluPheIleLysArgPheGlyAspGly AlaSerArgValValLeuSerLeuProPhe AlaGluGlySerSerSerValGluTyrIle AsnAsn
  CH5   CB6   CB7

ValGluLeuGluIleAsnPheGluThrArg GlyLysArgGlyGlnAspAlaMetTyrGlu TyrMetAlaGlnAlaCysAlaGlyAsnArg ValArg
  CB8   CH6   CH7

CysIleAsnLeuAspTrpAspValIleArg AspLysThrLysThrLysIleGluSerLeu LysGluHisGlyProIleLysAsnLysMet SerGlu
                  TH1   TH2

GluLysAlaLysGlnTyrLeuGluGluPhe HisGlnThrAlaLeuGluHisProGluLeu SerGluLeuLysThrValThrGlyThrAsn ProVal
        TH3   .   TH4

TrpAlaValAsnValAlaGlnValIleAsp SerGluThrAlaAspAsnLeuGluLysThr ThrAlaAlaLeuSerIleLeuProGlyIle GlySer
    TH5   TH6   TH7

ValHisHisAsnThrGluGluIleValAla GlnSerIleAlaLeuSerSerLeuMetVal AlaGlnAlaIleProLeuValGlyGluLeu ValAspIleGlyPhe
_____
           TH8
- - - - - - - - - - - - - - - - -

ValGluSerIleIleAsnLeuPheGlnVal ValHisAsnSerTyrAsnArgProAlaTyr SerProGlyHisLysThrGlnProPheLeu HisAspGlyTyrAla
_____
           TH9   RB1
- - - - - - - - - - - - - - - - - - - - - -

ValGluAspSerIleIleArgThrGlyPhe GlnGlyGluSerGlyHisAspIleLysIle ThrAlaGluAsnThrProLeuProIleAla GlyValLeuLeuPro
                                 _____   _____
                                           RB3   RB4

LeuAspValAsnLysSerLysThrHisIle SerValAsnGlyArgLysIleArgMetArg CysArgAlaIleAspGlyAspValThrPhe CysArgProLysSer
             _____   _____   _____
               RB5   RB6   RB7

AsnGlyValHisAlaAsnLeuHisValAla PheHisArgSerSerSerGluLysIleHis SerAsnGluIleSerSerAspSerIleGly ValLeuGlyTyrGln
_____
           RB8

ThrLysValAsnSerLysLeuSerLeuPhe PheGluIleLysSer
_____
                 RB10   535

Figure 1. Sequence of diphtheria toxin. The secondary structural elements identified in the crystal st sequence using the nomenclature of Choe et al., 1992. The solid bars highlight the regions containing hydrophobic residues. Sequences with alternating hydrophobic residues are shown by broken bars. I 186 and 201, and between Cys 461 and 471.

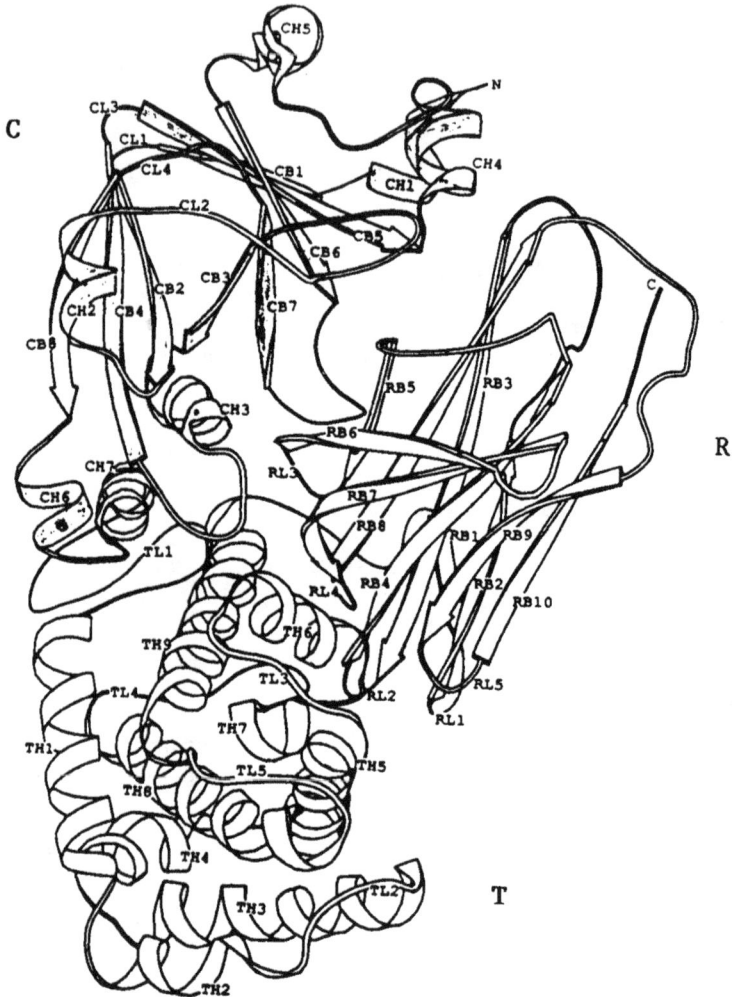

Figure 2. Crystal structure of diphtheria toxin. **C**, catalytic domain; **T**, "translocation" domain; **R**, receptor domain. Figure from Choe et al., 1992.

anionic ligands of the toxin, such as ATP, may also bind so as to interact with both the A chain and P-site.

III. THE DIPHTHERIA TOXIN RECEPTOR

Earlier observations of a complicated relationship between anion transport and toxicity in cells led to speculation that an anion transporter might be the receptor for the toxin (Olsnes and Sandvig, 1986). There are a number of factors that complicate interpretation of such experiments, and at present the molecular basis of this behavior is not understood (see London 1992a for details).

The best candidate for the functional cellular receptor for the toxin appears to be a plasma membrane protein that has been identified in Vero monkey cells. The receptor cDNA encodes a protein of 185 residues with a single transmembrane domain and an extracellular domain corresponding to a heparin-binding epidermal growth factor (Naglich et al., 1992). The receptor appears to interact or form a complex with a second membrane protein of unknown function, called DRAP 27 (Iwamoto et al., 1991). DRAP 27 is essentially identical to human CD 9 protein (Mitamura et al., 1992). Expression of the DRAP 27 protein increases the sensitivity of cells to diphtheria toxin (Mitamura et al., 1992).

IV. ENTRY OF DIPHTHERIA TOXIN INTO ENDOSOMES

The most important role for toxin binding to its receptor is probably to allow the toxin to be taken into the cell by receptor-mediated endocytosis. There is a great deal of evidence that upon endocytosis the toxin reaches the lumen of the endosomal vacuole system (see London, 1992a for details). Endosomes are acidic vacuoles, i.e. they have an aqueous lumen with a low pH. There appear to be, at least in some cell types, two categories of endosomes: early endosomes with a lumenal pH close to 6, and late endosomes which have a lumenal pH near 5.3 (Cain et al., 1989; Fuchs et al., 1989).

There is very strong evidence that at low pH the toxin encounters in the lumen of late endosomes is the trigger for membrane penetration by the toxin. At the cellular level, genetic studies show mutants which are unable to properly acidify endosomes are resistant to the toxin. Furthermore, both chemical inhibitors of the ATPase responsible for acidifying endosomes and pH-raising lysosomotropic amines also protect cells from the toxin (see London, 1992a for details). In fact, by decreasing external pH the toxin can even artificially translocate across plasma membranes. In this way protein synthesis inhibition can be obtained with the toxin bypassing passage through endosomes. The biochemical properties of the toxin also suggest that low pH is the physiological trigger for membrane insertion. Below pH 5.3 the toxin changes into a hydrophobic, membrane-penetrating form (see below). On the other hand, it should be noted that one group has proposed that toxin entry might only be completed after recycling of endocytic vesicles to the cell surface (Hudson et al., 1988). It is not known whether the toxin receptor or any additional plasma membrane or endosomal proteins play any additional roles in regulating translocation.

It is noteworthy that the triggering of toxin membrane penetration by low pH parallels the process whereby certain viruses invade cells. Most noteworthy is the well-characterized case of influenza virus, where exposure to the low pH within endosomes induces a change in the viral hemagglutinin protein which exposes a buried hydrophobic fusion peptide that appears to insert into the endosomal membrane (Wiley and Skehel, 1987).

V. THE EFFECTS OF LOW PH ON DIPHTHERIA TOXIN CONFORMATION AND MEMBRANE PENETRATION

Exposure to low pH induces a dramatic change in diphtheria toxin conformation. This change appears to be accompanied by increased exposure of Trp residues to solution (Blewitt et al., 1985), and increased accessibility to proteolysis (Hu and Holmes, 1984; Dumont and Richards, 1988). Overall, the changes at low pH involve a partial unfolding event in which tertiary structure is disrupted, but in which secondary structure remains largely intact (Zhao and London, 1986). These are the features associated with the molten globule protein conformation (London, 1992b).

This partial unfolding at low pH renders diphtheria toxin hydrophobic, as shown by its binding to micelles of mild non-ionic detergents, and its insertion into model and biological membranes (see London, 1992a for details). It appears that the exposure of the hydrophobic helices within the T domain due to partial unfolding is one central event triggering membrane insertion. However, as discussed below, unfolding within other domains may be involved as well. Low pH also seems to contribute to the development of hydrophobic behavior via the loss of charge on anionic residues (Montecucco et al., 1985; Kieleczawa et al., 1990). It has been suggested that the protonation of acidic residues renders the tips of pairs of helices in the T domain sufficiently hydrophobic to initiate membrane insertion (Choe et al., 1992).

Conformational changes induced by low pH are not restricted to the T domain. The isolated A chain also undergoes a partial unfolding event at low pH (Zhao and London, 1988). The opening of a hinge between two apparent "lobes" of the A chain may be involved (Choe et al., 1992). As in the case of whole toxin, unfolding of the A chain results in the formation of a hydrophobic structure capable of membrane insertion at low pH (Montecucco et al., 1985; Zhao and London, 1988). One difference between A chain and whole toxin behavior is that a somewhat lower pH is necessary to trigger conformational change within isolated A chains under most conditions. Another important difference between A chain and whole toxin behavior is that the changes within the A chain are readily reversed when pH is returned to 7. This supports the possibility that *in vivo* low pH induced unfolding and membrane insertion of the A chain could be followed by refolding and dissociation from membranes upon exposure to the neutral pH of the cytoplasm (Zhao and London, 1988).

Why does low pH alter toxin conformation? The reasons are probably the same as in any case of pH-induced unfolding. Alterations of electrostatic interactions due to protonation of His, Asp, and Glu residues could favor unfolding as a result in formation of buried charges, breakdown of salt bridges, and the strengthening of repulsions due to increased local or global positive charge. There has also been speculation of a role for Pro isomerization at low pH. (See London, 1992a for details.)

It should be noted that other bacterial toxins show behavior in which low-pH-induced partial unfolding is closely linked to membrane insertion. Recently, *Pseudomonas* exotoxin A (Jiang et al., 1990) and colicins (Merrill et al., 1990; van der Goot et al., 1991; Benedetti et al., 1992; Muga et al., 1993) have been shown to exhibit this type of behavior.

VI. CONFORMATION OF MEMBRANE-INSERTED TOXIN

At low pH the toxin efficiently inserts into model membranes (Hu and Holmes, 1984; Chung and London, 1988). Model membrane-inserted toxin can take on one of several distinct conformations. In all of these conformations some part of the B chain has unfolded. At lower temperatures the membrane-inserted toxin takes on conformations in which the A chain is folded, and at higher temperatures conformations in which the A chain is at least partially unfolded are observed (Jiang et al., 1991a). In addition, there is a change in toxin conformation after toxin exposed to low pH is returned to neutral pH. After pH neutralization the toxin remains membrane bound, but is inserted into the membrane to a lesser degree (Montecucco et al., 1985; Jiang et al., 1991a). Figure 3 summarizes these conformations. There has been one IR study of the secondary structure of membrane-inserted toxin (Cabiaux et al., 1989). This study suggested that there is an alteration in the type

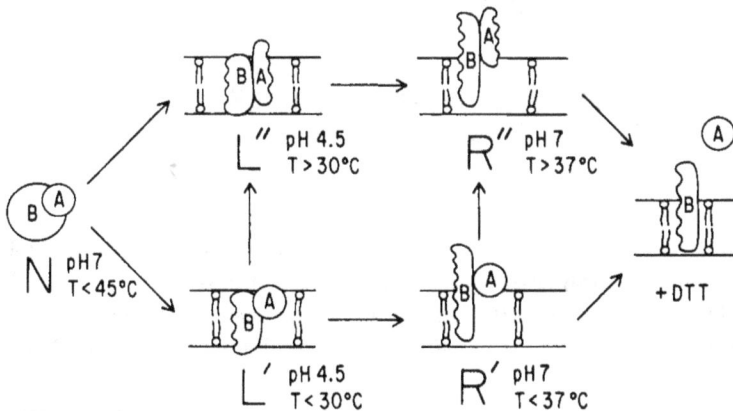

Figure 3. Conformations identified experimentally in membrane-inserted diphtheria toxin. From Jiang et al., 1991a.

of β-sheet structure present in model membrane-inserted whole toxin at low pH relative to that of native toxin in solution.

Interestingly, the various conformations described above are seen after low pH exposure both in solution and in model membrane-inserted toxin. A probable reason for the similarity in what should be two very different environments is that in solution after exposure to low pH the strong toxin–toxin interactions that form via hydrophobic interactions effectively substitute for toxin–lipid interactions (Blewitt et al., 1985; Jiang et al., 1991a).

Other model membrane experiments have yielded information concerning the topography of membrane-inserted toxin. Photolabeling experiments suggest parts of both the A and B chain contact lipid (Hu and Holmes, 1984; Zalman and Wisnieski, 1984; Montecucco et al., 1985). Vesicle-entrapped toxin has been used to show that some regions of membrane-inserted toxin cross all the way through the membrane (Gonzalez and Wisnieski, 1988; Jiang et al., 1991b). Experiments with biotinylated toxin suggest sites on both the A and B chains cross the membrane and reach the face of the membrane opposite the side of insertion (Jiang et al., 1991b). Bilayer lipid membrane studies suggest that the anion binding and perhaps even the entire ApUp binding sites reach the side opposite that of insertion (Donovan et al., 1982; Kagan et al., 1984).

The topography of the B chain inserted into the plasma membrane has been examined by proteolysis. It was found that in the membrane-inserted form, much of the R domain is insensitive to externally added proteases, and it was suggested that this domain translocates across the membrane. An alternate possibility is that it becomes protected from proteolysis due to approaching the bilayer surface closely and/or interaction with the receptor. Analysis of the protected regions was used to formulate a preliminary model of B chain topography (Moskaug et al., 1991). However, this model differs somewhat from that predicted from the position of α-helices in the crystal structure.

VII. PORE FORMATION BY DIPHTHERIA TOXIN AND ITS POSSIBLE ROLE IN TRANSLOCATION

Pore formation is one of the key properties associated with membrane insertion of the toxin (Donovan et al., 1981; Kagan et al., 1981). The fact that the isolated T domain is able to form pores suggests it is involved in pore formation (Kagan et al., 1981), although some sort of membrane lesion can even be formed by isolated A chain under the appropriate *in vitro* conditions (Jiang et al., 1989b; Jiang et al., 1991b).

The observation that the toxin makes pores in model membranes led to the proposal that the A chain passes into the cytoplasm via this pore, or that the pore is formed by a discarded hydrophobic B chain wrapper which acts to shield the membrane-inserted A chain from contact with the bilayer (Kagan et al., 1981; Misler, 1984).

A role for pore formation in translocation is supported by the close linkage between pore formation and translocation seen in a number of studies (see, for example, Falnes et al., 1992). In these cases it is found that the toxin is unable to form a pore and at the same time is unable to translocate and/or inhibit protein synthesis. However, establishing a cause and effect relationship between pore formation and translocation is difficult. For example, the simultaneous loss of pore formation and translocation may simply mean membrane insertion has been blocked.

There are a number of other reasons that the pore-forming activity of the toxin is hard to interpret. One is that pore properties vary from system to system. In model membrane vesicles, the pores observed seem relatively large (Jiang et al., 1989a, 1991b), and appear to be large enough to allow translocation of A chains (Jiang et al., 1991b). At the other extreme, the pores induced in plasma membranes by exposure of bound toxin to low pH seem very small (Papini et al., 1988; Sandvig and Olsnes, 1988). The explanation for this disparity may arise from the formation of toxin oligomers of various sizes within the membrane. For complement protein, the consequence of such behavior is that large pore size is correlated with large oligomer size (Malinski and Nelsesteun, 1989). If toxin behaves similarly, it would explain why in experimental systems in which toxin concentration is relatively high and oligomers might form (e.g., model membrane vesicle systems), pore size is large. In contrast, where toxin concentration is lower, or when binding to a receptor limits oligomer formation, as might occur in cells, pore size would be much smaller.

Another factor complicating interpretation of pore formation is that the quantitative relationship between pore formation and translocation is not clear. In particular, it is not known what percent of the time a membrane-inserted toxin molecule is engaged in pore formation. It is also possible that pore size shrinks after the A chain passes through. In that case the properties of the experimentally observed "empty" pore might not reflect that of the functioning pore.

VIII. RELATIONSHIP OF DISULFIDE AND PEPTIDE BOND CLEAVAGE TO TOXIN ENTRY

The functional roles and timing of the cleavage of toxin disulfide bonds and cleavage of the peptide link between the A and B chains is not yet completely understood. The cleavage of the peptide link between A and B chains is known to be necessary for toxicity (Sandvig and Olsnes, 1981). Most likely this is because the A chain cannot be released into the cytoplasm until this link is cleaved or because the linkage to the B chain sterically blocks elongation factor 2 binding. Given the sequence of the cleavage site, the most likely place for this cleavage to occur is in a vacuole subsequent to endocytosis, although extracellular cleavage is also possible (London, 1992a).

The behavior and roles of the two disulfide bonds of the toxin are even less well understood (see London, 1992a). Cleavage of the disulfide linking the A and B chains most likely involves reduction upon exposure of the disulfide to the cytoplasm (Moskaug et al., 1987), although reduction at the cytoplasmic surface cannot yet be ruled out (Ryser et al., 1991). For example, in Vero monkey kidney cells it appears that reduction occurs subsequent to membrane penetration (Papini et al., 1993). In contrast, in Chinese Hamster Ovary cells there is some evidence that there is inhibition of toxin action by antibodies against plasma membrane disulfide isomerase, a protein that could shuffle toxin disulfides (Mandel et al., 1993). It should be pointed out that the origin of antibody-induced inhibition of cytotoxicity is not clear, as these experiments involved extended incubations, and other components in the antibody-containing ascites fluid might influence toxicity.

The role of the intra-B chain disulfide is also unclear. Two mutants lacking this disulfide have different phenotypes. One mutant has greatly reduced cytotoxicity (Papini et al., 1987), whereas the second shows normal cytotoxicity (Stenmark et al., 1991a). In the latter case, the mutant exhibited both normal pore formation and low-pH-induced plasma membrane insertion. It has been suggested that a dependence of toxin behavior on the exact nature of the residues substituting for Cys might be responsible for this disparity (Stenmark et al., 1991a).

IX. TRANSLOCATION OF DIPHTHERIA TOXIN ACROSS REAL AND MODEL MEMBRANES

The analysis of the effects of low pH on model membrane inserted toxin has led to the proposal of various translocation mechanisms. What the membrane-inserted toxin would look like according to each of these models is shown in Figure 4. As

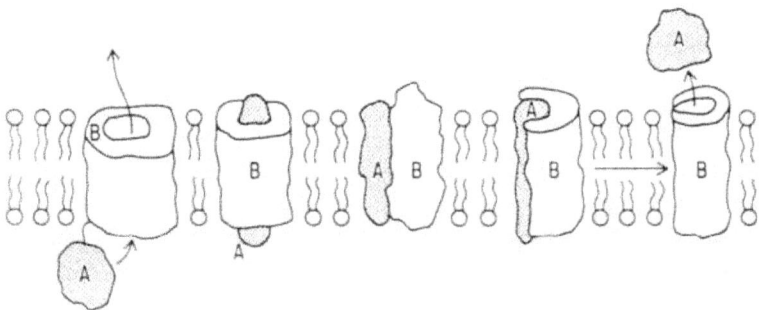

Figure 4. Proposed structures for membrane-inserted diphtheria toxin according to different translocation models. From London, 1992a.

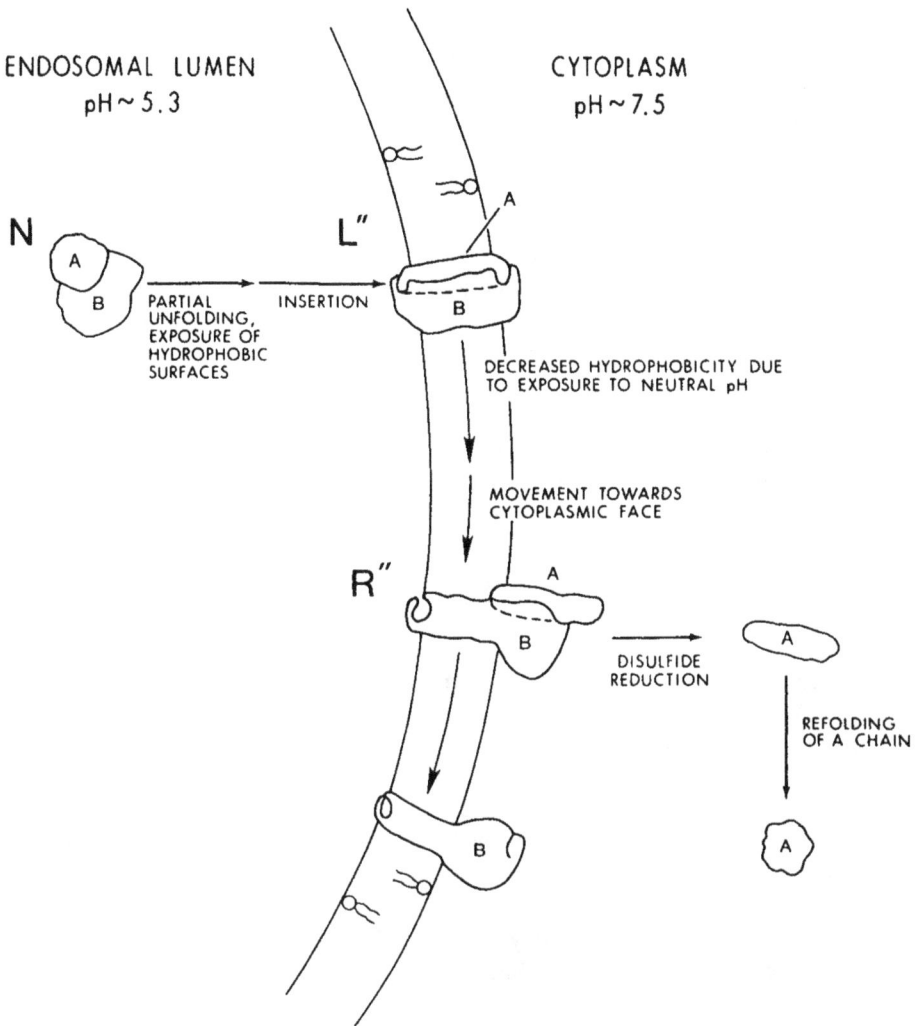

Figure 5. A proposed mechanism for toxin translocation across membranes. From London, 1992a.

noted above, one model predicts that the A chain passes into the cytoplasm via a pore, perhaps squeezing through in an unfolded form (Boquet et al., 1976; Kagan et al., 1981). Another model predicts that the pore is formed by a discarded hydrophobic B chain wrapper. This wrapper would shield the membrane-inserted A chain from contact with the bilayer (Misler, 1984).

Studies demonstrating that the A chain will insert into lipid bilayers led to proposals that the translocation process involves lipid contact with the A chain (Hu and Holmes, 1984; Zalman and Wisnieski, 1984; Montecucco et al., 1985). In more recent versions of this idea, it has been proposed that there is transmembrane insertion of the A chain with a partially surrounding B chain (Papini et al., 1987; Zhao and London, 1988), and that A chain unfolding is linked to proper A chain insertion (Zhao and London, 1988). The B chain may help properly orient the inserted A chain, and engage in pore formation when the A chain leaves the bilayer. Figure 5 illustrates the details of this model.

Another possibility is that the toxin can escape endosomes by inducing membrane lysis. This seems unlikely in view of the limited size of the lesions formed by toxins in model membrane vesicles (Jiang et al., 1989a; Jiang et al., 1991b).

One approach to studying translocation has been to examine the effect of environmental conditions on toxin-induced cytotoxicity (see London, 1992a). Using this approach, the effects of transmembrane pH gradients and membrane potential on cytotoxicity have been studied. It appears a transmembrane pH gradient, such as would occur *in vivo* across the endosomal membrane, promotes translocation, but there is disagreement whether there is a role for transmembrane potential (Sandvig et al., 1986; Hudson et al., 1988).

Another approach has been to study translocation across the plasma membrane induced by decreasing the pH outside the cell. In this way one starts with receptor-bound toxin and measures cytotoxicity, protection of the A chain from external pronase, and/or the release of the A chain into a soluble cell lysate (Moskaug et al., 1987, 1988). Studies using this system have shown that only a subpopulation of plasma membrane bound toxin molecules can translocate their A chains to the cytoplasm. It has also been shown that reduction of the A-B disulfide occurs upon membrane insertion. This is due to the increased exposure of the A-B disulfide in the low pH conformation and presumably to the exposure of the A-B disulfide to the reducing environment of the cytoplasm subsequent to membrane transport (Moskaug et al., 1987).

There has been one direct study of the translocation of diphtheria toxin trapped within isolated endosomes (Beaumelle et al., 1992). Translocation to external solution was observed, although it was slow (about 5% of trapped toxin per 40 min). Unfortunately, the effect of adding a reducing agent to the outside of the endosomes on translocation rate was not examined. Translocation was low pH-dependent, and dependent on a pH gradient across the membrane. Translocation of co-entrapped transferrin and horseradish peroxidase was not seen, but nonspecific pore formation cannot be ruled out as these markers are larger than the A chain. Many experiments were performed with a transferrin–toxin conjugate. In this system, the investigators were able to demonstrate it was the A chain being released from the endosomes. Overall, it appears that the endosomal system may develop into a useful method for looking at translocation.

It has also been shown that toxin trapped within the lumen of model membrane vesicles can be used to study translocation. By exposing the trapped toxin to low pH, reversing pH to 7, and reducing the disulfide linking the A and B chains translocation can be obtained, with release of about half of the A chains into the external solution (Jiang et al., 1991b). Under the conditions of these experiments (100–250 toxins trapped per vesicle) it could be shown that at least some of the toxin molecules form nonspecific pores that would efficiently release a 17-kDa dextran but not a 64-kDa dextran. A single large pore in a vesicle could be sufficient to allow nonspecific release of many A chains, and therefore not all of the trapped toxin molecules may have formed large pores. As a result, it is impossible to tell whether the similar level of A chain release that was observed with conformations having folded and unfolded A chain reflects a similar intrinsic efficiency of translocation for these conformations. It could also be shown that A chains trapped by themselves form enough large, nonspecific pores to allow their own release under some conditions (Jiang et al., 1989b; Jiang et al., 1991b). An unresolved question is whether translocation behavior would be different with low numbers of toxins trapped in model membrane vesicles. Until this is answered it is not possible to specify the exact roles of pore formation and A chain unfolding in translocation. There is even the possibility of different translocation mechanisms *in vivo* depending on toxin concentration within endosomes.

X. TRANSLOCATION OF MUTANTS WITH DELETIONS AND INSERTIONS

Use of natural toxin mutants (cross-reacting materials: CRMs) has provided useful information about the roles of different regions and residues in the toxin (London, 1992a). It is now possible to extend this approach by construction of any desired mutation. A number of recent studies have looked at the tolerance of the termini of the A and B chains to deletions and insertions in this way. In the case of the N-terminus of the A chain, it has been shown that peptide and protein extensions can often be tolerated such that the extended A chain can be translocated (Stenmark et al., 1991b; Madshus et al., 1992), although some N-terminal deletions may inhibit translocation (Chaudhary et al., 1991). Perhaps most intriguing, when the N-terminus of the A chain was fused to acidic fibroblast growth factor, it could be shown that only conditions resulting in tight folding of the growth factor prevented translocation of the A chain growth factor hybrid (Wiedlocha et al., 1992). The implication is that some degree of unfolding is necessary to maintain the translocation competent state, as in the case of ordinary protein translocation. This observation makes it seem more likely that the A chain must also maintain an unfolded state to be translocated.

A study of the role of the C-terminus of the A chain (C-terminal to Cys 186) also showed considerable tolerance for deletions and substitutions (Ariansen et al., 1993). To some degree there is a discrepancy between this behavior and that observed in an earlier study by the same group in which it appeared that a particular C-terminal sequence was important for translocation (Moskaug et al., 1989).

A study of the role of B chain structure upon translocation indicated that deletions of up to 12 residues at the N-terminus are more easily tolerated than smaller deletions at the C-terminus (Stenmark et al., 1992).

XI. PROPERTIES OF TOXIN MUTANTS WITH SINGLE RESIDUE SUBSTITUTIONS

Perhaps the most powerful method to identify the roles of individual amino acid residues in translocation will be to construct mutant toxins with single substitutions at specific residues. To date, there have been several studies in which residues in the B chain have been modified.

In one study, Glu 349 was changed to Lys (O'Keefe et al., 1992). This mutation was originally identified using the *E. coli* selection system of O'Keefe and Collier (1989). It inhibits ordinary cytotoxicity and the cytotoxicity obtained by exposing plasma membrane bound toxin to low pH, despite the fact that its receptor binding and *in vitro* enzymatic activity is intact. Whether this mutation would affect membrane interaction or the low-pH-induced conformational change was not examined. Nevertheless, in view of the location of residue 349 in the loop linking hydrophobic helices 8 and 9 of the T domain. It was proposed the substitution of a Lys may impede membrane insertion of these helices.

A nearby mutation in which Asp 352 is changed to Asn does seem to show significant effects on toxin pore forming properties (Mindell et al., 1992). It appears that this residue is in a position that influences pore formation, and may even be part of the pore. Furthermore, experiments by Mindell et al. altering pH on the both sides of the membrane imply that this residue crosses the membrane and ends up close to the side of the membrane opposite the side of insertion. In other words, in the conductive form of the toxin, helices 8 and 9 have inserted such that the loop between them reaches the opposite membrane face.

Another mutation that has been examined is the substitution of Ile 364 with Lys (Cabiaux et al., 1993). This mutation inhibits channel-forming activity and the cytotoxicity of plasma membrane-bound toxin after its exposure to low pH. It was not determined whether the loss of activity is due to the lack of a conformational change at low pH, lack of membrane insertion, or an alteration in another step in translocation.

The same limitations apply to the interpretation of mutations at Pro 345 (Johnson et al., 1993). Substitution of Glu or Gly at this position largely abolished cytotoxicity, while cell binding and *in vitro* ADP-ribosylation activity remained intact. Again, it was not determined what alteration leads to a loss of translocation.

All of the studies above involved residues in the helix 8 and 9 region of the T domain. Falnes et al. (1992) examined the roles of a wider set of acidic residues in the B chain by mutating them to Lys. A range of properties was examined. In each case the toxins produced seemed to fold into the native state at neutral pH. Mutation of Asp 392 to Lys decreased receptor binding. Mutation of Glu 362 to Lys appears to have destabilized the native state relative to the low pH conformation such that the change to the membrane-inserting form was shifted to higher pH by several tenths of a pH unit. This was reflected in the external pH necessary to induce cytotoxicity with plasma membrane bound toxin, and in the pH necessary to obtain pore formation within cells. Mutation of Asp 285 or 318 to Lys reduced pore formation and, as judged by pronase protection experiments, translocation. Mutation of Asp 403, Asp 442, or Glu 413 to Lys seemed to result in proteins with more or less normal properties, although the isolated B chain of the 403 mutant was very protease sensitive.

XII. PREDICTING THE STRUCTURE OF MEMBRANE-INSERTED DIPHTHERIA TOXIN BASED ON ITS CRYSTAL STRUCTURE

Our knowledge of the crystal structure of diphtheria toxin may be a great help in understanding its structure in membranes and thus translocation. One reason is that the toxin seems to undergo only relatively small changes in secondary structure at low pH, so in the membrane inserted form the toxin may preserve many of the sheets and helices present in its native state. This idea is supported by the observation that the most hydrophobic stretches in the T domain, which should show the most propensity to form transmembrane α-helices according to hydropathy calculations (Greenfield et al., 1983), are already packaged into α-helices in the native toxin (see Figure 1). Of course, it is likely that there are some changes in secondary structure upon membrane insertion. Nevertheless, the structure of the toxin in the native state is the first place to look when formulating hypotheses about the membrane-inserted conformation.

XIII. THE T DOMAIN AND TRANSMEMBRANE HELICES

The first structures that should be considered are transmembrane α-helices. The potential transmembrane helices in the native state are found in the T-domain. Helices 5–9 of the T domain most closely overlap the uncharged regions of the toxin (see Figure 1). The possibility that helices 5 and 6 and helices 8 and 9 form two inserting helical pairs was suggested by Choe et al. (1992). Helix 6 is a little short for a classical transmembrane helix, but could form a larger helix by combining with helix 7. Combining this idea and the location of the most hydrophobic residues in the T domain sequence allows us to propose a modified model showing where the helix boundaries might be in membrane-inserted toxin (Figure 6). In this model, helices 6 and 7 would combine into a single helix, helices 7 and 8 would be extended on their C-terminal sides, and helix 9 would be extended a few residues at its N-terminus. This arrangement would place helix-bending prolines within

```
                   .                              Ala
                   .                              Pro
        Lys***                                    Arg***
  TH4   Thr*                                      Tyr
        Val              Asn*Thr*                 Ser*
        Thr*        His***        Glu***          Asn*
        Gly         His***        Glu***          His***
  ─────────────────────────────────────────────────────
        Thr*        Val           Ile             Val
        Asn*        Ala           Val             Val
        Pro         Gly           Ala             Gln*
        Val         Asp***        Gln*            Phe
        Phe         Ala           Ser*            Leu
        Ala         Ile           Ile             Asn*
        Gly         Gly           Ala      TH9    Ile
        Ala         Met           Leu             Ile
        Asn*  TH7   Val           Ser*            Ser*
        Tyr         Ser*          Ser*            Glu***
        Ala         Gly           Leu             Val
        Ala         Ile           Met             Phe
        Trp         Gly           Val             Asn*
        Ala         Pro     TH8   Ala             Tyr
        Val         Leu           Gln*            Ala
  TH5   Asn*        Ile           Ala             Ala
        Val         Ser*          Ile             Phe
        Ala         Leu           Pro             Gly
        Gln*        Ala           Leu             Ile
        Val   TH6   Ala           Val             Asp***
        Ile         Thr*          Gly             Val
  ─────────────────────────────────────────────────────
        Asp***      Thr*             Glu***Leu
        Ser*        Lys***
        Glu***      Glu***
        Thr*        Leu
        Ala         Asn*
             Asp***
```

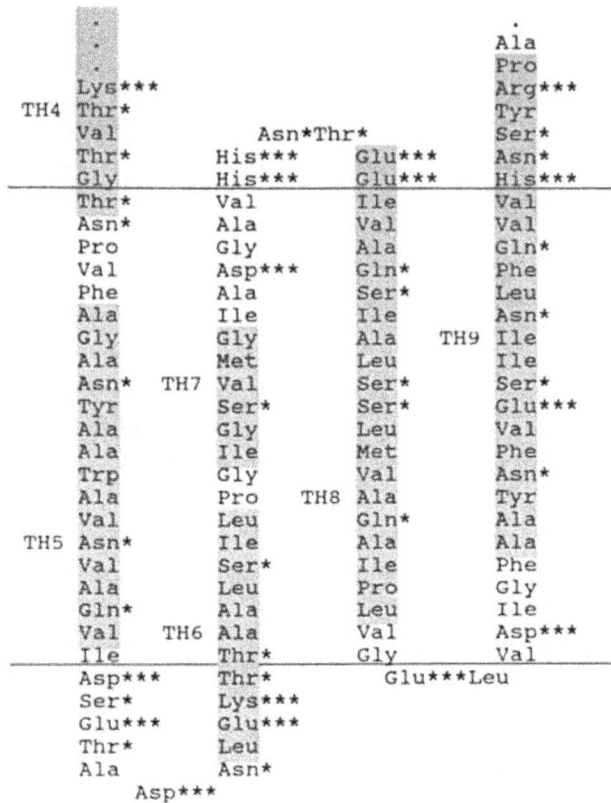

Figure 6. A hypothetical transmembrane orientation of the hydrophobic segment of the T domain. *Indicates a polar residue. ***Indicates a charged residue. Helices in the solution structure are indicated by bars. The boundary of the hydrophobic part of the bilayer is indicated by horizonal lines. The toxin sequences in this part of the membrane are likely to be helical.

transmembrane helices, but such behavior has already been seen for other membrane proteins.

The sequences in Figure 6 shown as being in the membrane are not totally hydrophobic. Some sort of amphipathic helix-like periodicity can be seen in helices 8 and 9. Any model of the membrane-inserted toxin structure must take the presence of polar and charged residues in the membrane into account. One possibility is that the more polar residues face toward the interior of the protein. Some of them could line the pore formed by the toxin, or hydrogen bond with other polar/charged residues. In this regard, it is interesting that a rough correspondence in depth can be proposed for the polar residues in helix 5 with those in helix 6 and 7 (Figure 6). Another possibility is that the polar residues make contact with transmembrane β-sheets. This idea is discussed in detail below. A third possibility is that some of these helices act as amphipathic helices and lie along the membrane surface, as suggested for colicins (Lakey et al., 1993). However, it does seem likely that these helices are transmembraneous under some conditions. They could take on different structures at different stages of the translocation process.

The inter-helical regions may also play an important role in membrane insertion. It has been proposed that the protonation of the Asp and Glu residues between helices 5 and 6 and those between helices 8 and 9 may render them more hydrophobic (Kieleczawa et al., 1990; Choe et al., 1992) and even initiate membrane insertion (Choe et al., 1992). A similar mechanism, in which hydrophobic tips at the ends of helices initiate membrane interaction, has been proposed for a lipoprotein (Breiter et al., 1991).

XIV. DOES THE C DOMAIN (A CHAIN) CONTAIN TRANSMEMBRANE β-SHEETS?

It is not absolutely known whether the A chain comes into contact with lipid during translocation. However, its hydrophobic behavior at low pH suggests it is likely to do so (see above). This creates a dilemma at first glance, because hydropathy analysis of A chain structure reveals no hydrophobic α-helices (Greenfield et al., 1983).

However, hydropathy analysis only searches for transmembrane helices. One other possibility is that the A chain forms transmembrane β-sheets. This is now a serious possibility for membrane proteins because crystallography of porins has shown that membrane proteins composed of transmembrane β-sheets do exist (Weiss et al., 1991). Transmembrane β-sheets are characterized by a shorter length than transmembrane helices (about 10 residues for sheets vs. 20 for helices). This is no surprise as they are relatively extended structures. They are also characterized by a tendency to have hydrophobic residues at alternating positions (Schlitz et al., 1991; Weiss et al., 1991).

In the A chain there are four 10-residue sequences that are dominated by hydrophobic residues at alternating positions and are coincident or substantially overlap β-sheets. These are at A chain sheets 4, 5, 6, and 8. These potentially transmembrane sheets are divided into two groups on either side of the potential hinge identified by crystallography, and might become more exposed upon partial unfolding of the A chain. It should also be noted there is one C-domain helix—helix 7—that might be converted into a transmembrane β-sheet upon membrane insertion.

The edges of a membrane-inserted β-sheet would have broken hydrogen bonds. In the porins this is solved by having the sheet curl back on itself to form a barrel, but in the A chain the number of strands is probably too low to do this. An alternate possibility is that hydrogen bonding groups on transmembrane helices might contact the edges of the A chain β-sheets to form the necessary hydrogen bonds. (If this is so, then the proper transmembrane structure of the toxin may not be formed by isolated domains.) Another possibility is that a β-sheet barrel is formed by an oligomer of toxin molecules.

XV. DOES THE β-SHEET RICH R DOMAIN PENETRATE MEMBRANES?

It is not clear whether the R domain actually penetrates membranes. As noted above, studies have shown its main function appears to be to bind to the receptor for the toxin. Unlike the T-domain it contains no helices, and the only long hydrophobic stretch revealed by hydropathy analysis involves a β-sheet with several Pro residues. However, the possibility of additional R-domain functions have not been ruled out and one group has proposed its translocation across membranes does occur (Moskaug et al., 1991).

Inspection of the R domain does show the presence of some potential transmembrane β-sheets. R domain sheets 8 and 10 have patterns of dominating hydrophobic residues at alternating positions over at least 10 residue stretches. Therefore, the possibility of the transmembrane stretches being formed by the R domain should be seriously considered. In this regard, it is noteworthy that the β-sheet rich receptor binding domain of *Pseudomonas* exotoxin A exhibits hydrophobic behavior at low pH (Idziorek, 1990).

XVI. THE POSSIBILITY OF NONCLASSICAL TRANSMEMBRANE STRUCTURES IN MEMBRANE-INSERTED DIPHTHERIA TOXIN

Up to now, we have only considered structures that have already been found in transmembrane proteins. However, the possibility that there are other transmembrane structures present should not be discounted. One possibility is hydrophilic transmembrane helices. These could be found in helices that are not in contact with lipid. Such helices could be present in the interior of the toxin, or at the boundary of toxin molecules in an oligomer. Hydrophilic helices in contact with lipid also should not be ruled out. Such structures could exist if the local lipid bilayer was distorted by the toxin so that the lipid polar groups came into contact with the helix rather than the acyl chain tails. The fact that the toxin has a tendency to destabilize bilayers, as indicated by its ability to induce membrane fusion at low pH (Cabiaux et al., 1984), lends some credence to this somewhat remote possibility.

Another structure that should not be ruled out is a hydrophobic surface formed by a cluster of hydrophobic residues that are noncontiguous. Such a hydrophobic cluster would exist only at the level of tertiary structure (Jiang and London, 1990).

This type of structure has recently been observed in what appears to be the lipid interacting surface of a colipase (van Tilbeurgh et al., 1992).

XVII. LESSONS FOR NORMAL PROTEIN TRANSLOCATION FROM TOXIN TRANSLOCATION: ROLE OF UNFOLDING OF THE TRANSLOCATING PROTEIN

It should be pointed out that the studies described above imply close parallels between the translocation of diphtheria toxin and that of ordinary cellular proteins. The concept that the translocated A domain of the toxin must unfold in order to be translocated is equivalent to the necessity of cellular proteins to remain in a partly unfolded "translocation-competent" state in order to cross membranes (Eilers and Schatz, 1988). It is interesting to speculate whether the possibility of A chain transmembrane β-sheets which come into contact with lipid might also apply to translocation of some cellular proteins.

XVIII. LESSONS FROM TOXIN TRANSLOCATION: THE TRANSLOCATION-PROMOTING PROTEIN SECA SHOWS TOXIN-LIKE BEHAVIOR

Since the linkage of membrane insertion to unfolding appears to partly involve the B domain of the toxin, there might also be a role for unfolding in the action of translocation-promoting proteins. This idea is borne out by the behavior of the SecA protein. SecA is a central component of the translocation machinery in *E. coli*. It has a role in the early events in protein translocation, interacting with the signal peptide of a precursor protein, and possessing an ATPase activity which probably plays an important role at some translocation step (Oliver et al., 1990).

The SecA protein is found both in the cytoplasm and in a dissociation-resistant form in the inner membrane (Cabelli et al., 1991). Early models envisioned that SecA would be a peripheral protein interacting with the membrane via other membrane proteins. However, more recent studies on the interaction of SecA with model membranes have shown that this protein is capable of taking on a membrane-inserted form in the absence of other proteins (Breukink et al., 1992; Ulbrandt et al., 1992). Furthermore, its membrane insertion is linked to the partial unfolding of at least one domain within the protein (Ulbrandt et al., 1992). In addition, as in the case of diphtheria toxin, anionic phospholipids promote the unfolding of SecA. This is probably one reason why such lipids promote SecA activity. Such properties imply that behavior similar to that of diphtheria toxin is likely to be found in other proteins involved in assisting translocation.

XIX.   LESSONS FROM TOXIN TRANSLOCATION: A STRUCTURAL MODEL FOR A TRANSLOCATOR THAT COULD BE THE SAME AS THAT INVOLVED IN ORDINARY CELLULAR TRANSLOCATION

Although this article is mainly concerned with diphtheria toxin, there is one aspect of the structure of cholera toxin that should not be overlooked when thinking of the lessons that may be from toxin proteins. Cholera toxin is familiar as a protein that binds to cells via the interaction of its B subunit with ganglioside GM_1 and acts by ADP-ribosylation of G_s proteins catalyzed by its A subunit. The crystal structure of a close relative of cholera toxin, heat-labile toxin of *E. coli*, shows that these toxins contain a ring of five B subunits. This ring surrounds a single extended A2 chain which is attached to a globular A1 domain (Sixma et al., 1991).

The crystal structure reveals the potential to form an "ideal" structure for the transmembrane section of a cellular protein translocation system (London, 1992b). A sliding motion of the B subunits of cholera toxin would open a channel within the ring of B subunits large enough to allow the A subunit to pass through in a partly unfolded form. The most exciting aspect of this model is that this opening would create a structure already adapted to binding of unfolded proteins. The interior of the open form would have five clefts, each bounded by a β-sheet outer wall and by two α-helical lateral walls. This is very similar to the structure found in the cleft of the human lymphocyte antigen (HLA) protein (Bjorkman et al., 1987) which binds antigens in the form of extended peptide chains held within its cleft (Madden et al., 1991). The size of the clefts within the B chain pentamer could be regulated by sliding between B subunits, and thus the binding and release of translocating peptide could be controlled through environmental changes. In this regard, it is fascinating that pH regulates binding and release of peptides by HLA (Reay et al., 1992). It is also provocative that there has been speculation the HLA structure will also be found in chaperones—soluble proteins that play a critical role in protein translocation through their interaction with unfolded proteins (Rippmann et al., 1991).

However, it should be cautioned that there is no direct experimental evidence for this model. In fact, the lack of a well-defined hydrophobic outer surface on the B subunits, and the observation of a cholera toxin-like structure in Shiga-like toxin B chains that are too small to span the bilayer (Stein et al., 1992) brings into question the idea that a transmembrane pore could be formed by the B chains. An alternate possibility is that the B subunits act as a chaperone for the unfolded A subunit at some stage in toxin entry.

In this regard, it should be pointed out that it is even possible that toxins interact with the cellular endoplasmic reticulum translocation machinery to use it in reverse! The endoplasmic reticulum retention signal KDEL is found at the C-terminus of the A subunit in cholera toxin, and might target the toxin to endoplasmic reticulum. The observation that a KDEL-like sequence has an critical role in *Pseudomonas* exotoxin A action (Chaudhary et al., 1990; Seetharam et al., 1991) certainly strengthens the possibility that some toxins travel to the endoplasmic reticulum.

XX. SUMMARY

Much progress has been made in understanding the translocation of diphtheria toxin. This progress has been made possible by the elucidation of the cellular entry pathway, in particular the role of low pH in translocation. Studies of the structure of membrane-inserted toxin may allow the formulation of a fairly detailed model for the translocation process in the not too-far distant future. For other toxins, we are still limited in our knowledge that at the present time we are restricted to speculation on how they may translocate across membranes. Nevertheless, it is clear that further studies of toxin translocation will be a rich source of information on protein translocation in general.

ACKNOWLEDGMENT

This work was supported by N.I.H. Grant GM 31986.

REFERENCES

Ariansen, S., Afanasiev, B.N., Moskaug, J.O., Stenmark, H., Madshus, I.H., & Olsnes, S. (1993). Membrane translocation of diphtheria toxin A fragment: Role of carboxy-terminal region. Biochemistry 32, 83–90.

Benedetti, H., Lloubes, R., Lazdunski, C., & Letellier, L. (1992). Colicin A unfolds during its translocation in *Escherichia coli* cells and spans the whole cell envelope when its pore has formed. EMBO J. 11, 441–447.

Beaumelle, B., Bensammar, L., & Bienvenue, A. (1992). Selective translocation of the A chain of diphtheria toxin across the membrane of purified endosomes. J. Biol. Chem. 267, 11525–11531.

Bjorkman, P.J., Saper, M.A., Samraoui, B., Bennett, W.S., Strominger, J.L., & Wiley, D.C. (1987). Structure of the human class I histocompatibility antigen, HLA-A2. Nature 329, 512–518.

Blewitt, M.G., Chung, L.A., & London, E. (1985). The effect of pH upon the conformation of diphtheria toxin and its implications for membrane penetration. Biochemistry 24, 5458–5464.

Boquet, P., Silverman, M.S., Pappenheimer, Jr., A.W., & Vernon, W.B. (1976). Binding of Triton X-100 to diphtheria toxin, cross reacting material 45 and their fragments. Proc. Natl. Acad. Sci. USA 73, 4449–4453.

Breiter, D.R., Kanost, M.R., Benning, M.M., Wesenberg, G., Law, J.H., Wells, M.A., Rayment, I., & Holden, H.M. (1991). Molecular structure of an apolipoprotein determined at 2.5 Å resolution. Biochemistry 30, 603–608.

Breukink, E., Demel, R.A., de Korte-Kool, G., & de Kruijff, B. (1992). SecA insertion into phospholipids is stimulated by negatively charged lipids and inhibited by ATP. Biochemistry 31, 1119–1124.

Cabelli, R.J., Dolan, K.M., Qian, L., & Oliver, D.B. (1991). Characterization of membrane-associated and soluble states of SecA protein from wild-type and *SecA51(TS)* mutant strains of *Escherichia coli*. J. Biol. Chem. 266, 24420–24427.

Cabiaux, V., Vandenbranden, M., Falmagne, P., & Ruysschaert, J.-M. (1984). Diphtheria toxin induces fusion of small unilamellar vesicles at low pH. Biochim. Biophys. Acta 775, 31–36.

Cabiaux, V., Brasseur, R., Wattiez, R., Falmagne, P., Ruysschaert, J.-M., & Goormagtigh, E. (1989). Secondary structure of diphtheria toxin and its fragments interacting with acidic liposomes studied by polarized infrared spectroscopy. J. Biol. Chem. 264, 4928–4938.

Cabiaux, V., Mindell, J., & Collier, R.J. (1993). Membrane translocation and channel forming properties of diphtheria toxin are blocked by replacing isoleucine 364 with lysine. Infect. Immun. 61, 2200–2202.

Cain, C.C., Sipe, D.M., & Murphy, R.F. (1989). Regulation of endocytic pH by the Na^+K^+ATPase in living cells. Proc. Natl. Acad. Sci. USA 86, 544–548.

Chang, M.P., Bramhall, J., Graves, S., Bonavida, B., & Wisnieski, B.J. (1989a). Internucleosomal DNA cleavage precedes diphtheria toxin-induced cytolysis. J. Biol. Chem. 264, 15261–15267.

Chang, M.P., Baldwin, R.L., Bruce, C., & Wisnieski B.J. (1989b). Second cytotoxic pathway of diphtheria toxin suggested by nuclease activity. Science 246, 1165–1168.

Chaudhary, V.K., Jinno, Y., FitzGerald, D., & Pastan, I. (1990). Pseudomonas exotoxin contains a specific sequence at the carboxyl terminus that is required for cytotoxicity. Proc. Natl. Acad. Sci. USA 87, 308–312.

Chaudhary, V.K., FitzGerald, D.J., & Pastan, I. (1991). A proper amino-terminus of diphtheria toxin is important for cytotoxicity. Biochem. Biophys. Res. Commun. 180, 545–551.

Choe, S., Bennett, M.J., Fujii, G., Curmi, P.M.G., Kantardjieff, K.A., Collier, R.J., & Eisenberg, D. (1992). The crystal structure of diphtheria toxin. Nature 357, 216–221.

Collier, R.J. (1982). Structure and activity of diphtheria toxin, in ADP-ribosylation reactions In: Biology and Medicine (Hayashi, O. & Ueda, K., Eds.), pp. 575–592. Academic Press, New York.

Collins, C.M., Barbieri, J.T., & Collier, R.J. (1984). Interaction of diphtheria toxin with adenylyl-(3′,5′)-uridine-3′-monophosphate. J. Biol. Chem. 259, 15154–15158.

Donovan, J.J., Simon, M.I., Draper, R.K., & Montal, M. (1981). Diphtheria toxin forms transmembrane channels in bilayers. Proc. Natl. Acad. Sci. USA 78, 172–176.

Donovan, J.J., Simon, M.I., & Montal, M. (1982). Insertion of diphtheria toxin into and across membranes: role of phosphoinositide asymmetry. Nature 298, 669–672.

Dumont, M.E. & Richards, F.M. (1988). The pH-dependent conformation change of diphtheria toxin. J. Biol. Chem. 263, 2087–2097.

Eilers, M. & Schatz, G. (1988). Protein unfolding and the energetics of protein translocation across biological membranes. Cell 52, 481–483.

Falnes, P.O., Madshus, I.H., Sandvig, K., & Olsnes, S. (1992). Replacement of negative by positive charges in the presumed membrane-inserted part of diphtheria toxin B fragment. J. Biol. Chem. 267, 12284–12290.

Fuchs, R., Schmid, S., & Mellman, I. (1989). A possible role for Na^+K^+ ATPase in regulating ATP-dependent endosome acidification. Proc. Natl. Acad. Sci. USA 86, 539–543.

Greenfield, L., Bjorn, M.J., Horn, G., Fong, D., Buck, G.A., Collier, R.J., & Kaplan, D.A. (1983). Nucleotide sequence of the structural gene for diphtheria toxin carried by corynebacteriophage β. Proc. Natl. Acad. Sci. USA 80, 6853–6857.

Gonzalez, J.E. & Wisnieski, B.J. (1988). An endosomal model for acid triggering of diphtheria toxin translocation. J. Biol. Chem. 263, 15257–15259.

Hu, V.W. & Holmes, R.K. (1984). Evidence for direct insertion of fragments A and B of diphtheria toxin into model membranes. J. Biol. Chem. 259, 12226–12233.

Hudson, T.H., Scharff, J., Kimak, M.A.G., & Neville, D.M., Jr. (1988). Energy requirements for diphtheria toxin translocation are coupled to the maintenance of a plasma membrane potential and a proton gradient. J. Biol. Chem. 263, 4773–4781.

Idziorek, T.D., FitzGerald, D., & Pastan, I. (1990). Low pH induced changes in Pseudomonas exotoxin and its domains: increased binding of Triton X-114. Infect. Immun. 58, 1415–1420.

Iwamoto, R., Senoh, H., Okada, Y., Uchida, T., & Mekada, E. (1991). An antibody that inhibits the binding of diphtheria toxin to cells revealed the association of a 27-kDa membrane protein with the diphtheria toxin receptor. J. Biol. Chem. 266, 20463–20469.

Jiang, G.-s., Solow, R., & Hu, V.W. (1989). Characterization of diphtheria toxin-induced lesions in liposomal membranes. J. Biol. Chem. 264, 13424–13429.

Jiang, G.-s., Solow, R., & Hu, V.W. (1989). Fragment A of diphtheria toxin causes pH-dependent lesions in model membranes. J. Biol. Chem. 264, 17170–17173.

Jiang, J.X., Abrams, F.S., & London, E. (1991a). Folding changes in membrane-inserted diphtheria toxin that may play important roles in its translocation. Biochemistry 30, 3857–3864.

Jiang, J.X. & London, E. (1990). Involvement of denaturation-like changes in pseudomonas exotoxin A hydrophobicity and membrane penetration determined by characterization of pH and thermal transitions. J. Biol. Chem. 265, 8636–8641.

Jiang, J.X., Chung, L.A., & London, E. (1991b). Self-translocation of diphtheria toxin across model membranes. J. Biol. Chem. 266, 24003–24010.

Johnson, V.G. (1990). Does diphtheria toxin have nuclease activity? Science 250, 832–834.

Johnson, V.G., Nicholls, P.J., Habig, W.H., & Youle, R.J. (1993). The role of proline 345 in diphtheria toxin translocation. J. Biol. Chem. 268, 3514–3519.

Kagan, B.L. (1991). Inositol 1,4,5-trisphosphate directly opens diphtheria toxin channels. Biochim. Biophys. Acta 1069, 145–150.

Kagan, B.L., Finkelstein, A., & Colombini, M. (1981). Diphtheria toxin fragment forms large pores in phospholipid bilayer membranes. Proc. Natl. Acad. Sci. USA 78, 4950–4954.

Kagan, B.L., Reich, K.A., & Collier, R.J. (1984). Orientation of the diphtheria toxin channel in lipid bilayers. Biophys. J. 45, 102–104.

Kieleczawa, J., Zhao, J.-M., Luongo, C.L., Dong, L.-Y.D., & London, E. (1990). Effect of high pH upon diphtheria toxin conformation and membrane penetration. Arch. Biochem. Biophys. 282, 214–220.

Kochi, S.K. & Collier, R.J. (1993). DNA fragmentation and cytolysis in U(#& cells treated with diphtheria toxin or other inhibitors of protein synthesis. Exp. Cell Res. 208, 296–302.

Lakey, J.H., Duche, D., Gonzalez-Manas, J.-M., Baty, D., & Pattus, F. (1993). Fluorescence energy transfer measurements. The hydrophobic helical hairpin of colicin A in the membrane bound state. J. Mol. Biol. 230, 1055–1067.

Lessnick, S.L., Lyczak, J.B., Bruce, C., Lewis, D.G., Kim. P.S., Stolowitz, M.L., Hood, L., & Wisnieski, B.J. (1992). Localization of diphtheria toxin nuclease activity to fragment A. J. Bacteriology 174, 2032–2038.

London, E. (1992a). Diphtheria toxin: Membrane interaction and membrane translocation. Biochim. Biophys. Acta 1113, 25–51.

London, E. (1992b). How bacterial toxins enter cells. Mol. Microbiology 6, 3277–3282.

Madden, D.R., Gorga, J.C., Strominger, J.L., & Wiley, D.C. (1991). The structure of HLA-B27 reveals nonamer self-peptides bound in an extended conformation. Nature 353, 321–325.

Mandel, R., Ryser, H.J.-P., Ghani, F., Wu, M., & Peak, D. (1993). Inhibition of a reductive function of the plasma membrane by bacitracin and antibodies against protein disulfide-isomerase. Proc. Natl. Acad. Sci. USA 90, 4112–4116.

Madshus, I.H., Olsnes, S., & Stenmark. H. (1992). Membrane translocation of diphtheria toxin carrying passenger protein domains. Infect. Immun. 60, 3296–3302.

Malinski, J.A. & Nelsestuen, G.L. (1989). Membrane permeability to macromolecules mediated by the membrane attack complex. Biochemistry 28, 61–70.

Merrill, A.R., Cohen, F.S., & Cramer, W.A. (1990). On the nature of the structural change of the colicin E1 channel peptide necessary for its translocation-competent state. Biochemistry 29, 5829–5836.

Mindell, J.A., Silverman, J.A., Collier, R.J., & Finkelstein, A. (1992). Locating a residue in the diphtheria toxin channel. Biophys. J. 62, 41–44.

Misler, S. (1984). Diphtheria toxin fragment channels in lipid bilayer membranes: Selective sieves or discarded wrappers? Biophys. J. 45, 107–109.

Mitamura, T., Iwamoto, R., Umata, T., Yomo, T., Urabe, I., Tsuneoka, M., & Mekada, E. (1992). The 27-kDa diphtheria toxin receptor-associated protein (DRAP 27) from vero cells is the monkey homologue of human CD9 antigen. J. Cell. Biol. 118, 1389–1399.

Montecucco, C., Schiavo, G., & Tomasi, M. (1985). pH-dependence of the phospholipid interaction of diphtheria-toxin fragments. Biochem. J. 231, 123–128.

Morimoto, H. & Bonavida, B. (1992). Diphtheria toxin and *Pseudomonas* A toxin-mediated apoptosis. J. Immunol. 149, 2089–2094.

Moskaug, J.O., Sandvig, K., & Olsnes, S. (1987). Cell-mediated reduction of the interfragment disulfide in nicked diphtheria toxin. J. Biol. Chem. 262, 10339–10345.

Moskaug, J.O., Sandvig, K., & Olsnes, S. (1988). Low pH-induced release of diphtheria toxin A-fragment in vero cells. J. Biol. Chem. 263, 2518–2525.

Moskaug, J.O., Sletten, K., Sandvig, K., & Olsnes, S. (1989). Translocation of diphtheria toxin A-fragment to the cytosol. J. Biol. Chem. 264, 15709–15713.

Moskaug, J.O., Stenmark, H., & Olsnes, S. (1991). Insertion of diphtheria toxin B-fragment into the plasma membrane at low pH. J. Biol. Chem. 266, 2652–2659.

Muga, A., Gonzalez-Manas, J.M., Lakey, J.H., Pattus, F., & Surewicz, W.K. (1993). pH-dependent stability and membrane insertion of the pore-forming domain of colicin A. J. Biol. Chem. 268, 1553–1557.

Naglich, J.G., Metherall, J.E., Russell, D.W., & Eidels, L. (1992). Expression cloning of a diphtheria toxin receptor: Identity with a heparin-binding EGF-like growth factor receptor. Cell 69, 1051–1061.

O'Keefe, D.O., Cabiaux, V., Choe, S., Eisenberg, D., & Collier, R.J. (1992). pH-dependent insertion of proteins into membranes: B-chain mutation of diphtheria toxin that inhibits membrane transloca- tion, Glu 349-Lys. Proc. Natl. Acad. Sci. USA 89, 6202–6206.

O'Keefe, D.O. & Collier, R.J. (1989). Cloned diphtheria toxin within the periplasm of *Escherichia coli* causes lethal membrane damage at low pH. Proc. Natl. Acad. Sci. USA 86, 343–346.

Oliver, D.B., Cabelli, R.B., & Jarosik, G.P. (1990). SecA protein: Autoregulated initiator of secretory precursor protein translocation across the *E. coli* plasma membrane. J. Bioenerg. Biomemb. 22, 311–336.

Olsnes, S. & Sandvig, K. (1986). Interactions between diphtheria toxin entry and anion transport in vero cells. Inhibition of anion antiport by diphtheria toxin. J. Biol. Chem. 261, 1553–1561.

Papini, E., Schiavo, G., Tomasi, M., Colombatti, M., Rappuoli, R., & Montecucco, C. (1987). Lipid interaction of diphtheria toxin and mutants with altered fragment B. Eur. J. Biochem. 169, 637–644.

Papini, E., Rappuoli, R., Murgia, M., & Montecucco, C. (1993). Cell penetration of diphtheria toxin. J. Biol. Chem. 268, 1567–1574.

Papini, E., Sandona, D., Rappuoli, R., & Montecucco, C. (1988). On the membrane translocation of diphtheria toxin: at low pH the toxin induces ion channels in cells. EMBO J. 7, 3353–3359.

Proia, R.L., Wray, S.K., & Eidels, L. (1980). Characterization and affinity labeling of the cationic phosphate-binding (nucleotide binding) peptide located in the receptor-binding region of the B-fragment of diphtheria toxin. J. Biol. Chem. 255, 12025–12033.

Reay, P.A., Wettstein, D.A., & Davis, M.M. (1992). pH dependence and exchange of high and low responder peptides binding to a class II MHC molecule. EMBO J. 11, 2829–2839.

Rippmann, F., Taylor, W.R., Rothbard, J.B., & Green, N.M. (1991). A hypothetical model for the peptide binding domain of hsp70 based on the peptide binding domain of HLA. EMBO J. 10, 1052–1059.

Rolf, J.M. & Eidels, L. (1993). Characterization of the diphtheria toxin receptor-binding domain. Mol. Microbiol. 7, 585–591.

Ryser, H.J.-P., Mandel, R., & Ghani, F. (1991). Cell surface sulfhydryls are required for the cytotoxicity of diphtheria toxin but not of ricin in chinese hamster ovary cells. J. Biol. Chem. 266, 18439–18442.

Sandvig, K. & Olsnes, S. (1981). Rapid entry of nicked diphtheria toxin into cells at low pH. J. Biol. Chem. 256, 9068–9076.

Sandvig, K., Tonessen, T.I., Sand, O., & Olsnes, S. (1986). Requirement of a transmembrane pH gradient for the entry of diphtheria toxin into cells at low pH. J. Biol. Chem. 261, 11639–11644.

Sandvig, K. & Olsnes, S. (1988). Diphtheria toxin-induced channels in vero cells selective for monovalent cations. J. Biol. Chem. 263, 12352–12359.

Schlitz, E., Kreusch, A., Nestel, U., & Schulz, G.E. (1991). Primary structure of porin from *Rhodobacter capsulatus*. Eur. J. Biochem. 199, 587–594.

Seetharam, S., Chaudhary, V.K., FitzGerald, D., & Pastan, I. (1991). Increased cytotoxic activity of *Pseudomonas* exotoxin and two chimeric toxins ending in KDEL. J. Biol. Chem. 266, 17376–17381.

Sixma, T.K., Pronk, S.E., Kalk, K. H., Wartna, E.S., van Zanten, B.A.M., Witholt, B., & Hol, W.G.J. (1991). Crystal structure of a cholera toxin-related heat-labile enterotoxin form *E. coli*. Nature 351, 371–377.

Stein, P.E., Boodhoo, A., Tyrrell, G.J., Brunton, J.L., & Read, R.J. (1992). Crystal structure of the cell-binding B oligomer of verotoxin-1 from *E. coli*. Nature 355, 748–760.

Stenmark, H., Olsnes, S., & Madshus, I.H. (1991a). Elimination of the disulphide bridge in fragment B of diphtheria toxin. Mol. Microbiol. 5, 595–606.

Stenmark, H., Ariansen, S., Afanasiev, B.N., & Olsnes, S. (1992). Interaction of diphtheria toxin B-fragment with cells. J. Biol. Chem. 267, 8957–8962.

Stenmark, H., Moskaug, J.O., Madshus, I.H., Sandvig, K., & Olsnes, S. (1991b). Peptides fused to the amino-terminal end of diphtheria toxin are translocated to the cytosol. J. Cell. Biol. 113, 1025–1032.

Ulbrandt, N.D., London, E., & Oliver, D.B. (1992). Deep penetration of a portion of the *Escherichia coli* SecA protein into model membranes is promoted by anionic phospholipids and by partial unfolding. J. Biol. Chem. 267, 15184–15192.

van der Goot, F.G., Gonzalez-Manas, J.M., Lakey, J.H., & Pattus, F. (1991). A 'molten-globule' membrane-insertion intermediate of the pore forming domain of colicin A. Nature 354, 408–410.

van Tilbeurgh, H., Sarda, L., Verger, R., & Cambillau, C. (1992). Structure of the pancreatic lipase-prolipase complex. Nature 359, 159–162.

Weiss, M.S., Abele, U., Weckesser, J., Welte, W., Schiltz, E., & Schulz, G.E. (1991). Molecular architecture and electrostatic properties of a bacterial porin. Science 254, 1627–1630.

Wiedlocha, A., Madshus., I.H., Mach, H., Middaugh, C.R., & Olsnes, S. (1992). Tight folding of acidic fibroblast growth factor prevents its translocation to the cytosol with diphteria toxin as vector. EMBO J. 11, 4835–4842.

Wiley, D.C. & Skehel, J.J. (1987). The structure and function of the hemagglutinin membrane glycoprotein of influenza virus. Annu. Rev. Biochem. 56, 365–394.

Wilson, B.A., Blanke, S.R., Murphy, J.R., Pappenheimer, Jr., A.M., & Collier, R.J. (1990). Does diphtheria toxin have nuclease activity? Science 258, 834–836.

Zalman, L.S. & Wisnieski, B.J. (1984). Mechanism of insertion of diphtheria toxin: Peptide entry and pore size determinations. Proc. Natl. Acad. Sci. USA 81, 3341–3345.

Zhao, J.-M. & London, E. (1988). Conformation and model membrane interactions of diphtheria toxin fragment A. J. Biol. Chem. 263, 15369–15377.

Zhao, J.-M. & London, E. (1986). Similarity of the conformation of diphtheria toxin at high temperature to that in the membrane-penetrating low pH state. Proc. Natl. Acad. Sci. USA 83, 2002–2006.

PROTEIN TRANSLOCATION INTO CHLOROPLASTS

Marinus Pilon, Twan America, Ron van't Hof,

Ben de Kruijff, and Peter Weisbeek

Membrane Protein Transport
Volume 1, pages 229–256.
Copyright © 1995 by JAI Press Inc.
All rights of reproduction in any form reserved.
ISBN: 1-55938-907-9

I. INTRODUCTION

Like all eukaryotic cells the plant cell contains different organelles which are separated from the cytosol by one or more membranes. In each organelle a specific set of biochemical reactions takes place. In order to fulfill their special tasks each organelle has a unique protein composition. Depending on the cell type, different plastids occur in plant cells. Chloroplasts are characteristic organelles of green tissue. The plastids as well as mitochondria contain their own genetic system. These organelles have their own DNA and protein synthesizing machinery but the size of the DNA is too small to account for all the gene products found in them. Therefore, mitochondria and plastids as well as all other organelles depend on cytosolic protein synthesis. The presence of an intracellular targeting system ensures the correct delivery of proteins synthesized in the cytosol to each organelle in the plant cell.

Three membranes divide the chloroplast into compartments (see Figure 1): a two-membrane envelope and an internal membrane system, called the thylakoids (Kirk and Tilney-Bassett, 1978). At least five localizations within the organelle are known for nuclear-encoded precursors (de Boer and Weisbeek, 1991). The final localization of a protein depends on the presence of topogenic information (targeting information) present in the protein at the time of synthesis. Two types of topogenic information can be distinguished: routing and sorting information. After synthesis in the cytosol a protein can either remain there or, depending on the

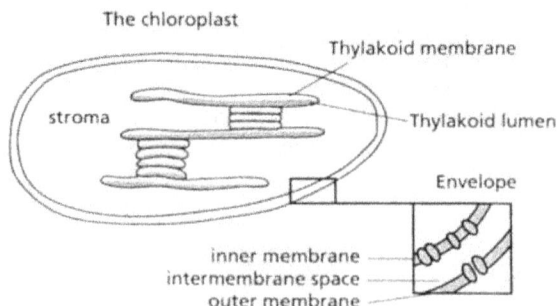

Figure 1. Schematic representation of a chloroplast.

presence of a routing sequence, be transported (routed) to a specific organelle. Routing thus is defined as the export of proteins from the cytosol (Pugsley, 1989). Proteins routed to plastids might undergo further relocation (termed sorting) to reach their final destination. Sorting can be defined as transport within an organelle and depends on the presence of further topogenic information termed sorting signals (Pugsley, 1989).

Using a combination of *in vitro* and *in vivo* approaches, our laboratory as well as many others have first focused on the routing and sorting signals that direct proteins to their specific localization within the chloroplast [see Keegstra (1989) for a concise review, and de Boer and Weisbeek (1991) for a more comprehensive review]. We then analyzed the import step into the chloroplast more in detail. We focused on the targeting to the chloroplast, the folding of the precursor, the secondary structure of the transit sequence, the role of lipid-protein interactions, and the conformation during translocation across the envelope. We will discuss these topics and emphasize the experimental set up that is used to investigate each subject.

II. THE ANALYSIS OF CHLOROPLAST TARGETING INFORMATION

A. Methods to Study Targeting

Methods to reconstitute the posttranslational import into isolated organelles *in vitro* were first developed for the chloroplast. Most of our present knowledge on the import of proteins into chloroplasts (de Boer and Weisbeek, 1991) comes from such analysis performed *in vitro*. In this approach radioactively labeled precursor proteins are added to isolated chloroplasts. The precursors are synthesized from labeled amino acids in lysates derived from wheat germ or reticulocytes. The lysates are programmed with either poly-A mRNA isolated from plants or with synthetic mRNA obtained by *in vitro* transcription of a cloned gene (which encodes a chloroplast protein of known function/localization) from a plasmid. The lysates containing the labeled precursors are subsequently added to isolated chloroplasts. The organelles used for such studies are isolated from crude homogenates by a combination of differential and density gradient centrifugation. It is important to verify the purity and intactness of the isolated organelles. Following the incubation the chloroplasts are recovered by centrifugation and the associated proteins are analyzed by SDS-PAGE and fluorography. To distinguish proteins associated with the outside of the organelle from imported protein a protease treatment can be performed. To further investigate the localization inside the organelle, a gentle lysis followed by differential or density gradient centrifugation can be used.

The expression of recombinant genes in transgenic plants is another way to study protein transport in plant cells. This method relies on the possibility of plant transformation and on methods to detect the protein of interest; for instance with a specific antibody or by measuring enzyme activities in subcellular fractions. De Boer et al. (1991) obtained different results when comparing the topogenic information required for sorting in chloroplasts when the localization was determined both *in vitro* and *in vivo*.

A schematic overview of the topogenic (routing and sorting) information present in imported chloroplast proteins is shown in Figure 2.

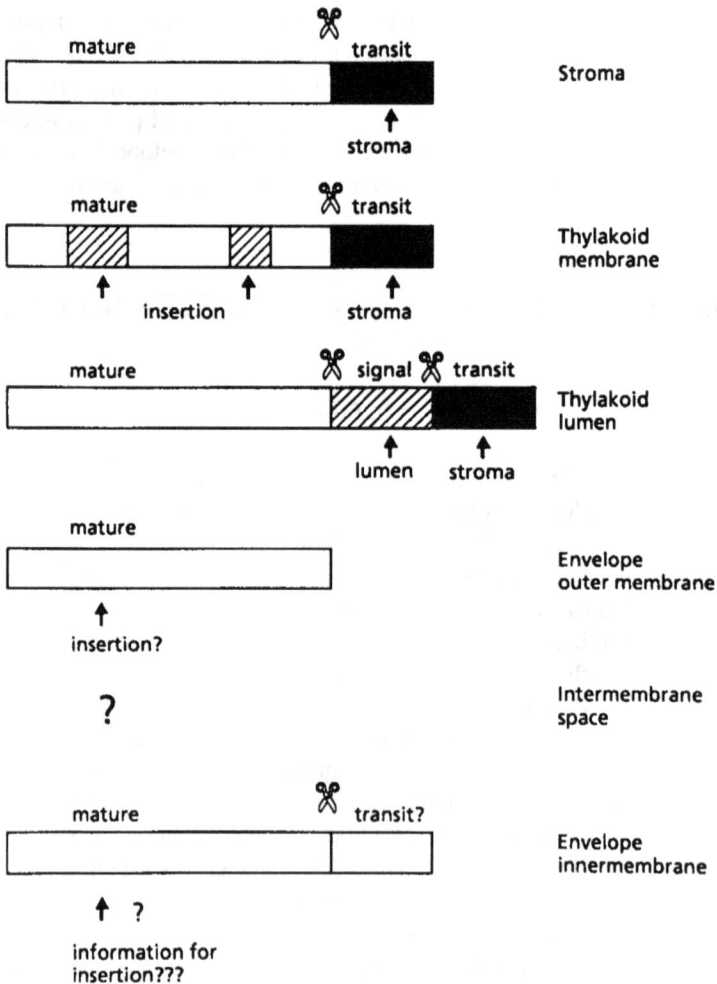

Figure 2. Organization of chloroplast routing and sorting signals. Precursors are drawn schematically with their N-terminus to the right. Topogenic information is indicated with arrows. Scissors indicate cleavage sites. See text for details.

B. The Transit Sequence

Nuclear-encoded chloroplast proteins of the stroma and thylakoids all have to be translocated across the envelope. These proteins have within a cleavable N-terminal extension a routing signal that is called the transit sequence. Because the transit sequences of different precursors share the same function, it can be expected that they are structurally similar and different from other topogenic sequences like mitochondrial presequences and secretory signal sequences. The latter two types of topogenic sequences indeed both have their own structural motifs (Roise and Schatz, 1988; Gierasch, 1989). For chloroplast transit sequences a common motif is not easily described; they differ greatly in length and have little primary sequence homology (von Heijne et al., 1989), and only a loosely defined processing site (Gavel and von Heijne, 1990) is conserved. Transit sequences as a whole are characterized by a low content of acidic residues and the high frequency of ser, thr, and ala. They can be divided into three regions, based on the frequency of specific amino acids. The N-terminal region is more hydrophobic in nature and is characterized by the absence of residues with a charged side chain. The middle region is highly variable in length and composition; it lacks acidic residues. The third, most C-terminal region contains the most positive charges, with a preference for positions -2, -6 and -10, relative to the processing site; this region could be amphiphilic if it is a β-strand (Von Heijne et al., 1989). The low degree of primary sequence similarity and the absence of a consistent secondary structure prediction of transit sequences contrast with the functional specificity.

Deletion- and gene-fusion experiments have indicated that the transit sequence is necessary (Reiss et al., 1987; Smeekens et al., 1989) and in principle sufficient for translocation into the chloroplast, both *in vitro* (Smeekens et al., 1987) and *in vivo* (van den Broeck et al., 1985). Deletions in the C- and N-terminal parts of transit sequences interfere with the function of the transit, whereas deletions in the central region are more tolerated. However, large deletions in this region also disturb the function (Reiss et al., 1987; Hageman et al., 1990). A deletion in the C-terminal part of the transit sequence of the small subunit of ribulose bisphosphate carboxylase/oxygenase (Rubisco) only affected the processing and not the import (Ostrem et al., 1989). In contrast, a small deletion in the C-terminal part of the ferredoxin transit sequence interfered with both import and binding (Smeekens et al., 1989).

C. Sorting in the Chloroplast

Proteins that insert into the thylakoid membrane, like the light-harvesting complex proteins, depend on a transit sequence for their import into the chloroplast, but in the stroma topogenic information inside the mature sequence is responsible for insertion into the thylakoid membrane (See de Boer and Weisbeek, 1991).

Precursors of lumenal proteins, like plastocyanin, have a bipartite, cleavable topogenic sequence; in addition to the routing signal, a prokaryotic-like signal sequence is present which contains information for thylakoid membrane transfer (Smeekens et al., 1985). These two parts of the topogenic sequence are functionally independent (Smeekens et al., 1986; Hageman et al., 1990). The routing signals of the precursors of lumenal-, thylakoid membrane-, and stromal proteins are functionally interchangeable and all serve as a stromal targeting sequence (Hageman et al., 1990, Smeekens et al., 1990) and, therefore, all deserve the name transit sequence.

According to the theory of endosymbiotic origin of organelles, chloroplasts are derived from an ancestor of present day blue-green algae (Raven, 1970; Margulis, 1970). In these photosynthetic bacteria proteins of the thylakoid lumen are synthesized with a signal sequence which mediates the uptake into the thylakoids. It is hypothesized that this transport route is conserved in the chloroplast of plants (Smeekens et al., 1986). When the gene of a thylakoid lumen protein moved to the nucleus, the addition of a transit sequence was necessary to ensure import into the organelle, after which the existing sorting pathway to the thylakoid can be used. This hypothesis is known as the conservative sorting theory.

Much less is known about envelope proteins. Two proteins localized in the outer envelope are synthesized in the cytosol without a cleavable extension. The import route of these proteins also differs from the other import routes in that they do not depend on protease-sensitive components, nor on ATP hydrolysis (Salomon et al., 1990; Li et al., 1991). Sequence information is not available for proteins located in the envelope inter-membrane space. Proteins of the envelope inner membrane contain a cleavable extension of the mature sequence (Flügge et al., 1989; Dreses-Werringloer et al., 1991), but it is not yet clear whether it can be functionally exchanged with a transit sequence. Additional insertion information might be present in the mature parts of these proteins.

An interesting aspect of plants is the occurrence of different plastid types in different tissues. Using an elegant *in vivo* approach de Boer et al.(1988) demonstrated that different plastids all have the capability to take up and process different precursors. Probably all plastids have a similar import apparatus. This excludes the presence of more than one plastid type per cell (de Boer and Weisbeek, 1991).

D. Comparison to Protein Targeting to Other Organelles in the Plant Cell

Most of the knowledge on the biogenesis of mitochondria comes from studies performed in fungi (especially yeast) and mammalian cells. Protein import into mitochondria has been reviewed recently (Hartl and Neupert, 1990; Beasley et al., 1992). Import into the matrix usually depends on the presence of a presequence, which is proteolytically removed after the import. Import requires energy in the form of ATP and a membrane potential across the inner membrane.

Although relatively few studies have been performed on plant mitochondria they seem to use similar mechanisms as mentioned above. Plant precursors also have presequences. Plant precursors can be imported by yeast mitochondria and vice versa emphasizing the evolutionary conservation of the mechanisms involved (Whelan et al., 1988; Bowler et al., 1989).

The supposed similarities in routing signals and uptake mechanisms for mitochondria and chloroplasts have raised the question whether mis-routing can occur in plant cells. Using hybrid proteins that contained parts of chloroplast transit sequences some import into fungal mitochondria was observed, although at low efficiency (Hurt et al., 1986; Pfaller et al., 1989). Mechanisms to avoid mis-routing, that do not have to be present in fungi, might be present in plant cells, where both mitochondria and chloroplasts occur. Therefore, the two observations described above might be of limited value. However, Huang et al. (1990) observed that a yeast mitochondrial presequence can target a bacterial enzyme to both mitochondria and chloroplasts in transgenic plants. The experiments showed that the possibility of mis-routing exists in plant cells. One should keep in mind that there are no plastids in yeast, and therefore the non-homologous system used might just lack additional information that prevents mis-routing. In addition it should be mentioned that the presequence used is structurally atypical and might follow an alternative import pathway in yeast (Miller and Cumsky, 1991). When homologous routing signals are used no mis-routing is observed both *in vivo* (Boutry et al., 1987) and *in vitro* (Whelan et al., 1990).

The thylakoid insertion or transfer, sorting, signal of thylakoid proteins could in principle be recognized by the endoplasmic reticulum (ER) translocation machinery. Until now no attempts have been reported that investigate whether these signals can indeed be recognized by the ER machinery *in vitro*. In general prokaryotic signal sequences can be recognized by the eukaryotic secretion pathway. Also an intermediate form of plastocyanin containing the thylakoid transfer domain can be used as a signal to efficiently export plastocyanin to the periplasmic space of *E. coli* (A.D. de Boer and P.J. Weisbeek., unpublished observations). In any case we must assume that the presence of a transit sequence provides a dominant signal directing these precursors to the chloroplast.

II. PROTEIN TRANSLOCATION ACROSS THE CHLOROPLAST ENVELOPE

A. General Characteristics of Import

In vitro systems have greatly increased our understanding of protein translocation across the envelope. Protein import is posttranslational (Chua and Schmidt, 1978). The envelope most likely contains the major part of the protein import machinery. The precursor first binds to the outer surface of the envelope (Cline et al., 1985). This process requires low levels of nucleotide triphosphate hydrolysis in the intermembrane space (Olsen and Keegstra, 1992). Evidence for the involvement of proteinaceous components involved in the import is present (See: de Boer and Weisbeek, 1991). ATP hydrolysis in the stroma is required for translocation across the envelope (Theg et al., 1989). During or shortly after translocation, the transit sequence is cleaved off by a stromal protease (Oblong and Lamppa, 1992). Some imported proteins are transiently associated with the chloroplast chaperonin 60, (Lubben et al., 1989). Then enzyme assembly or, depending on additional topogenic information, further sorting to the thylakoid membrane or lumen takes place.

B. Purified Ferredoxin Precursor is Translocation Competent by Itself

Transport into plastids as well as peroxisomes, the endoplasmic reticulum and mitochondria involves the insertion into and/or translocation across membranes. Although the presence of topogenic information is a prerequisite for a protein to be translocated across its target membrane, it is not always enough. It is now generally accepted that a protein has to be in a translocation competent state (Meyer, 1988).

In the case where a precursor is not translocation-competent by itself, cytosolic proteins might help to ensure a productive pathway. In principle, two functions can be fulfilled by cytosolic factors. This can be either a chaperon role, prevention of misfolding or aggregation (Ellis and Hemmingsen, 1989), or this can be a role in targeting. A combination of both functions is possible. Examples of chaperons are the HSP70 proteins that function in yeast; these proteins aid in the translocation of precursors across the endoplasmic reticulum or mitochondrial membranes (Deshaies et al., 1988). From *in vitro* studies a role for HSP70 has also been implied in protein targeting in plant cells for proteins destined to the ER (Miernyk et al., 1992) and for the import of the LHCP precursor into chloroplasts (Waegeman et al., 1990). When cytosolic factors function in targeting, this implies that topogenic information is recognized in the cytosol.

Until recently it was not known which role cytosolic factors can play in the import and also little was known about the secondary structure and folding of chloroplast precursors. The reason for the lack of information on these points is the lack of sufficient amounts of purified precursor protein. Most *in vitro* studies so far have been performed with precursors synthesized in wheat-germ or reticulocyte lysates. These lysates are complex in composition and can contain several factors that influence the import. Furthermore, the amount of precursor that can be synthesized in such a system is limited.

To address the involvement of cytosolic factors in the import we took a new approach. We have chosen to express the ferredoxin precursor (prefd) in *E. coli* and to purify it (Pilon et al., 1990). Several characteristics, like the need for cytosolic factors, the energy requirements, the kinetics of import and the structure of the precursor were analyzed.

Purified prefd diluted out of 8 M urea was imported by isolated chloroplasts although with a relatively low efficiency (see Figure 3). Import was greatly stimulated by the addition of reducing agents like dithiothreitol (DTT) or glu-

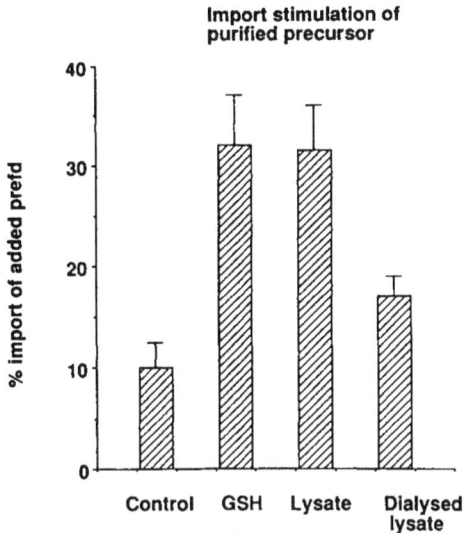

Figure 3. Stimulation of prefd translocation into chloroplasts. Purified prefd was incubated in the presence of chloroplasts without (*control*), or with the indicated additions. GSH: 1 mM reduced glutathione. *Lysate*: 6 µl of a 100.000 × g supernatant of wheat-germ lysate, containing 30 µg protein. *Dialysed lysate*: 7.1 µl of lysate (30 µg protein) that had been dialyzed for 4 hr against buffer without DTT. See Pilon et al. (1992a) for experimental details.

tathione (GSH). Also wheat germ lysate stimulated the import but the stimulation was largely lost after dialysis of the lysate, indicating that a compound of low molecular weight was responsible for this stimulation. The stimulation was not lost after dialysis of the lysate in the presence of DTT. Together these experiments indicate that prefd is translocation-competent by itself and that thiol reducing agents stimulate the import.

Since prefd was added from urea, a need for cytosolic factors could have been masked. A possible function of cytosolic factors could be to preserve translocation competence. However prefd did not lose its translocation competency for at least 1 h after dilution from urea. Prefd even remained import competent after complete removal of urea from the precursor solution by gelfiltration (Pilon et al., 1992a).

Also by gelfiltration it was shown that *in vitro* (in a wheat germ system) synthesized translocation-competent prefd is also not associated with a macro-molecular factor (Pilon et al., 1992a)

C. A Role for Cytosolic Factors in Import?

Cytosolic factors involved in protein translocation can have a function in target-ing or a chaperonin function. Ferredoxin does not need cytosolic factors and thus is targeted efficiently to the import machinery by itself. However, our finding cannot be generalized for all chloroplastic precursors since another protein, the precursor of a light-harvesting complex protein, apparently needs cytosolic factors to obtain import competence (Waegemann et al., 1990).

All the topogenic information for chloroplast targeting in a precursor of a stromal or thylakoidal protein is present in the transit sequence. Furthermore chloroplast transit sequences are functionally interchangeable. Taken together this means that a general role in targeting by a cytosolic signal recognition factor seems excluded. Therefore, we propose that if cytosolic factors are needed this will be a chaperon-like function, prevention of improper interactions: misfolding or aggregation (Ellis and Hemmingsen, 1989).

A requirement for cytosolic factors in posttranslational protein transport in a eukaryotic cell *in vivo* has so far only been shown in yeast (Deshaies et al., 1988). The chaperons involved in protein translocation in yeast belong to the HSP-70 family (Chirico et al., 1988; Murakami et al., 1988). These proteins have ATPase activity, which is used to release the bound protein from the chaperon (Gething and Sambrook, 1992). Theg et al. (1989) have concluded that ATP was not required outside the chloroplast for the import of the precursor of plastocyanin (prepc), the small subunit of Rubisco (pressu) and prefd. Thus it seems unlikely that these HSP-70 proteins are involved in the import of prepc, pressu, and prefd precursors. If one of the cytosolic factors in the leaf extract that stimulated the import of the light harvesting complex precursor indeed is a HSP-70 protein (Waegeman et al., 1990) then we have no reason to assume that this is a general aspect of chloroplast import.

It is now generally recognized that precursors have to be in a translocation competent state in order to be translocated across a membrane (Meyer, 1988). There is not an exact description of what a translocation-competent state of a precursor is. Clearly, proteolysis, aggregation and interactions of either mature sequence or topogenic signal at nonproductive sites in the cell have to be avoided. Experiments with yeast mitochondria and *E. coli* cells have indicated that proteins can only be transported when in a not-tightly folded conformation (Eilers and Schatz, 1986; Randall and Hardy, 1986). This principle also seems to be valid for the ER (Sanders and Schekman, 1992) and the chloroplast system (Della-Cioppa and Kishore, 1988). However, it should be mentioned that up until now no direct proof has been presented to indicate that a tightly folded precursor cannot be imported into chloroplasts (see below). In the case of protein import into mitochondria, the role of cytosolic factors, when present, also seems to be solely a chaperon function (Glick and Schatz, 1991). Both mitochondrial and chloroplastic precursors have a rather hydrophilic targeting sequence (von Heijne et al., 1989). In contrast, the signal sequence of secreted proteins is rather hydrophobic (von Heijne, 1988). In this case topogenic sequence recognition in the cytosol could be necessary more often to ensure a productive pathway.

D. Thiol Reducing Agents Stimulate Activity of the Import Machinery

The specificity of the stimulation by thiol reducing agents was further analyzed. A modification of the cysteines of the purified precursor with iodoacetamide was used to show that DTT and GSH have their principal effect on the chloroplast (Pilon et al., 1992a). In addition, because this modified precursor was imported while at the same time iron–sulphur cluster insertion was blocked, these experiments showed that import takes place independently from holo–enzyme formation.

To investigate whether DTT stimulates the import of other precursors, we synthesized prepc and pressu and as a control prefd *in vitro. In vitro* translation systems contain DTT and therefore we had to remove this by gelfiltration on a fast desalting column. As a control, samples were also desalted in the presence of DTT. Import experiments (2 mM ATP) and binding experiments (0.05 mM ATP) were performed. For all three precursors tested, the import in the presence of DTT was much better. DTT influenced the rate of import, as was determined from the time courses of import (M. Pilon, unpublished results). Import of the samples where DTT was removed could be stimulated by addition of extra DTT to the original levels. Interestingly, the effect of DTT was also seen on the level of binding (M. Pilon, unpublished results). We therefore conclude that DTT stimulates the import machinery involved in the import of all these precursors, most likely at an early stage.

Our results show that DTT not only stimulates the import of prefd, but in addition it also stimulates the import of preplastocyanin and pressu. It is important to note that DTT addition only stimulates the import: there is no absolute requirement. The DTT in our *in vitro* experiments probably substitutes for glutathione present *in vivo*. Despite the presence of different amounts of cysteines in each precursor (only one cysteine is present in plastocyanin, 3 in pressu and 5 in prefd) the effect of DTT is quite similar for all precursors and this points to an effect of DTT directly on the chloroplast. We propose that a reduced cysteine residue of a protein in or nearby the import machinery is required for an efficient translocation activity.

Protease pretreatment of chloroplasts has shown that proteinaceous components exposed at the chloroplast surface are involved in the import and binding of precursors (Cline et al., 1985). Stable binding of precursors to the outer envelope requires ATP hydrolysis in the intermembrane space (Olsen and Keegstra, 1992). This means that a putative proteinaceous component of the import machinery has its ATP binding site exposed to the intermembrane space. It is not known whether this component is also exposed to the surface of the chloroplast. Import and also binding are stimulated by reducing agents like glutathione. Possibly one of the protease-sensitive components or the ATP consuming component involved in binding are effected by the thiol reducing compounds. It might well be worthwhile to further investigate this, because it might lead to the identification of a component involved in the import process.

E. Physiological Relevance of the *In Vitro* Import System

To further characterize and to investigate the physiological relevance of the import pathway of the purified ferredoxin precursor *in vitro* we performed several experiments. First, the localization of the imported ferredoxin was determined. An import experiment was performed. One aliquot did not receive protease treatment, the rest was treated with the protease thermolysin. Part of the protease-treated chloroplasts were fractionated by centrifugation (see methods) into stroma and membranes (thylakoids, containing a large part of the envelope fraction). The results presented in Figure 4 indicate that the imported and processed prefd accumulated in the stromal fraction. We next investigated whether the cofactor was inserted in this protein. We made use of the different elution patterns of holofd and apofd on a mono-Q 5/5 ion exchange column. As is shown in Figure 5 (upper panel), holofd from *Silene pratensis* elutes from the mono-Q 5/5 column as a single peak with a defined position in the salt gradient. In contrast, the apofd is more spread out over the column and elutes in two peaks, possibly due to oligomer formation on the column. Fractions of 0.5 ml were collected and used for holofd to determine

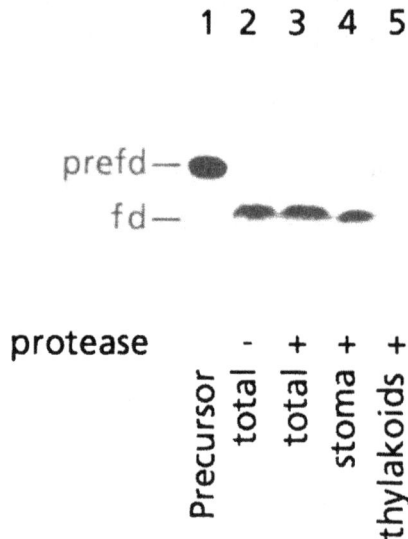

Figure 4. Example of an *in vitro* chloroplast import and fractionation experiment using the purified ferredoxin precursor. *Precursor*: 20% of the amount of prefd added to each of the lanes 2-5; *Total*: whole chloroplasts; *Stroma*: stromal fraction; *Thylakoids*: thylakoid fraction. Protease treatment prior to chloroplast lysis is indicated. Samples were analyzed by SDS-PAGE and fluorography. The positions of precursor (*prefd*) and mature form of ferredoxin (*fd*) are indicated. Proteins were separated by SDS-PAGE, radioactively labeled proteins are visualized by fluorography.

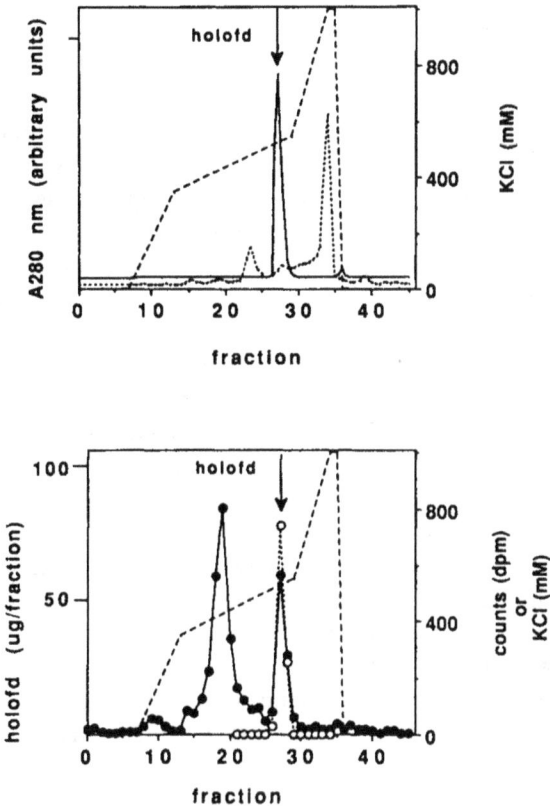

Figure 5. Elution patterns from Mono-Q 5/5 column. (*Upper panel*): elution patterns detected by A280 nm. (*Solid line*): purified holofd (*Dotted line*): purified apofd. (*Lower panel*): proteins recovered in fractions. (*Open circles*): purified holofd recovered from the column in fractions and quantified by A_{420} nm. (*Closed circles*): stromal fraction after import experiment, detection by scintillation counting of fractions. The dashed line indicates the salt (KCl) gradient.

the exact elution volume and recovery. Based on the A_{420nm}, 100% of the ferredoxin applied to the column was recovered, of which 70% was collected in fraction 27 (Figure 5, lower panel). Stromal fractions from import experiments were now injected onto the column and the elution profile was determined by scintillation counting of fractions (Figure 5, lower panel). Three peaks of radioactivity could be distinguished. None of the radioactivity peaks coeluted with one of the major peaks detected by A_{280nm} of the stromal fraction. Two peaks eluted before and one peak exactly at the elution volume of the purified *S. pratensis* holofd. This peak probably represents holofd. We thus tentatively conclude that part of the imported and processed purified ferredoxin is converted to the holoenzyme. To investigate the formation of holoenzyme in time, relative to the import process, we performed a time-course of import. The stromal fractions were prepared and analyzed. We did not observe a tight coupling of import and holoenzyme formation (not shown).

We showed that the 2Fe–2S cofactor is inserted in part of the imported purified prefd. This is an important conclusion, because it shows the physiological relevance of the import. We have used a column system to separate the holoprotein. Previously, Takahashi et al. (1986) used a nondenaturing gelelectrophoresis system to show the formation of iron–sulphur clusters in ferredoxin in chloroplasts, independent from import. Later, Li et al. (1990) and Suzuki et al. (1991) used a similar gel system to show cofactor assembly to take place after import of *in vitro* synthesized precursor. In all these cases the purified apoprotein migrated differently from the apoprotein present in the stroma fractions. We observed a similar difference with our column system. For this reason, both types of analysis are not fully conclusive on the exact nature of the imported proteins, although the identification of the holoprotein looks trustworthy in both cases. The column system offers the advantage of possible scale-up and subsequent analysis by, for instance, protease-sensitivity assays or even a second column type. We did not observe tight coupling of holoenzyme formation with import and this leads us to conclude (in agreement with the findings obtained for the acetamide modified precursor) that import takes place independently from cofactor insertion.

Next we investigated the energy requirements of import. Import and binding of the purified prefd are ATP dependent (Pilon et al., 1992a). Finally a kinetic analysis

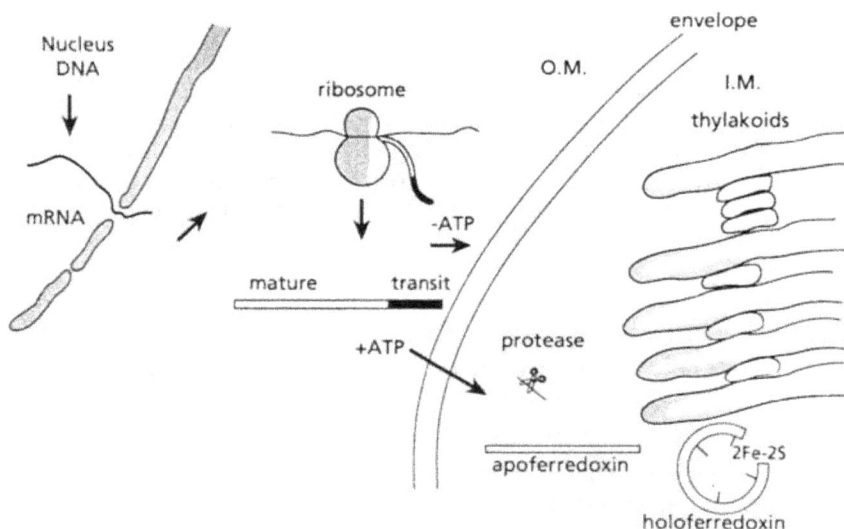

Figure 6. The biogenesis of chloroplast ferredoxin. O.M.: outer membrane, I.M.: inner membrane. Low ATP levels stimulate binding to the chloroplast surface (–ATP), while high ATP levels stimulate translocation to the stroma (+ATP). See text for details.

indicated that the import proceeds via a saturable high-affinity import system. An apparent K_m of 100 nM and a capacity of 2.5×10^4 (precursors/min/chloroplast) was found (Pilon et al., 1992c). This value is in proportion to the uptake rate that would be needed to sustain chloroplast division at rates observed *in vivo* (Pilon et al., 1992c). This conclusion is of importance as it validates the use of *in vitro* systems to analyze chloroplast protein uptake. The biogenesis of chloroplast ferredoxin is schematically depicted in Figure 6.

III. STRUCTURAL ANALYSIS OF A FUNCTIONAL PRECURSOR

A. Overall Structural Characteristics of the Ferredoxin Precursor

When trying to unravel the molecular mechanism of protein import into the chloroplast, one needs to understand the structural features of the precursor in its translocation-competent form: especially the transit sequence is of interest. The unique properties of preferredoxin, a soluble protein that requires no cytosolic factors, make it possible to perform structural studies on this precursor in the translocation-competent state. In addition, it is a relatively small protein of which the transit sequence is a relatively large domain. Therefore we can say something about the structure of individual domains.

We analyzed functional and structural properties of the translocation-competent ferredoxin precursor as well as its isolated constituent domains (see Figure 7). Only unlabeled prefd and transit peptide were able to compete with labeled prefd in a

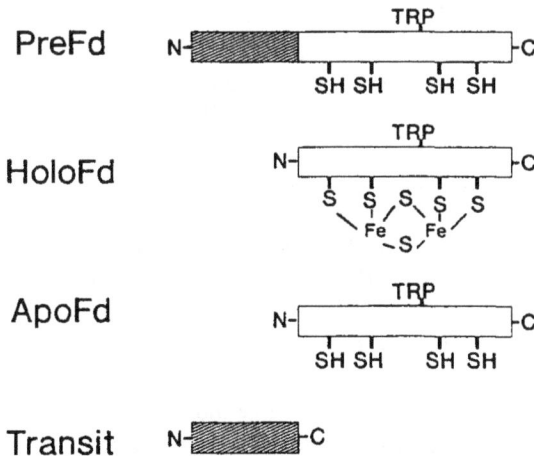

Figure 7. Proteins used for structural analysis. The full precursor, the mature protein both in holo- and apo-form (without iron–sulfur cluster) and the isolated transit peptide were analyzed separately. See Pilon et al. (1992b) for details.

Table 1. Secondary Structure of the Prefd Transit Sequence as Determined by Circular Dichroism[a]

	Alpha-Helix	Beta-Strand
Calculated from prefd and apofd:	11%	19%
Measured in isolated peptide in:		
buffer (10 mM Tris/HCL, pH 8)	5%	19%
25 mM SDS in buffer	32%	6%
Trifluoroethanol	39%	19%

Note: [a]For methods see Pilon et al., 1992b.

chloroplast import experiment; apofd and holofd did not compete (Pilon et al., 1992b).

We analyzed the secondary structure by circular dichroism (CD). Prefd and apofd have only a very low content of secondary structure in buffer (not shown). The contribution of the transit sequence to the secondary structure of prefd was estimated after subtraction of the obtained apofd spectrum (see Table 1). Interestingly, a very similar low content of α-helix and β-strand was obtained by measuring the transit sequence in buffer directly. We tentatively conclude that this is the structure of the prefd transit sequence prior to import.

The overall folding state was analyzed by tryptophan fluorescence quenching, protease sensitivity, and gelfiltration experiments. These measurements indicated that the monomeric prefd, like apofd, has an open and flexible conformation, whereas in contrast holofd is highly structured. The folding properties of preferredoxin are due to both mature part and transit sequence (Pilon et al., 1992b).

These properties of preferredoxin might well explain the lack of cytosolic factor requirements of this precursor. We can try to explain the folding state of prefd from the properties of both mature and transit peptide part. From model studies we know that hydrophobic interactions are most probably the dominant forces in protein folding (Dill, 1990). The rather low overall hydrophobicity and the overall negative charge of −15 of the mature part of prefd might act as sufficient "antifolding" properties. Because the transit sequence itself is also quite hydrophilic, this might explain the unfolded state of the precursor without a high tendency to aggregate, and thus the lack of cytosolic and stromal chaperon requirement.

B. Lipid-Protein Interactions

To investigate the lipid inserting capacity monolayer experiments can be performed. In such an experiment a monomolecular lipid layer is formed at the air–water interface and its lateral pressure is measured. The insertion of a protein, that is injected into the subfase, is detected by an increase in the lateral pressure of the lipid monolayer. Using this set up van't Hof et al. (1991) investigated the

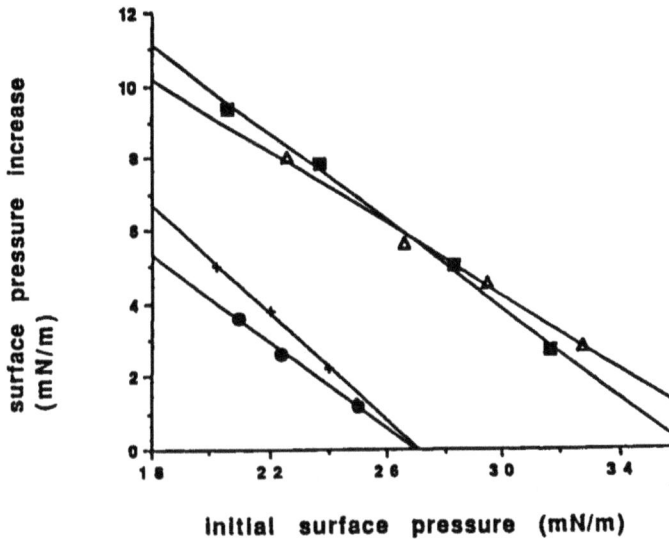

Figure 8. Surface pressure increases, in milli Newtons per meter (mN/m), measured after the injection of different peptides underneath monolayers consisting of SQDG. See van't Hof et al. (1993) for experimental details. (*Squares*): precursor ferredoxin. (*Triangles*): transit peptide. (*Crosses*): apo-ferredoxin. (*Circles*): holo-ferredoxin.

insertion of fragments of the transit peptide of the small subunit of Rubisco into a monolayer prepared from lipids found in the chloroplast envelope. The fragments showed a differential interaction with the monolayers and had a preference to insert into the chloroplast galacto- and sulfolipid and phosphatidylglycerol. Even more interesting, the full ferredoxin precursor was shown to specifically interact with these lipids due to its transit sequence (van't Hof et al., 1993) An example of such an experiment, where the insertion into the chloroplast-specific sulfolipid (SQDG) is measured, is presented in Figure 8. At lateral pressures, which are relevant for biological membranes (30–35 mN/m) (Demel et al., 1975), only prefd and the isolated transit peptide induce a pressure increase, whereas apofd and holofd have no effect (van't Hof et al., 1993). The preferential interaction with lipids found in the chloroplast was observed in this way. We conclude that the transit sequence provides prefd with a specific lipid inserting capacity.

Interactions with components of the import machinery might modulate the structure of the transit sequence. Conformational flexibility might be of importance. We therefore also estimated by CD analysis the secondary structure of the transit peptide in the presence of micelles and in an apolar environment. The neutral detergent octylglucoside did not influence the structure of the transit peptide; just as in buffer it remains mainly random coiled. In contrast, an increase in α-helix was induced by the negatively charged detergent SDS and by an apolar solvent (Table

1), but always a high content of random coil is maintained. The isolated transit peptide thus can undergo large conformational changes when placed in different environments. The conformational behavior of the ferredoxin transit peptide in the presence of bilayer-forming lipids found in membranes is from a biological point of view more interesting. We found that α-helix induction in the transit peptide of ferredoxin requires a surface containing negative charges. Up to 40% α-helix is induced by vesicles containing negatively charged lipids (L. Horniak, 1993).

C. The Role of Lipids in Import

The possibility of a direct role for the envelope membrane lipids in the import process has not yet received much attention. In other protein translocation systems, evidence for a direct involvement of lipids has been obtained. In *E. coli*, the negatively charged phosphatidyl glycerol (PG) is involved in translocation of preproteins across the inner membrane (De Vrije et al., 1988; Kusters et al., 1991). Protein import into mitochondria is probably most similar to chloroplast protein import: both organelles are surrounded by an envelope of two membranes (De Boer and Weisbeek, 1991) and both are thought to be the result of independent endosymbiotic events in the evolution of eukaryotic cells (Raven, 1970). Import receptors are identified on the mitochondrial surface, but these are functionally redundant (Baker and Schatz, 1991).The import of mitochondrial precursors with a cleavable presequence has, however, an absolute requirement for an integral outer membrane protein (Baker and Schatz, 1991). In addition, the presequence has a high capacity to interact with lipids, and its physical properties in model membrane studies can be interpreted to support a role for lipids in the import of these precursors (Roise and Schatz, 1988; De Kroon et al., 1991; Maduke and Roise, 1993). The mitochondrial intermembrane space protein cytochrome *c* is a special case. At present, no receptors are known (Stuart and Neupert, 1990) and the precursor protein, apocytochrome *c*, has special lipid-penetrating properties that enable it to pass a lipid bilayer at least partially (Rietveld and de Kruijff, 1984). The important role of negatively charged phospholipids in this import process was demonstrated by a competition assay where pure lipid vesicles inhibited the import of apocytochrome *c* into purified mitochondria (Jordi et al., 1992).

While irreversible binding to the chloroplast envelope requires ATP (Olsen and Keegstra 1992) binding can also be observed in the absence of ATP (Flugge, 1990). The lipids of the chloroplast outer envelope membrane deserve special attention with respect to the import process for several reasons. (1) Most important, fragments of a chloroplast transit peptide and the full ferredoxin precursor can insert into lipid monolayers with a preference for the lipids that are found in the chloroplast outer membrane (van't Hof et al., 1991, 1993). Also the ferredoxin precursor and its complete transit peptide were found to bind, with an affinity in the micromolar range, to large unilamellar lipid vesicles prepared from chloroplast outer membranes (van 't Hof et al., unpublished results). (2) The lipid composition of this membrane is unique; it contains galactolipids and sulfolipids. These classes of lipids are absent from any other membrane facing the cytosol (Douce and Joyard, 1990) and could therefore potentially be used for targeting. (3) This membrane has a very low protein content, providing more space for precursors to insert into the lipids (Douce et al., 1984).

A way to assess to what extent the lipid–protein interactions alone are used for targeting is by doing a chloroplast import/vesicle binding competition assay. In such an assay, small unilamellar lipid vesicles are included in an *in vitro* import assay, allowing precursor proteins the possibility to either bind to the vesicles or to be imported. Only a rough estimate for the amount of lipid present in the chloroplast outer membrane can be made, based on literature data. The outer membrane contains approximately 1% of the total chloroplast protein (Douce et al., 1984; Flügge, 1990). The lipid-to-protein ratio of the chloroplast outer membrane is 3:1 (Douce et al., 1984). Our chloroplast preparations contain 10–20 μg protein per μg chlorophyll. We now estimate, assuming an average M_r of 800 for the lipids, that at maximum 0.75 nmol outer membrane lipid is present in chloroplasts equivalent to 1 μg chlorophyll. We have chosen to use lipid quantities up to 5 nmol per μg chlorophyll in our competition assays.

Chloroplast outer envelope membranes were isolated and a lipid extract was obtained as described (van't Hof, 1991). Small unilamellar vesicles were prepared by sonication. In the presence of vesicles, 5 nmol lipid/μg chlorophyll, 24% of the added prefd was imported. In the control without vesicles, 25% was imported (average of two experiments). During the import assay the vesicles might for some reason have aggregated, rendering their exposed surface much smaller. As a control for this, the supernatant from one of these experiments with chloroplast outer membrane lipids was analyzed on a Sepharose 4B size exclusion column equilibrated in the import buffer. The recovery of the vesicles from the import experiment was 91%. These vesicles, which had been marked by the incorporation of tritiated cholesteryl–ether, were quantitatively recovered from the column in an elution volume expected for small unilamellar vesicles (not shown). From these results, we conclude that recognition of the precursor takes more then is accomplished by lipids alone.

If we combine all the available data we can speculate and come to the following role of lipids in the import process. We propose that the transit sequence has the capability to insert into the lipid bilayer of the chloroplast outermembrane. Participation of the transit sequence, which is rich in hydroxylated amino acids, in the hydrogen bonding network between the lipids might be involved in this process (van't Hof et al., 1993). Lipid-mediated conformational changes of the transit sequence or lateral diffusion in the plane of the membrane can then contribute to more efficient delivery of the precursor to the proteinaceous translocation machinery (Cline et al., 1985). During interaction with protein components, the ATP hydrolysis necessary for stable and irreversible binding (Olsen and Keegstra, 1992) takes place.

Should we now assume that the unique lipid composition of plastids is not used for the recognition of precursors? Recently two different clones encoding chloroplast outer membrane proteins were isolated (Salomon et al., 1990; Li et al., 1991). These precursors that do not contain a cleavable topogenic sequence insert into the chloroplast outer membrane without making use of a protease-sensitive component in the outer membrane, or an ATP requirement. It is very probable that these proteins directly recognize the specific chloroplast lipids.

D. What are the Functional Properties of Transit Sequences?

What are the implications of our findings with respect to the functioning of the transit sequence? Statistical analysis of transit sequences from different precursors indicated three regions within the sequence (von Heijne et al., 1989) and revealed a loosely defined processing site (Gavel and von Heijne, 1990). However, only a very low conservation of primary structure was found (von Heijne et al., 1989). It can therefore be expected that the essential features of transit sequences reside in their secondary or even tertiary structure. Von Heijne and Nishikawa (1991) proposed that transit sequences are designed to be random-coiled. Such design would make transit sequences susceptible to recognition by chaperon-like proteins.

Our experiments indicate that the ferredoxin transit sequence possesses strikingly low contents of secondary structure. However, no interaction with a chaperon is observed in the cytosol in this case. If we do not favor the hypothesis of von Heijne and Nishikawa, then what is the special property of transit sequences that can make them recognizable? Based on our current knowledge we could speculate on this. In the cytosol, the transit sequence might be quite unstructured. Its hydrophilic nature and position at the N-terminus could help to ensure the exposure on the outside of the precursor. It can penetrate into a suitable lipid interface, and CD analysis has indicated that it might then adopt a defined secondary structure. Certain parts of the transit sequence in this state will be important for recognition (possibly the ATP dependent step), while others only serve as connectors that might remain unstructured. It should be emphasized that this model has similarities with the 'framework of primary structure' model proposed by Karlin-Neuman and Tobin (1986), but in contrast it states that a framework of secondary or tertiary structure elements is important. The model can explain how non-homologous transit sequences can fulfill the same specific function despite their lack of primary sequence similarity.

E. An Evolutionary Context for the Structural Design of Transit Sequences?

It is easy to recognize a signal sequence of a secretory protein or a presequence of a mitochondrial protein by statistical analysis due to their relatively simple design, but in the case of transit sequences this is not the case. Why are the functional properties of transit sequences so cryptic as compared to signal sequences and presequences? The simple physicochemical properties of signal sequences and presequences might be rationalized in an evolutionary context. Signal sequences are probably the evolutionary oldest cleavable topogenic sequences. Blobel (1980) has proposed that signal sequences, which contain a hydrophobic stretch that can form an α-helix (Gierasch, 1989) might have evolved from transmembrane domains (hydrophobic α-helices). Presequences possibly evolved to give optimal response to the membrane potential across the inner membrane, which is important to translocate the presequence across the inner membrane (Martin et al., 1991). The positively charged amphiphilic nature of presequences is a good functional design to do this (Roise and Schatz, 1988; de Kroon et al., 1991).

The endosymbiosis event leading to the chloroplast occurred at a later stage in evolution (Raven, 1970). Chloroplast import therefore most likely evolved later then secretion and mitochondrial import. Recognition processes had to be different from the existing ones to avoid large-scale mistargeting. Therefore another type of design for chloroplast topogenic sequences (see above) might have been necessary.

IV. CONFORMATION OF THE PRECURSOR DURING ENVELOPE PASSAGE

To analyze the conformation of a precursor protein during envelope translocation we needed to follow a different approach. We have chosen to analyze the import of a set of fusion proteins consisting of the full sequence of the murine cytosolic enzyme dihydrofolate reductase (DHFR) fused to the plastocyanin chloroplast lumen targeting sequence. Most of the experiments were performed with the precursor outlined in Figure 9 (PC-DHFR). Other constructs containing shorter segments of the plastocyanin targeting sequence or the full ferredoxin transit sequence were also used and gave similar results as described below.

DHFR fusion proteins have been used to show that unfolding is required for membrane translocation into isolated mitochondria. To this aim the substrate analog methotrexate, which tightly binds to DHFR and keeps it in a stably folded conformation, was used (Eilers and Schatz, 1986). We tried to make use of this property of the DHFR fusion proteins. To this aim we required a radiolabeled precursor. First, the PC-DHFR fusion protein was expressed in *E. coli*. The protein could be partially purified from the crude cell homogenate by binding it to a methotrexate agarose affinity column and subsequent elution with dihydrofolate. These experiments strongly indicate that the fusion protein can specifically bind methotrexate. However a purification to homogeneity was difficult to achieve due to the presence of mature sized degradation products that were already present during the expression in *E. coli* and could not be efficiently removed later. We therefore turned to the use of an *in vitro* transcription/translation system to express the radiochemically pure precursor in a lysate.

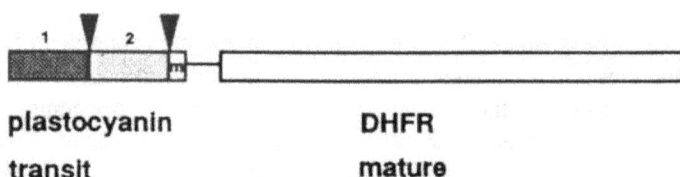

plastocyanin　　　　　　　DHFR

transit　　　　　　　　　mature

Figure 9. Schematic presentation of the PC-DHFR fusion protein. The fusion protein contains the full plastocyanin transit sequence (1) and thylakoid transfer domain (2) and the first amino acid of mature plastocyanin from Silene pratensis (m). A small spacer is used to fuse mature DHFR to the transit sequence. Details of this construct can be found in Hageman et al. (1990).

Upon addition of the PC-DHFR fusion protein to intact isolated chloroplasts in the absence of methotrexate the protein was readily imported. The plastocyanin sorting sequence does not target a non-chloroplast lumen protein to the thylakoid lumen efficiently (Hageman et al., 1990), and therefore the imported protein became localized in the stromal compartment. A complication of the analysis is the appearance of a characteristic set of labeled peptides smaller than mature DHFR within the stroma derived from the imported precursor. This phenomenon is seen with all the DHFR fusions that we tested and is probably due to a degradation of mature DHFR in the stroma.

The presence of methotrexate had no effect on an import reaction of the parent plastocyanin or ferredoxin precursor. In contrast, upon addition of methotrexate to the fusion protein two differences were observed in the chloroplast import. First, the appearance of labeled protein inside the chloroplast was delayed when time courses were compared from import experiments both in the presence and in the absence of methotrexate. Methotrexate thus decreased the rate of import but did not block import. Similar results were reported for the precursor of 5-enol-pyruvate shikimate phosphate synthase (Della-Cioppa and Kishore, 1988). As a second effect of methotrexate, the degradation products in the stroma were not observed and instead the mature sized DHFR accumulated. The observed import of precursor in the presence of methotrexate and the altered processing pattern inside the organelle led us to investigate whether methotrexate itself was imported together with the precursor. For this purpose, tritiated methotrexate was used in combination with ^{35}S-labeled precursor and both molecules were now quantified simultaneously in chloroplast fractions by liquid scintillation counting. In addition we quantified the imported precursor by excision of gel slices after SDS-PAGE. No accumulation of methotrexate inside the chloroplasts was observed when the full plastocyanin precursor was imported. When a DHFR fusion protein was imported a near stoichiometric accumulation of methotrexate was observed. However, when proteins were allowed to import first, followed by washing of the chloroplasts, and then incubated with methotrexate, a near stoichiometric accumulation of the drug was also seen. These data strongly suggest that methotrexate can enter the chloroplast independently from protein import and becomes trapped inside the organelle by the DHFR moiety.

The question still remained whether the fusion protein was loosely folded during the import or whether it was imported as a stably folded protein. In order to shed some light on this issue we performed protease accessibility experiments. The rationale of using protease to probe the folding is the assumption that stably folded proteins have a low degree of chain flexibility compared to more loosely folded proteins and thus a lower degree of accessibility of the polypeptide chain to the active site of the protease. PC-DHFR fusion protein is completely digested by the protease thermolysin in the absence of methotrexate. In the presence of methotrexate the fusion protein was processed to a size similar to that of mature DHFR. This suggests that in this case the transit sequence was accessible to protease while the mature part became protected, thus stably folded, due to the binding of methotrexate. The fusion protein can be stably bound to the outer envelope membrane of the chloroplast at 4 °C. This early translocation intermediate can subsequently be chased to imported protein by incubating the chloroplasts at 25 °C in the presence of ATP. We observed that the bound precursor was sensitive to protease, even at low concentrations, both in the absence and presence of methotrexate. The change in protease accessibility strongly suggests that the fusion protein which is bound to the chloroplast envelope has undergone a conformational change (see Figure 10). The chloroplast thus contains at the envelope a machinery that has the capacity to destabilize the complexed fusion protein. We tentatively

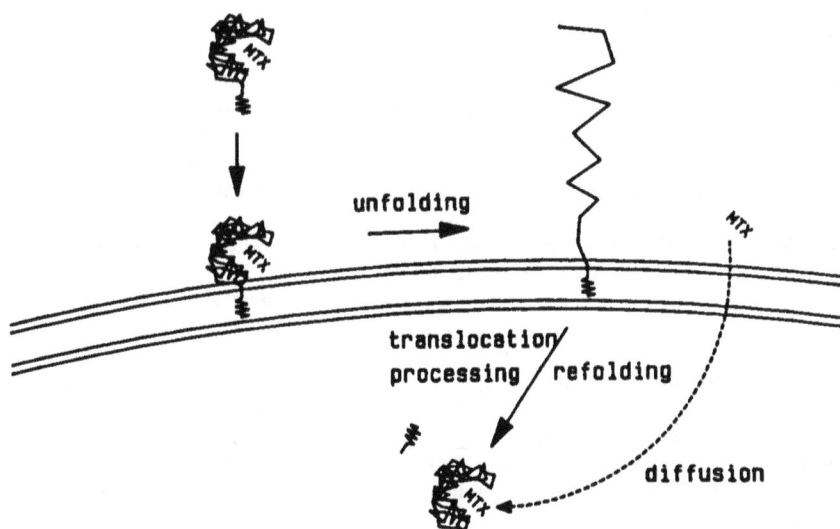

Figure 10. Model for the import into chloroplasts of DHFR fusion protein in the presence of methotrexate (MTX).

speculate that this conformational change is rate-limiting causing the fusion protein to be imported more slowly in the presence of methotrexate.

V. CONCLUSION AND FUTURE PROSPECTS

Our knowledge of the transport of proteins into chloroplasts is now in the stage that we know which targeting signals direct proteins to each of the compartments inside the organelle. The use of purified precursors of which the import characteristics are well defined will make it possible to identify components of the import and sorting machineries. At present our laboratory is using the purified precursor of ferredoxin in a chemical cross-linking approach to this aim. It is to be expected that in the near future several protein components of the import and sorting machinery will be identified. When sequence information becomes available it will be interesting to see whether the chloroplast components bare any similarity to components of translocation machinery known from other organelles or prokaryotes. Then the task will become to analyze exactly how identified components mediate each of the specific steps involved in the translocation. Another major task is to unravel how degenerate signals like transit sequences can fulfill their specific function.

REFERENCES

Baker, K.P. & Schatz, G. (1991). Mitochondrial proteins essential for viability mediate protein import into yeast mitochondria. Nature 349, 205–208.

Beasley, E.M., Wachter, C., & Schatz, G. (1992). Putting energy into mitochondrial protein import. Curr. Opin. Cell Biol. 4, 646–651.

Blobel, G. (1980). Intracellular protein topogenesis. Proc. Natl. Acad. Sci. USA 77, 1496–1500.

Boutry, M., Nagy, F., Poulsen, C., Aoyagi, K., & Chua, N-H. (1987). Targeting of bacterial chloramphenicol acetyltransferase to mitochondria in transgenic plants. Nature 328, 340–342.

Bowler, C., Alliotte, T., Van den Bulke, M., Bauw, G., Vanderkerkhove, J., Van Montagu, M., & Inze, D. (1989). A plant manganese superoxide dismutase is efficiently imported and correctly processed by yeast mitochondria. Proc. Natl. Acad. Sci. USA 86, 3237–3241.

Chirico, W.J., Waters, M.G., & Blobel, G. (1988). 70-Kda heat-shock related proteins stimulate protein translocation into microsomes. Nature 332, 805–809.

Chua, N.H. & Schmidt, G.W. (1978). Post-translational transport into intact chloroplasts of a precursor to the small subunit of ribulose-1,5-bisphosphate carboxylase. Proc. Natl. Acad. Sci. USA 75, 6110–6114.

Cline, K., Werner-Washburne, M., Lubben, T.H., & Keegstra, K. (1985). Precursors to two nuclear-encoded chloroplast proteins bind to the outer envelope membrane before being imported into chloroplasts. J. Biol. Chem. 260, 3691–3696.

de Boer, D., Cremers, F., Teertstra, R., Smits, L., Hille, J., Smeekens, S., & Weisbeek, P. (1988). *In vivo* import of plastocyanin and a fusion protein into developmentally different plastids of transgenic plants. EMBO J. 7, 2631–2635.

de Boer, D., Bakker, H., Lever, A., Bouma, T., Salentijn , E., & Weisbeek, P. (1991). Protein targetting towards the thylakoid lumen of chloroplasts: proper localization of fusion proteins is only observed *in vivo*. EMBO J. 10, 2765–2772.

de Boer, A.D. & Weisbeek, P.J. (1991). Chloroplast topogenesis: Protein import, sorting and assembly. Biochim. Biophys. Acta 1071, 221–253.

de Kroon, A.I.P.M., de Gier, J., & de Kruijff, B. (1991). The effect of a membrane potential on the interaction of mastoparan X, a mitochondrial presequence, and several regulatory peptides with phopholipid vesicles. Biochim. Biophys. Acta 1068, 111–124.

Della-Cioppa, G. & Kishore, G.M. (1988). Import of a precursor protein into chloroplasts is inhibited by the herbicide glyphosate. EMBO J. 7, 1299–1305.

Demel, R.A., Geurts van Kessel, W.S.M., Zwaal, R.F.A., Roelofsen, B., and Van Deenen, L.L.M. (1975). Relation between various phospholipase actions on human red cell membranes and the interfacial pressure in monolayers. Biochim. Biophys. Acta 406, 97–107.

Deshaies, R.J., Koch, B.D., Werner-Washburne, M., Craig, E.A., & Schekman, R. (1988). A subfamily of stress proteins facilitates translocation of secretory and mitochondrial precursor polypeptides. Nature 332, 800–805.

De Vrije, T., de Swart, R.L., Dowhan, W., Tommassen, J., & de Kruijff, B. (1988). Phosphatidylglycerol is involved in protein translocation across *Escherichia coli* inner membranes. Nature, 334, 173–175.

Dill, K.A. (1990). Dominant forces in protein folding. Biochemistry 29, 7133–7155.

Dreses-Werringloer, U., Fisher, K., Wachter, E., Link, T.A., & Flügge, U.I. (1991). cDNA sequence and deduced aminoacid sequence of the precursor of the 37-kDa inner envelope membrane polypeptide from spinach chloroplasts. Eur. J. Biochem. 195, 361–368.

Douce, R., Block, M.A., Dorne, A.J., & Joyard, J. (1984). The plastid envelope membranes: their structure, composition and role in chloroplast biogenesis. Subcell. Biochem. 10, 1–84.

Douce, R. & Joyard, J. (1990). Biochemistry and function of the platid envelope. Annu. Rev. Cell Biol. 6, 173–216.

Eilers, M. & Schatz, G. (1986). Binding of a specific ligand inhibits import of a purified precursor protein into mitochondria. Nature 322, 228–232.

Ellis, R.J. & Hemmingsen, S.M. (1989). Molecular chaperones: proteins essential for the biogenesis of some macromolecular structures. Trends Biochem. Sci. 14, 339–342.

Flügge, U.I., Fischer, K., Gross, A., Sebald, W., Lottspeich, F., & Eckershorn, C. (1989). The triose phosphate-3-phosphoglycerate-phosphate translocator from spinach chloroplasts: nucleotide sequence of a full length cDNA clone and import of the in vitro synthesized precursor protein into chloroplasts. EMBO J. 8, 39–46.

Flügge, U.I. (1990). On the translocation of proteins across the chloroplast envelope. J. Bioenerg. Biomem. 22, 769–787.

Gavel, Y. & von Heijne, G. (1990). A conserved cleavage-site motif in chloroplast transit peptides. FEBS Lett., 261, 455–548.

Gething, M.J. & Sambrook, J. (1992). Protein folding in the cell. Nature 355, 33–45.

Gierasch, L.M. (1989). Signal sequences. Biochemistry 28, 923–930.

Glick, B. & Schatz, G. (1991). Import of proteins into mitochondria. Annu. Rev. Genet. 25, 21–44.

Hageman, J., Baecke, C., Ebskamp, M., Pilon, R., Smeekens, S., & Weisbeek, P. (1990). Protein import into and sorting inside the chloroplast are independent processes. Plant Cell 2, 479–494.

Hartl, F-U. & Neupert, W. (1990). Protein sorting to mitochondria: evolutionary conservations of folding and assembly. Science 247, 930–938.

Horniak, L., Pilon, M., Van't Hof, R., & De Kruijff, B. (1993). The secondary structure of the ferredoxin transit sequence is modulated by its interaction with negatively charged lipids. FEBS Lett. 334, 241–246.

Huang, J., Hack, E., Thornburg, R.W., & Myers, A.M. (1990). A yeast mitochondrial leader peptide functions *in vivo* as a dual targeting signal for both chloroplasts and mitochondria. Plant Cell 2, 1249–1260.

Hurt, E.C., Soltanifar, N., Goldschmidt-Clermont, N., Rochaix, J.D., & Schatz, G. (1986). A cleavable presequence of an imported protein directs attached polypeptides into yeast mitochondria. EMBO J. 5, 1343–1350.

Jordi, W., Hergersberg, C., and de Kruijff, B. (1992). Bilayer penetrating properties enable apocytochrome c to follow a special import pathway into mitochondria. Eur. J. Biochem. 204, 841–846.

Karlin-Neuman, G.A. & Tobin, E.M. (1986). Transit peptides of nuclear-encoded chloroplast proteins share a common amino acid framework. EMBO J. 5, 9–13.

Keegstra, K. (1989). Transport and routing of proteins into chloroplasts. Cell 56, 247–253.

Kirk, J.T.O. & Tilney-Basset, R.A.E. (1978). The Plastids: Their Chemistry, Structure, Growth and Inheritance. Elsevier Biomedical Press, Amsterdam, pp. 787–872.

Kusters, R., Dowhan, W., & de Kruijff, B. (1991). Negatively charged phospholipids restore prePhoE translocation across phosphatidylglycerol-depleted *Escherichia coli* inner membranes. J. Biol. Chem. 266, 8659–8662.

Li, H., Theg, S.M., Bauerle, C.M., & Keegstra, K. (1990). Metal-ion-center assembly of ferredoxin and plastocyanin in isolated chloroplasts. Proc. Natl. Acad. Sci. USA 87, 6748–6752.

Li, H., Moore, T., & Keegstra, K. (1991). Targeting of proteins to the outer envelope membrane uses a different pathway than transport into chloroplasts. Plant Cell 3, 709–717.

Lubben, T.H., Donaldson, G.K., Viitanen, P., & Gatenby, A.A. (1989). Several proteins imported into chloroplasts form stable complexes with the GroEL-related chloroplast molecular chaperone. Plant Cell 1, 1223–1230.

Maduke, M. & Roise, D. (1993). Import of a mitochondrial presequence into protein-free phospholipid vesicles. Science 260, 364–367.

Margulis, L. (1970). Origin of Eukaryotic Cells. Yale University Press, New Haven, CT.

Martin, J., Mahlke, K., & Pfanner, N. (1991). Role of an energized inner membrane in mitochondrial protein import. J. Biol. Chem. 266, 18051–18057.

Miernyk, J.A., Duck, N.B., Shatter, R.G., & Folk, W.R. (1992). The 70-kilodalton heat-shock cognate can act as a molecular chaperone during the membrane translocation of a plant secretory protein precursor. Plant Cell 4, 821–829.

Meyer, D.I. (1988). Preprotein conformation: the year's major theme in translocation studies. Trends Biochem. Sci. 13, 471–474.

Miller, B. & Cumsky, M. (1991). An unusual mitochondrial import pathway for the precursor to yeast cytochrome c oxidase subunit Va. J. Cell. Biol. 112, 883–841.

Murakami, H., Pain, D., & Blobel, G. (1988). 70-kDa heat shock-related protein is one of at least two distinct cytosolic factors stimulating protein import into mitochondria. J. Cell Biol. 107, 2051–2057.

Oblong, J.E. & Lamppa, G.K. (1992). Identification of two structurally related proteins involved in proteolytic processing of precursors targeted to the chloroplast. EMBO J. 11, 4401–4409.

Olsen, L.J. & Keegstra, K. (1992). The binding of precursor proteins to chloroplasts requires nucleotide triphosphates in the intermembrane space. J. Biol. Chem. 267, 433–439.

Ostrem, J.A., Ramage, R.T., Bohnert, H.J., & Wasmann, C.C. (1989). Deletion of the carboxyl-terminal portion of the transit peptide affects processing but not import or assembly of the small subunit of ribulose-1,5-bisphosphate carboxylase. J. Biol. Chem. 264, 3662–3665.

Pfaller, R., Pfanner, N., & Neupert, W. (1989). Mitochondrial protein import: Bypass of proteinaceous surface receptors can occur with low specificity and efficiency. J. Biol. Chem. 264, 34–39.

Pilon, M., de Boer, A.D., Knols, S.L., Koppelman, M.H.G.M., van der Graaf, R.M., de Kruijff, B., & Weisbeek, P. (1990). Expression in Escherichia coli and purification of a translocation-competent precursor of the chloroplast protein ferredoxin. J. Biol. Chem. 265, 3358–3361.

Pilon, M., de Kruijff, B., & Weisbeek, P.J. (1992a). New insights into the import mechanism of the feredoxin precursor into chloroplasts. J. Biol. Chem. 267, 2548–2556.

Pilon, M., Rietveld, A.G., Weisbeek, P.J., & de Kruijff, B. (1992b). Secondary structure and folding of a functional chloroplast precursor protein. J. Biol. Chem. 267, 19907–19913.

Pilon, M., Weisbeek, P.J., & de Kruijff, B. (1992c). Kinetic analysis of translocation into isolated chloroplasts of the purified ferredoxin precursor. FEBS Lett. 302, 65–68.

Pugsley, A.P. (1989). Protein Targeting. Academic Press, San Diego, CA.

Randall, L.L. & Hardy, S.J.S. (1986). Correlation of competence for export with lack of tertiary structure of the mature species: a study *in vivo* of maltose-binding protein in *E. coli*. Cell 46, 921–928.

Raven, P.H. (1970). A multiple origin for plastids and mitochondria. Science 169, 641–646.

Reiss, B., Wasmann, C.C., & Bohnert, H.J. (1987). Regions in the transit peptide of SSU essential for transport into chloroplasts. Mol. Gen. Genet. 209, 116–121.

Rietveld, A. & de Kruijff, B. (1984). Is the mitochondrial precursor protein apocytochrome c able to pass a lipid bilayer? J. Biol. Chem. 259, 2548–2556.

Roise, D. & Schatz, G. (1988). Mitochondrial presequences. J. Biol. Chem. 263, 4509–4511.

Salomon, M., Fischer, K., Flügge, U.I., & Soll, J. (1990). Sequence analysis and protein import studies of an outer chloroplast envelope polypeptide. Proc. Natl. Acad. Sci. USA 87, 5778–5782.

Sanders, S. & Schekman, R. (1992). Polypeptide translocation across the endoplasmic reticulum membrane. J. Biol. Chem. 267, 13791–13794.

Smeekens, S., de Groot, M., van Binsbergen, J., & Weisbeek, P. (1985). Sequence of the precursor of the chloroplast thylakoid lumen protein plastocyanin. Nature 317, 456–458.

Smeekens, S., van Steeg, H., Baurle, C., Bettenbroek, H., Keegstra, K., & Weisbeek, P. (1987). Import into chloroplasts of a yeast mitochondrial protein directed by ferredoxin and plastocyanin transit peptides. Plant. Mol. Biol. 9, 377–388.

Smeekens, S., Bauerle, C., Hageman, J., Keegstra, K., & Weisbeek, P. (1986). The role of the transit peptide in the routing of precursors toward different chloroplast compartments. Cell 46, 365–375.

Smeekens, S., Geerts, D., Bauerle, C., & Weisbeek, P. (1989). Essential function in chloroplast binding and import of the ferredoxin transit peptide processing region. Mol. Gen. Genet. 216, 178–182.

Smeekens, S., Weisbeek, P., & Robinson, C. (1990). Protein transport into and within chloroplasts. Trends Biochem. Sci. 15, 73–76.

Stuart, R.A. & Neupert, W. (1990). Apocytochrome c: an exceptional mitochondrial precursor protein using an exceptional import pathway. Biochimie 72, 115–121.

Suzuki, S., Izumihara, K., & Hase, T. (1991). Plastid import and iron-sulfur cluster assembly of photosynthetic and non-photosynthetic ferredoxin isoproteins in maize. Plant Physiol. 97, 375–380.

Takahashi, Y., Mitsui, A., Hase, T., & Matsubara, H. (1986). Formation of the iron-sulfur cluster of ferredoxin in isolated chloroplasts. Proc. Natl. Acad. Sci. USA 83, 2434–2437.

Theg, S.M., Bauerle, C., Olsen, L.J., Selman, B.R., & Keegstra, K. (1989). Internal ATP is the only energy requirement for the translocation of precursor proteins across chloroplastic membranes. J. Biol. Chem. 264, 6730–6736.

van den Broeck, G., Timko, M.P., Kausch, A.P., Cashmore, A.R., Van Montagu, M., & Herrera-Estrella, L. (1985). Targeting of a foreign protein to chloroplasts by fusion to the transit peptide from the small subunit of ribulose-1,5-bisphosphate carboxylase. Nature 313, 358–363.

van 't Hof, R., Demel, R.A., Keegstra, K., & de Kruijff, B. (1991). Lipid-peptide interactions between fragments of the transit peptide of ribulose-1,5-bisphosphate carboxylase/oxygenase and chloroplast membrane lipids. Febs Lett. 291, 350–354.

van 't Hof, R., van Klompenburg, W., Pilon, M., Kozubec, A., de Korte-Kool, G., Demel, R.A., Weisbeek, P., & de Kruijff, B. (1993). The transit sequence mediates the specific interaction of the precursor of ferredoxin with chloroplast envelope membrane lipids. J. Biol. Chem. 268, 4037–4042.

von Heijne, G., Steppuhn, J., & Herrmann, R.G. (1989). Domain structure of mitochondrial and chloroplast targeting peptides. Eur. J. Biochem. 180, 535–545.

von Heijne, G. & Nishikawa, K. (1991). Chloroplast transit peptides. The perfect random coil? FEBS Lett. 278, 1–3.

Waegemann, K., Paulsen, H., & Soll, J. (1990). Translocation of proteins into isolated chloroplasts requires cytosolic factors to obtain import competence. FEBS Lett. 261, 89–92.

Whelan, J., Dolan, L., & Harmey, M.A. (1988). Import of precursor proteins into vicia faba mitochondria. Febs Lett. 236, 217–220.

Whelan, J., Knorpp, C., & Glazer, E. (1990). Sorting of precursor proteins between isolated spinach leaf mitochondria and chloroplasts. Plant Mol. Biol. 14, 977–982.

INTERNALIZATION OF PSEUDOMONAS EXOTOXIN A UTILIZES THE α₂MACROGLOBULIN RECEPTOR/LOW DENSITY LIPOPROTEIN RECEPTOR RELATED PROTEIN

Randal E. Morris and Catharine B. Saelinger

Membrane Protein Transport
Volume 1, pages 257–274.
Copyright © 1995 by JAI Press Inc.
All rights of reproduction in any form reserved.
ISBN: 1-55938-907-9

ABSTRACT

Pseudomonas exotoxin A (PE) is a three-domain bacterial toxin which enters sensitive mammalian cells by receptor-mediated endocytosis. Inside the cell, PE is converted to an enzyme active form, enters the cytosol, possibly from the Golgi, and inactivates elongation factor 2. The first step in this process is binding of PE to a receptor on the cell surface. We have isolated a high molecular weight glycoprotein from mouse LM cells and from mouse liver which has the properties of a PE receptor. We have evidence that this receptor is the same as the $\alpha_2$macroglobulin receptor/low-density lipoprotein receptor related protein (α_2MR/LRP). The α_2MR/LRP is a large glycoprotein (600 kDa) which functions to clear activated $\alpha_2$macroglobulin, apoE-enriched β-VLDL, and a variety of other ligands from the blood. We have shown that the PE receptor and the α_2MR/LRP have identical molecular weights when run on a polyacrylamide gel, bind PE specifically, and are immunologically indistinguishable. Proteolytic digestion of both receptors generates similar toxin binding fragments. In addition, a 39-kDa receptor associated protein (RAP), which blocks ligand binding to the α_2MR/LRP, also blocks binding of PE to and its subsequent toxicity for LM cells. Interestingly, several cell lines which are highly resistant to PE have low levels of PE receptor, but express high levels of RAP. RAP is found in very low levels in toxin sensitive cells. We hypothesize that the α_2MR/LRP is involved in receptor-mediated endocytosis of PE and that RAP plays a role in the regulation of ligand binding to cells.

I. INTRODUCTION: PSEUDOMONAS EXOTOXIN A

Mammalian cells are susceptible to a myriad of microbial products that have varying tissue and cell specificities. Some of these products are toxic and usurp existing cellular machinery to enter and destroy a target cell. Pseudomonas exotoxin A (PE) is one of several virulence factors produced by *Pseudomonas aeruginosa*. It is a three-domain bacterial toxin that kills mammalian cells by gaining entry to the cytosol and inhibiting protein synthesis (Iglewski and Kabat, 1975; Allured et al., 1986). This toxin binds to a surface receptor on susceptible cells, is internalized via coated pits and endosomes, and eventually enters the cytosol (FitzGerald et al., 1980). During its journey the toxin is cleaved to generate an enzymatically active fragment which ADP-ribosylates cytoplasmic elongation factor 2 (Leppla et al., 1978; Morris, 1990; Ogata et al., 1990; Fryling et al., 1992).

Crystallographic and genetic studies have shown that PE is composed of three distinct structural domains each with a unique function (Allured et al., 1986; Hwang et al., 1987; Siegall et al., 1989; Wick et al., 1990). Domain Ia (amino-terminus) contains sequences which bind to the cell surface receptor for PE, Domain II is involved in processing and translocation into the cytosol, and Domain III (carboxyl-terminus) contains the catalytic subunit responsible for ADP-ribosylation of elongation factor 2, which halts protein synthesis (reviewed in Wick et al., 1990; Guidi-Rontani and Collier, 1987; Hwang et al., 1987; Siegall et al., 1989; Chaudhary et al., 1990a,b; Seetharam et al., 1991). Native toxin is a proenzyme and must be activated to exhibit enzyme activity in cell free extracts. *In vitro* pretreatment with urea and dithiothreitol activates the toxin by unfolding the protein (Leppla et al., 1978; Lory and Collier, 1980; Farahbakhsh et al., 1986). In a sensitive cell it is believed that the toxin is internalized, and proteolytically cleaved and reduced in endosomes; this generates 37- and 28-kDa fragments. The 28-kDa fragment contains the binding domain of the toxin, and the 37-kDa fragment contains the ADP-ribosyltransferase (Ogata et al., 1990; Fryling et al., 1992). The site of escape of the enzyme active domain into the cytosol has not been fully resolved.

Mammalian cells exhibit a spectrum of susceptibility to pseudomonas exotoxin. This is most evident in an animal model following infection with viable *Pseudomonas aeruginosa* or injection of purified toxin. In this situation, liver is the primary target of the toxin, with protein synthesis and functional elongation factor 2 depleted by over 90%. Other organs are variably affected (Pavlovskis and Shackelford, 1974; Iglewski et al., 1977; Saelinger et al., 1977; Snell et al., 1978). A spectrum of sensitivity to toxin action is also displayed by established tissue culture cell lines. The mouse LM fibroblast is an example of a cell line which is exquisitely sensitive to PE (Middlebrook and Dorland, 1977; Michael and Saelinger, 1979). In contrast, Ovcar cells, a human ovarian carcinoma cell line, exhibit three logs greater resistance to the toxin.

We have shown that PE enters LM fibroblasts by receptor mediated endocytosis (FitzGerald et al., 1980; Morris and Saelinger, 1986; Morris, 1990). The first step in this process is binding to a specific receptor on the cell surface (Manhart et al., 1984). Several lines of evidence suggest that a receptor for PE exists, and would be present in reasonable numbers on LM and liver cells. First, LM cells have high-affinity PE binding sites (Manhart et al., 1984). Second, treatment of LM cells with trypsin transiently ablates toxin sensitivity. Third, PE is routed into LM cells via clathrin coated pits, structures typically associated with receptor-mediated endocytosis (FitzGerald et al., 1980). Fourth, PE, which is produced locally in small amounts in a murine burn infection, stops protein synthesis in liver; recognition by a receptor would facilitate internalization at this distant site (Iglewski et al., 1977; Saelinger et al., 1977; Snell et al., 1978). We have isolated and partially characterized a toxin binding protein which functions as the cell surface receptor for this process (Forristal et al., 1991; Thompson et al., 1991). This PE receptor is immunologically and functionally similar to $\alpha_2$macroglobulin receptor/low-density lipoprotein receptor related protein (α_2MR/LRP; Kounnas et al., 1992b). The following discussion will focus on this receptor and its characterization.

II. CHARACTERIZATION OF PSEUDOMONAS EXOTOXIN A RECEPTOR

Because of their high level of PE sensitivity, our initial studies focused on LM cells and mouse liver. Receptor for PE is readily purified from detergent extracts of LM cells (Thompson et al., 1991) or mouse liver (Forrsital et al., 1991) by PE Sepharose affinity chromatography. The receptor from either source is a high molecular weight glycoprotein (greater than 500 kDa) which is not reduced by treatment with reducing agent. Binding of PE to both liver and LM cell receptor is specific. Increasing concentrations of native PE progressively inhibit binding of biotinyl-PE to purified receptor; however, neither diphtheria toxin nor cholera toxin displaces biotinyl-PE (Thompson et al., 1991). In addition, PEglu57, which is altered in the binding domain and has diminished cytotoxicity (Jinno et al., 1988), does not block binding of biotinyl-PE to LM cell receptor in a toxin blot assay.

Although 10% of the molecular weight of the LM cell receptor is carbohydrate, there are several lines of evidence to show that glycosylation is not essential for receptor function. First, toxin-induced inhibition of protein synthesis in cells treated with tunicamycin (^{3}H-mannose incorporation reduced by approximately 85–90%) and control cells is identical (Figure 1). Receptor which has been affinity-purified

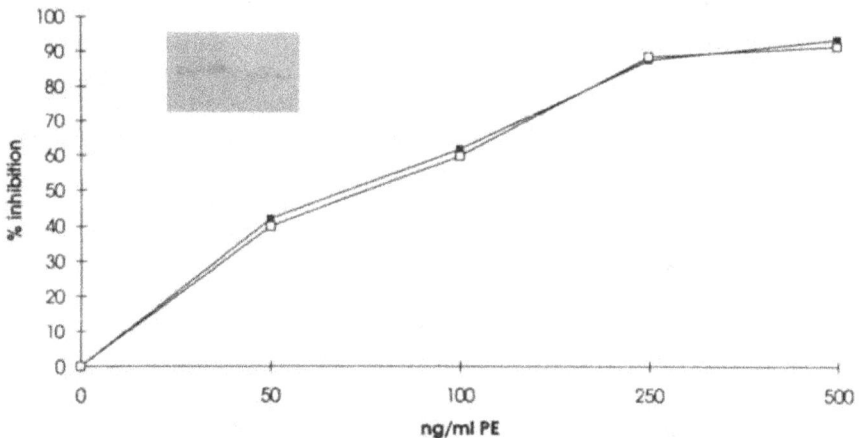

Figure 1. Tunicamycin treatment does not alter toxicity of PE for LM cells. Mouse LM fibroblasts are incubated with tunicamycin (150 ng/ml) for four days prior to initiation of experiment. At this time, tritiated mannose incorporation is reduced by approximately 90%. Cells are cooled to 4 °C, incubated with various concentrations of PE for 1 hr, washed and reincubated at 37 °C to allow expression of toxicity. Tritiated leucine is added to measure protein synthesis (FitzGerald et al., 1980). ■ no tunicamycin; □ with tunicamycin. *Insert*: Affinity purified receptor from tunicamycin treated (*right lane*) and control cells (*left lane*), subjected to SDS-PAGE, transferred to nitrocellulose paper, and probed for ability to bind PE, as described in Thompson et al., 1991.

Table 1. Reactivity of Purified Receptor with Digoxigen Labeled Lectins

Source	Lectin	GNA	SNA	MAA	DSA	PNA
Liver		+	+++	+	+	–
Kidney		–	–	–	+++	+++
LM cell		+	+++	+++	+++	–
Chang		+	+/–	+/–	+/–	–
OVCAR		+/–	+++	+/–	+/–	–

Note: Affinity purified receptors were electrophoresed on a 6% SDS gel and transferred to nitrocellulose paper. The blot was probed with the digoxigenin-labeld lectins indicated. GNA, *Galanthus nivalis* agglutinin; SNA, *Sambucus nigra* agglutinin; MAA, *Maackia amurensis* agglutinin; DSA, *Datura stramonium* agglutinin; PNA, peanut agglutinin.

from tunicamycin treated cells has a slightly lower molecular weight than receptor from cells not treated with tunicamycin, but binds PE normally when assessed by ELISA (data not shown) or ligand blot assay (Figure 1, insert). Second, treatment of purified LM or liver receptor with PNGase F may reduce, but does not ablate PE binding when assessed by ELISA or ligand blot assay (Thompson et al., 1991). In addition, treatment of affinity-purified receptor with endoH has no effect on ability to bind toxin (Thompson et al., 1991). Taken together the data suggest that deglycosylated receptor is able to function effectively in the internalization of pseudomonas exotoxin.

We have purified receptor from a variety of tissue culture cell lines and mouse tissues. In all cases, the affinity-purified receptors have a molecular weight and binding characteristics similar to receptor from LM cells or liver. The sole differences among receptors from different sources is their glycosylation patterns (see Table 1). The significance of these differences is unknown at the present time. It is not known if carbohydrate can modify routing of receptor or affinity of receptor for ligand.

III. CELLULAR LOCALIZATION OF RECEPTOR

To have a role in the intoxication process, at least some receptor must be localized on the cell surface. Treatment of LM cells with trypsin (37 C) renders them transiently resistant to PE; sensitivity is restored after reincubation in tissue culture medium (Figure 2). These results would suggest that a proteinaceous surface-located material is required for toxicity. To further localize receptor, LM cells were homogenized and subjected to sucrose density gradient fractionation; fractions were analyzed by SDS-PAGE, and receptor was identified by toxin binding activity. While a small amount of receptor was found in fractions co-migrating with plasma membrane markers, the majority was seen associated with galactosyltransferase, a Golgi marker enzyme. Our results suggest that there is a large intracellular pool of

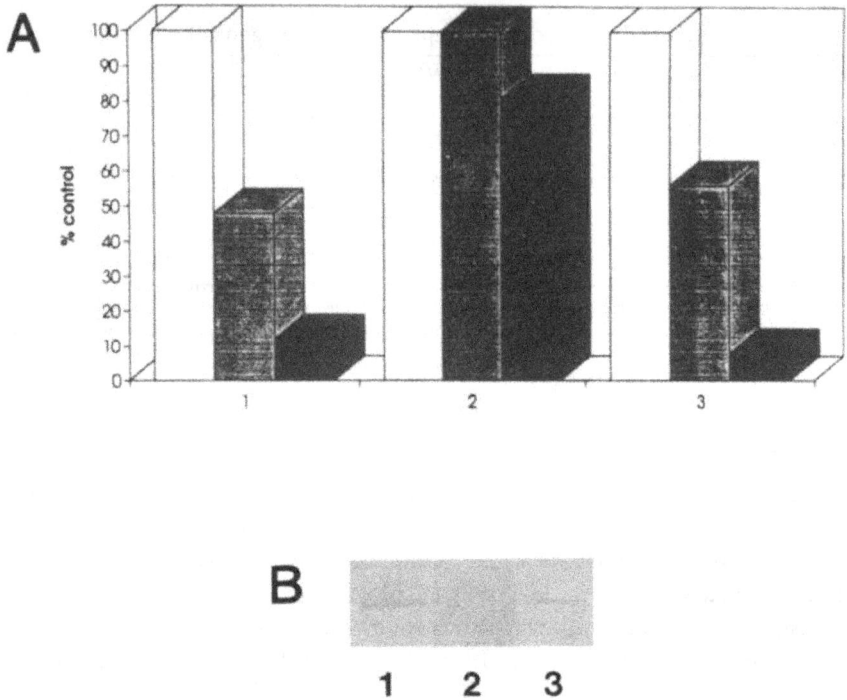

Figure 2. Presence of proteinaceous material is essential for toxicity. LM cell mono-layers are treated with 2% trypsin—10 mM EDTA for 20 min at 37 °C; cells are collected, incubated with soy bean trypsin inhibitor, and washed. Cells are replated in fresh tissue culture medium with 10% fetal calf serum for 0 hr (Expt 2) or 5 hr (Expt 3), prior to addition of toxin. Toxin biological activity is determined as described in Figure 1. (Expt 1), no trypsin treatment prior to toxin addition. Open bars = no PE, stippled bars = 10 ng PE, black bars = 25 ng PE. Insert: SDS-PAGE followed by toxin blot assay. Lane 1, no trypsin treatment; Lane 2, cells processed immediately after trypsin treatment; Lane 3, cells reincubated for 5 hr at 37 °C prior to collection.

receptor which could recycle rapidly between the cell interior and surface. How-ever, at any time a small percentage of the receptor is present on the cell surface.

IV. IDENTIFICATION OF FUNCTIONAL RECEPTOR

A question that has plagued researchers is what is the receptor that a toxin usurps to enter a cell and exert a toxic effect. We believe that we have answered that question for pseudomonas exotoxin. All evidence suggests that PE uses the α₂macroglobulin receptor/low-density lipoprotein receptor-related protein (α₂MR/LRP; Kounnas et al., 1992b) to enter mammalian cells. α₂MR/LRP belongs to a group of receptors which are internalized constitutively, i.e. ligand binding is not required for entry. This may be an advantage for toxin in that it need not trigger an "internalization button" to be taken into the cell.

α₂MR/LRP is a multifunctional receptor. It is responsible for the rapid hepatic clearance of α₂macroglobulin–proteinase complexes from blood, followed by delivery of protease to lysosomes for degradation (Ashcom et al., 1990; Moestrup and Gliemann, 1991). This receptor also has been proposed to mediate hepatic clearance of chylomicron remnants. This hypothesis is supported by *in vitro* studies showing the interaction of receptor with apoE and with apoE-enriched β-VLDL (Kowal et al., 1989, 1990; Lund et al., 1989). Amino acid sequencing studies of peptides derived from chemical and proteolytic cleavage of the α₂macroglobulin receptor and the LRP demonstrate that these two proteins are identical (Herz et al., 1988; Kristensen et al., 1990; Strickland et al., 1990). This multifunctional receptor also binds several other ligands including lactoferrin, plasminogen activator–inhibitor complexes, and lipoprotein lipase; because of the numerous ligands for α₂MR/LRP, it can be considered to be a true scavenger receptor (Chappell et al., 1992; Herz et al., 1992; Huettinger et al., 1992; Orth et al., 1992; Willnow et al., 1992; Choi et al., 1993; Nykjaer et al., 1993).

The receptor is synthesized as a 600-kDa single-chain precursor molecule that is cleaved in the *trans*-Golgi to a 515-kDa polypeptide and an 85-kDa polypeptide (Herz et al., 1990). The 515-kDa protein contains the ligand binding domain, and the 85-kDa protein contains a transmembrane domain and cytoplasmic tail. These two polypeptides remain noncovalently, yet tightly associated on the cell surface and during purification. In addition, a 39-kDa polypeptide, called receptor associated protein (RAP) copurifies with α₂MR/LRP (Herz et al., 1991; Strickland et al., 1991; Kounnas et al., 1992a). This receptor associated protein is able to block the interaction of ligands with receptor. We will discuss possible roles for RAP in the regulation of ligand binding later.

The α₂MR/LRP is closely related to the low-density lipoprotein (LDL) receptor. It is much larger than the LDL receptor. Like the LDL receptor, the 515-kDa peptide contains complement-like repeats and growth factor repeats (Willnow et al., 1992). In addition, there are two NPXY consensus sequences in the cytoplasmic tail which may signal the internalization of the receptor via clathrin-coated pits. Unlike the LDL receptor, α₂MR/LRP does not bind LDL (Kowal et al., 1989, 1990).

There are several pieces of evidence that suggest PE receptor affinity purified from LM cells or mouse liver is similar if not identical to α₂MR/LRP purified from human placenta (Kounnas et al., 1992b). The PE receptor and the heavy chain (515 kDa) of α₂MR/LRP have identical mobility on SDS-PAGE. Antibodies against purified α₂MR/LRP react with PE receptor from both sources. Similarly, antibodies

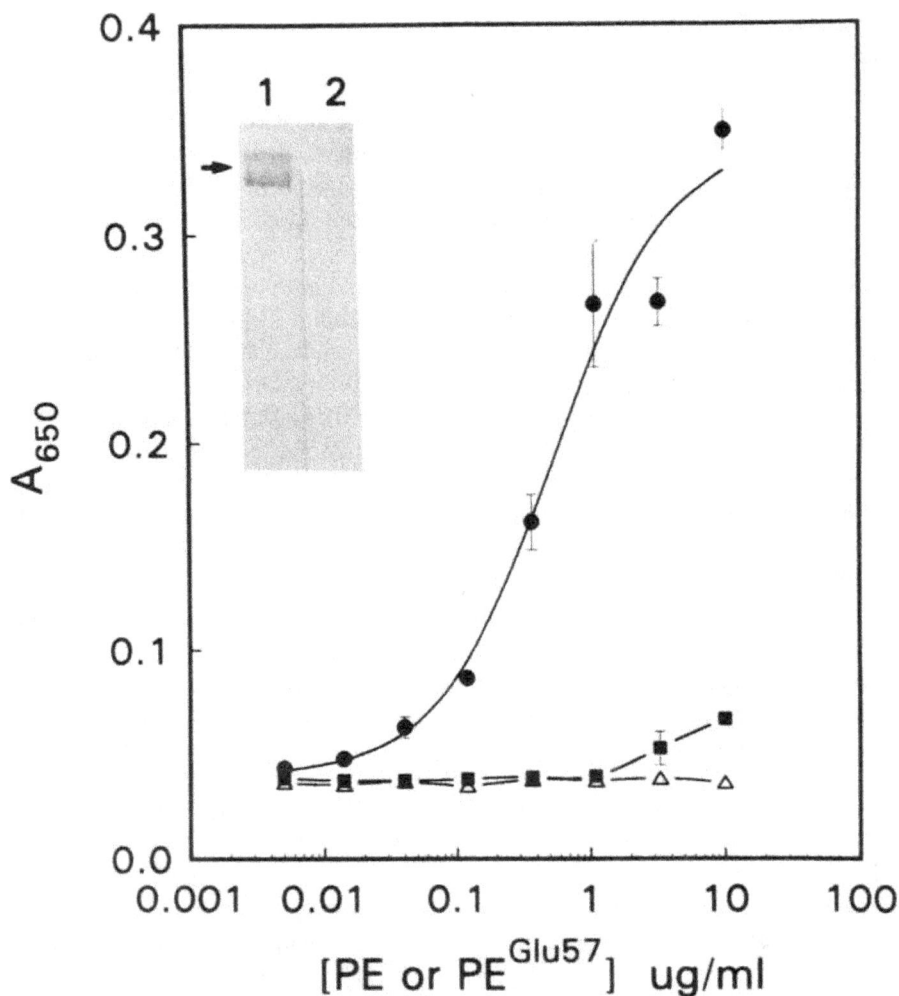

Figure 3. The α₂MR/LRP binds PE in a dose dependent manner. Microtiter plates were coated with affinity purified human placental α₂MR/LRP, wells were washed and increasing concentrations of PE (●) or PEglu57 (■) added. Following incubation and washing the amount of toxin bound to receptor was determined by sequential incubations with goat antitoxin and anti-goat IgG-HRP conjugate as described in Kounnas et al., 1992b. (△) = binding of PE and PEglu 57 to BSA-coated wells. *Insert*: Ligand blotting of purified α₂MR/LRP with PE (lane **1**) or PEglu57 (lane **2**). From Kounnas et al., J. Biol. Chem. 267, 12422. Used with permission.

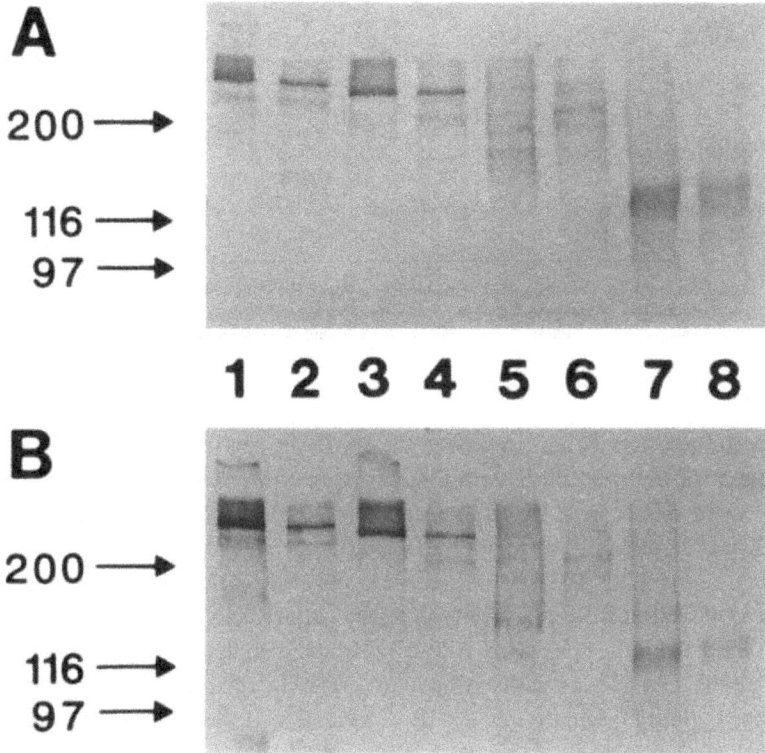

Figure 4. Protease digestion of the PE receptor and the α_2MR/LRP yields similar toxin binding fragments. Affinity purified PE receptor or α_2MR/LRP were incubated with PNGase F for 60 min at 37 °C to remove carbohydrate, then digested with staph V 8 protease (endoproteinase glu-C) or endoproteinase lys-C for 60 min at 25 °C or 37 °C respectively. Samples were run on SDS-PAGE, transferred to nitrocellulose paper, and bands identified by toxin blot assay (Panel **A**) or by incubation with antibody to α_2MR/LRP (Panel **B**). Lanes **1**, **3**, **5**, and **7** are α_2MR/LRP. Lanes **2**, **4**, **6**, and **8** are LM cell PE receptor. Lanes **1** and **2**, receptors received no treatment. Lanes **3** and **4**, receptors were incubated with PNGase F alone. Lanes **5** and **6**, receptors were incubated with PNGase F and Staph V8 protease, and Lanes **7** and **8**, receptors incubated with PNGase F and lys C.

against the PE receptor recognize human α_2MR/LRP. In addition both receptor preparations contain the 85-kDa light chain. Thus these two receptors are immunologically related. Lastly, PE binds to α_2MR/LRP in a dose-dependent manner (Figure 3). In contrast, PEglu57, a nonbinding form of toxin, does not bind to this receptor.

To further confirm the coidentity of the PE receptor and α_2MR/LRP, we digested both receptors with the endoproteinase Glu-C (Staph V8 protease) or with the

Table 2. Effect of Ligands on Toxin Binding to Intact LM Cells in Culture

Ligand (μg/ml)	% Control
None	100
apo E (20 μg)	82–95
Lactoferrin (3 mg)	75
α₂Macroglobulin* (10 μg)	65–92
Lipoprotein lipase (10 μg)	92

Note: LM cell monolyers were incubated with competing ligand for 18 hr at 4 °C, PE (1 μg/ml) was added and
incubation in the cold continued for 5 hr. Cells were collected and processed as in Figure 5. Results expressed
as percentage of PE bound in presence of competing ligand as compared to cells incubated with PE alone.

endoproteinase Lys C and compared the resulting peptide fragments (Figure 4).
Treatment of both receptors with Glu-C resulted in generation of several bands
which were able to bind PE and to react with antibody to α₂MR/LRP. Digestion
with Lys C resulted in one major toxin binding fragment (~120 kDa). These data
support the hypothesis that the PE receptor and α₂MR/LRP are the same protein.

Detergent extracts from a variety of tissue culture cells and mouse tissues have
been screened for the presence of a high molecular weight glycoprotein which binds
toxin and is able to react with antibody to α₂MR. In all situations in which the PE
receptor is present, one detects α₂MR. Conversely, if there is no PE receptor, no
α₂MR is detected.

V. MULTIPLE LIGAND BINDING SITES

The α₂MR/LRP is a scavenger receptor, capable of binding a variety of ligands.
Available evidence suggests that α₂MR/LRP has several independent ligand binding sites, which do not cross compete (Nykjaer et al., 1992; Willow et al., 1992),
e.g. plasminogen activator–inhibitor type 1-plasminogen activator complexes and
α₂M* do not cross-compete. Moestrup et al. (1993) have recently localized the
binding of these two ligands and the receptor-associated protein to a specific
624-residue region on the receptor. To further address this question, we examined
the ability of several ligands to block the binding of pseudomonas exotoxin to intact
LM cells. As shown in Table 2, none of the ligands tested consistently reduced
binding of PE to the surface of LM cells by more than 20–25%. Based on this
preliminary information, it is suggested that PE does not share binding sites with
other ligands internalized by the α₂MR/LRP. This is in agreement with recent
studies by Willnow et al. (1992) which suggest that the multiple ligands of
α₂MR/LRP each bind to different sites on the receptor.

VI. RECEPTOR ASSOCIATED PROTEIN ALTERS PE BINDING TO RECEPTOR

As already discussed, a 39-kDa receptor-associated protein copurifies with α_2MR/LRP (Herz et al., 1991; Moestrup and Gliemann, 1991; Williams et al., 1992). At present, the physiological role of RAP is not known. However, a RAP-glutathione S-transferase fusion protein (RAP-GST) blocks binding of activated α_2M* and apo-E enriched β-VLDL to α_2MR/LRP (Moestrup and Gilemann, 1989; Herz et al., 1991; Williams et al., 1992). In fact, the binding of all ligands shown to interact with α_2MR/LRP is blocked by RAP. It is not known if this is due to the fact that the binding of RAP to receptor induces a conformational change, which occludes the binding sites, or if the 39-kDa protein binds to all ligand binding sites directly. Support for the first model (indirect) comes from the data of Williams et al. (1992), which suggest that one single α_2MR/LRP binds only two molecules of RAP. Alternatively, if RAP directly blocks binding of ligands, one would predict that there is a high multiplicity of RAP molecules which bind to receptor (Willnow et al., 1992).

Because of the regulatory effect of RAP in other ligand systems, we asked if it was able to interfere with the interaction of PE with mammalian cells. Preincubation of the fusion protein RAP-GST with LM cells blocks binding of PE in a dose-dependent manner (Figure 5). Binding is reduced to 50% at a RAP-GST concentration

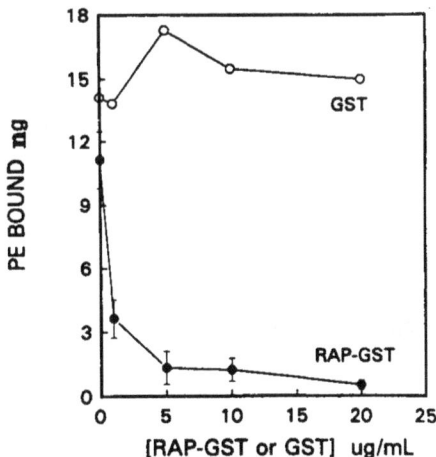

Figure 5. RAP blocks binding of PE to LM cells. LM cell monolayers were incubated with tissue culture medium containing RAP-GST (●) or GST (○) for 18 hours at 4 °C; 2 µg toxin was added for an additional 5 hours at 4 °C. The cells were washed, harvested, homogenized and the concentration of cell associated toxin assayed by ELISA. From Kounnas et al., J. Biol. Chem. 267, 12422. Used with permission.

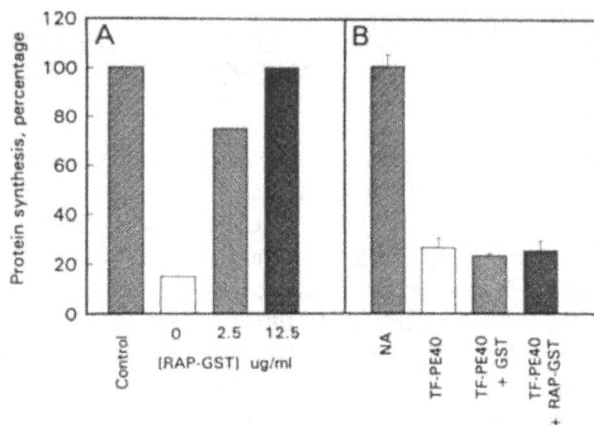

Figure 6. RAP protects LM cells from PE induced toxicity. LM cell monolayers were incubated with RAP-GST for 90 min at 37 °C. Cells were cooled and 40 ng PE or 500 ng transferrin-PE40 added. Incubation was continued for 45 min at 4 °C, cells were washed, reincubated for 4 hr at 37 °C in tissue culture fluid, and protein synthesis assayed. Control has no addition of toxin or RAP-GST. From Kounnas et al., J. Biol. Chem. 267, 12422. Used with permission.

of approximately 14 nM, a value virtually identical to the K_d of 4–20 nM measured for the interaction of RAP with purified α_2MR/LRP (Williams et al., 1992). GST alone does not alter PE binding. RAP also inhibits the biological activity of toxin (Figure 6). Preincubation of LM cells with RAP-GST completely abolishes PE toxicity. In control experiments, RAP-GST has no effect on the toxicity of a chimeric toxin, transferrin-PE40, which is internalized via the transferrin receptor. In data not shown here, RAP reduces but does not ablate the binding of PE to affinity-purified LM cell or liver PE receptor.

These results demonstrate that human placental α_2MR/LRP and murine PE receptor are similar polypeptides, are immunologically related, and are functionally identical. That RAP abolishes both toxin binding and toxicity strongly suggests that α_2MR/LRP is responsible for toxin internalization. Thus we believe that α_2MR/LRP is the functional receptor for PE in mouse liver and mouse LM fibroblasts.

VII. BASIS FOR CELLULAR SENSITIVITY TO TOXIN

Cells exhibit differences in their susceptibility to PE, as indicated in Table 3 (Middlebrook and Dorland, 1977; Vasil and Iglewski, 1978; Michael and Saelinger, 1979). Mouse LM and Swiss 3T3 cells are highly sensitive to PE, while Ovcar (human ovarian carcinoma) and DU145 (prostate carcinoma) cells are highly

Table 3. Differential Sensitivity of Cells to PE

	TCD$_{50}$ (ng/ml)	
Cell Line	18 hr[*]	5 hr[#]
LM	0.01	11
Swiss 3T3	0.90	375
Chang	3.5	~1000
Vero	11	~1000
Hela	32	ND
Ovcar	50	>1000
DU145	~35	>1000

Notes: [*]Cell monolayers were incubated for 18 hr with various concentrations of PE, toxin was removed, and protein synthesis measured by a 60 min pulse with tritiated leucine.
[#]Cell monolayers cooled to 4 °C, PE added for 60 min (4 °C). Monolayers were washed, and reincubated in medium for 5 hr prior to measurement of protein synthesis.

resistant. Short-term incubation of resistant cells with toxin at 4 °C does not afford sufficient binding of PE to inhibit protein synthesis when cells are brought to 37 °C. While some inhibition of protein synthesis is observed when resistant cells are continuously incubated with toxin for long time periods at 37 °C, the levels of toxin required are high, and it is presumed that toxin is being internalized by an inefficient route, perhaps by fluid-phase pinocytosis. We have shown previously that a similar toxin, diphtheria toxin, is internalized by resistant cells via a non-receptor-mediated route, and suggest that a similar process occurs for PE in resistant cells (Morris and Saelinger, 1983).

Toxin resistance could be explained by an inability of elongation factor 2 to be ADP-ribosylated by PE, to inefficient processing of toxin, to the absence of toxin receptor, or to the presence of a nonfunctional receptor. We have ruled out the first hypothesis by showing that PE is able to ADP-ribosylate elongation factor 2 from toxin resistant cells in an *in vitro* assay system. We are presently testing the last two explanations.

First presence/absence of receptor was correlated with toxin sensitivity/resistance. All cell lines and tissues which exhibit sensitivity to PE have receptor as detected by toxin blot assay or by reactivity with antibody to α_2MR/LRP. (Representative cell lines are shown in Figure 7.) In contrast, Hela cells which are resistant to PE, have no detectable receptor. Thus far, the data fit the hypothesis. However three cell lines (Ovcar, DU145, and LnCap) have been examined which are highly resistant to the biologic action of PE but which have moderate levels of PE receptor. Thus another explanation for resistance was needed, and we began to look at the contribution of RAP to toxin binding.

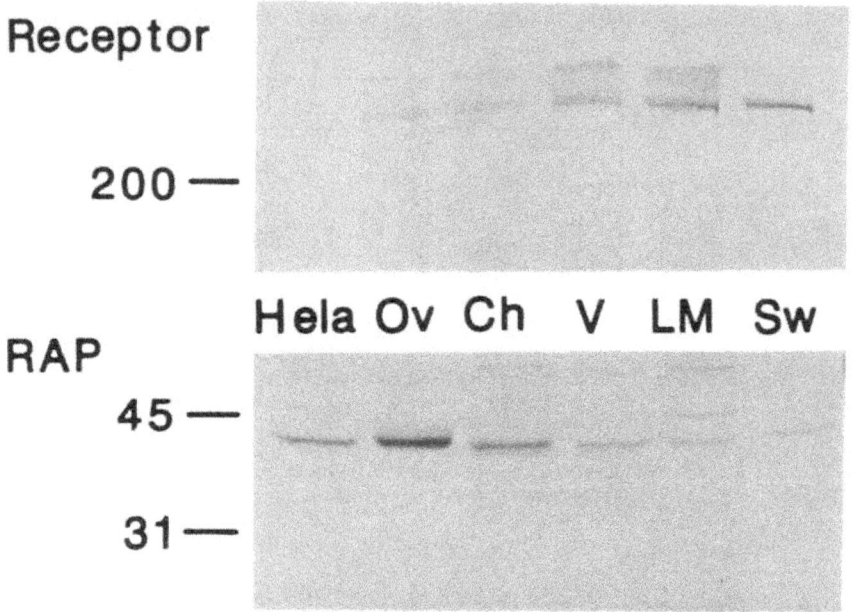

Figure 7. Presence of α₂MR/LRP and RAP in several cell lines. Detergent extracts of a variety of cells (Hela, Ovcar, Chang, Vero, mouse LM fibroblasts, and NIH Swiss 3T3) were subjected to SDS-PAGE, transferred to nitrocellulose and probed with antibody to α₂MR/LRP (*top panel*) or to RAP (*bottom panel*). ~25 μg protein applied in each lane.

One possible role for RAP is to regulate ligand binding to receptor. Therefore it would be expected that toxin-sensitive cells would have low levels of RAP, while resistant cells could have high levels of this protein. When detergent extracts of a variety of cells and tissues were examined, all sensitive cell lines had minimal RAP. The three highly resistant cell lines which have PE receptor have abundant levels of RAP (for representative data, see Figure 7).

In order to block the first step in the receptor-mediated internalization process, and thus render a cell resistant to toxin, RAP and receptor must be associated on the cell surface. We have begun to examine the colocalization of these two proteins. In initial experiments detergent extracts of Ovcar or DU 145 cells are subjected to sucrose density gradient fractionation, and the resulting fractions assayed by SDS-PAGE followed by transfer to nitrocellulose paper and probing with antibody to RAP or to receptor. RAP and receptor comigrate in a fraction which also contains plasma membrane marker enzymes. While this does not prove that receptor and RAP are colocalized on the surface of resistant cells, it does support our hypothesis that one explanation of cellular resistance to PE is the presence of sufficient surface-associated RAP to block—either directly or indirectly—the binding of ligand. A second explanation for resistance, as seen in Hela cells, is the virtual absence of receptor, and thus an inability to bind toxin and internalize it by the normal receptor mediated pathway.

If indeed the amount of RAP relative to receptor concentration is important in determining receptor activity, then it would be predicted that the amount of RAP in liver would be low since this tissue has high levels of receptor activity involved in clearing ligands from the circulation. In addition, liver is the primary target of toxin in an experimental burned-mouse model (Saelinger et al., 1977; Snell et al., 1978). As already described, receptor is readily purified from mouse liver, and is identical to receptor from other sources except for glycosylation pattern. When detergent extracts of liver were screened for RAP, only low levels were detected. Other tissues have been examined for RAP, and it is present in moderate amounts. The balance between receptor and RAP might explain the variable toxin sensitivity of organs other than liver described in various publications.

VIII. CONCLUSION

In conclusion, pseudomonas exotoxin A appears to have usurped the α_2MR/LRP to find a way to enter mammalian cells. This large receptor is able to internalize a variety of ligands, some of which follow different routes after internalization. Entry via this scavenger receptor allows delivery of pseudomonas exotoxin to the appropriate compartments for processing to an ADP-ribosyltransferase competent form and for escape into the cell cytoplasm. Not all cells are susceptible to PE, and this resistance can be attributed to either absence of receptor which mediates toxin internalization or to the presence of receptor-associated protein which is able to regulate binding of ligand to receptor. Further studies are required to determine whether RAP binds directly to the ligand (toxin) binding site on the α_2MR/LRP, or if RAP binding induces a conformational change in the receptor molecule, thus rendering it unable to bind other ligands.

ACKNOWLEDGMENTS

The studies on the relationship of toxin receptor and the α_2MR/LRP have been carried out in collaboration with Dr. David FitzGerald, National Cancer Institute, NIH, and Dr. Dudley Strickland, American Red Cross. I would like to acknowledge the work of Judith Forristal and Diane Mucci in studies carried out in my (CBS) laboratory. The work was supported by RO1 AI 17529, NIAID.

ABBREVIATIONS

α_2MR/LRP: $\alpha_2$macroglobulin receptor/low density lipoprotein like receptor protein

α_2Macroglobulin*:   activated form of $\alpha_2$macroglobulin
LDL: low density lipoprotein
PE: pseudomonas exotoxin A
RAP: receptor associated protein
RAP-GST: fusion protein of RAP and glutathione-S-transferase
SDS-PAGE: sodium dodecyl sulfate-polyacrylamide gel electrophoresis

REFERENCES

Allured, V. S., Collier, R. J., Carroll, S. F., & McKay, D. B. (1986). Structure of exotoxin A of *Pseudomonas aeruginosa* at 3.0-Angstrom resolution. Proc. Natl. Acad. Sci. USA 83, 1320–1324.

Ashcom, J., Tiller, S., Dickerson, K., Cravens, J., Argraves, W. S., & Strickland, D. K. (1990). The human $\alpha_2$macroglobulin receptor: Identification of a 420-kD cell surface glycoprotein specific for the activated conformation of $\alpha_2$macroglobulin. J. Cell Biol. 110, 1041–1048.

Chappell, D., Fry, G., Waknitz, M., Iverius, P., Williams, S., & Strickland, D. (1992). The low density lipoprotein receptor-related protein /$\alpha_2$macroglobulin receptor binds and mediates catabolism of bovine milk lipoprotein lipase. J. Biol. Chem. 267, 25764–25767.

Chaudhary, V., Jinno, Y., FitzGerald, D., & Pastan, I. (1990a). *Pseudomonas* exotoxin contains a specific sequence at the carboxyl terminus that is required for cytotoxicity. Proc. Natl. Acad. Sci. USA 87, 308–312.

Chaudhary, V., Jinno, Y., Gallo, M., FitzGerald, D., & Pastan I. (1990b). Mutagenesis of *Pseudomonas* exotoxin in identification of sequences responsible for the animal toxicity. J. Biol. Chem. 265, 16306–16310.

Choi, S. & Cooper, A. (1993). A comparison of the roles of the low density lipoprotein (LDL) receptor and the LDL receptor-related protein /$\alpha_2$macroglobulin receptor in chylomicron remnant removal in the mouse *in vivo*. J. Biol. Chem. 268, 15804–15811.

Farahbakhsh, Z. T., Baldwin, R. L., & Wisnieski, B. J. (1986). *Pseudomonas* exotoxin A: membrane binding, insertion, and traversal. J. Biol. Chem. 261, 11404–11408.

FitzGerald, D., Morris, R.E., & Saelinger, C. B. (1980). Receptor-mediated internalization of *Pseudomonas* toxin by mouse fibroblasts. Cell 21, 867–873.

Forristal, J. J., Thompson, M. R., Morris, R. E., & Saelinger, C. B. (1991). Mouse liver contains a *Pseudomonas aeruginosa* exotoxin A-binding protein. Infect. Immun. 59, 2880–2884.

Fryling, C., Ogata, M., & FitzGerald, D. (1992). Characterization of a cellular protease that cleaves *Pseudomonas* exotoxin. Infect. Immun. 60, 497–502.

Guidi-Rontani, C. & Collier, R. J. (1987). Exotoxin A of *Pseudomonas aeruginosa*: evidence that Domain I functions in receptor binding. Mol. Micro. 1, 67–72.

Herz, J., Clouthier, D. E., & Hammer. R. E. (1992). LDL receptor-related protein internalizes and degrades uPA-PAI-1 complexes and is essential for embryo implantation. Cell 71, 411–421.

Herz, J., Hamann, U., Rogne, S., Myklebost, O., Gausepohl, H., & Stanley, K. (1988). Surface location and high affinity for calcium of a 500-kDa liver membrane protein closely related to the LDL-receptor suggest a physiological role as lipoprotein receptor. EMBO J. 7, 4119–4127.

Herz, J., Goldstein, J., Strickland, D., Ho, Y., & Brown, M. (1991). 39-kDa protein modulates binding of ligands to low density lipoprotein receptor-related protein/α_2-macroglobulin receptor. 266, 21232–21238.

Herz, J., Kowal, R. C., Goldstein, J. L., & Brown, M. S. (1990). Proteolytic processing of the 600 kd low density lipoprotein receptor-related protein (LRP) occurs in a *trans*-Golgi compartment. EMBO J. 9, 1769–1776.

Huettinger, M., Retzek, H., Hermann, M., & Goldenberg, S. (1992). Lactoferrin specifically inhibits endocytosis of chylomicron remnants but not α-macroglobulin. J. Biol. Chem. 267, 18551–18557.

Hwang, J., FitzGerald, D. J., Adhya, S., & Pastan, I. (1987). Functional domains of *Pseudomonas* exotoxin identified by deletion analysis of the gene expressed in *E. coli*. Cell 48, 129–136.

Iglewski, B. H., Liu, P. V., & Kabat, D. (1977). Mechanism of action of *Pseudomonas aeruginosa* exotoxin A: Adenosine diphosphate-ribosylation of mammalian elongation factor 2 *in vitro* and *in vivo*. Infect. Immun. 15, 138–144.

Iglewski, B. H. & Kabat, D. (1975). NAD-dependent inhibition of protein synthesis by *Pseudomonas aeruginosa* toxin. Proc. Natl. Acad. Sci. USA 72, 2284–2288.

Jinno, Y., Chaudary, V., Kondo, T., Adhya, S., FitzGerald, D., & Pastan, I. (1988). Mutational analysis of Domain I of *Pseudomonas* exotoxin: Mutations in Domain I of *Pseudomonas* exotoxin which reduce cell binding and animal toxicity. J. Biol. Chem. 263, 13203–13207.

Kounnas, M., Argraves, W. S., & Strickland, D. K. (1992a). The 39-kDa receptor-associated protein interacts with two members of the low density lipoprotein receptor family, $\alpha_2$macroglobulin receptor and glycoprotein 330. J. Biol. Chem. 267, 21162–21166.

Kounnas, M., Morris, R., Thompson, M., FitzGerald, D., Strickland, D., & Saelinger, C. B. (1992b). The $\alpha_2$macroglobulin receptor/low density lipoprotein receptor-related protein binds and internalizes *Pseudomonas* exotoxin A. J. Biol. Chem. 267, 21232–21238.

Kowal, R., Herz, J., Goldstein, J., Esser, V., & Brown, M. (1989). Low density lipoprotein receptor-related protein mediates uptake of cholesteryl esters derived from apoprotein E-enriched lipoproteins. Proc. Natl. Acad. Sci. USA 86, 5810–5814.

Kowal, R., Herz, J., Weisgraber, K., Mahley, R., Brown, M., & Goldstein, J. (1990). Opposing effects of apolipoproteins E and C on lipoprotein binding to low density lipoprotein receptor-related protein. J. Biol. Chem. 265, 10771–10779.

Kristensen, T., Moestrup, S., Gliemann, J., Bendtsen, L., Sand, O., & Sottrup-Jensen, L. (1990). Evidence that the newly cloned low-density lipoprotein receptor-related protein (LRP) is the $\alpha_2$macroglobulin receptor. FEBS 276, 151–155.

Leppla, S. H., Martin, O. C., & Muehl, L. A. (1978). The exotoxin of *P. aeruginosa*: A proenzyme having an unusual mode of activation. Biochem. Biophys. Res. Commun. 81, 532–538.

Lory, S. & Collier, R. J. (1980). Expression of enzymic activity by exotoxin A from *Pseudomonas aeruginosa*. Infect. Immun. 28, 494–501.

Lund, H., Takahashi, K., Hamilton, R., & Havel, R. J. (1989). Lipoprotein binding and endosomal itinerary of the low density lipoprotein receptor-related protein in rat liver. Proc. Natl. Acad. Sci USA 86, 9318–9322.

Manhart, M., Morris, R. E., Bonventre, P. F., Leppla, S., & Saelinger, C. B. (1984). Evidence for *Pseudomonas* exotoxin A receptors on plasma membrane of toxin-sensitive LM fibroblasts. Infect. Immun. 45, 596–603.

Michael, M. & Saelinger, C. B. (1979). Toxicity of pseudomonas toxin for mouse LM cell fibroblasts. Curr. Microbiol. 2, 103–108.

Middlebrook, J. & Dorland, R. (1977). Response of cultured mammalian cells to the exotoxins of *Pseudomonas aeruginosa* and *Corynebacterium diphtheriae*: differential cytotoxicity. Can. J. Microbiol. 23, 175–182.

Moestrup, S. & Gliemann, J. (1991). Analysis of ligand recognition by the purified $\alpha_2$macroglobulin receptor (low density lipoprotein receptor-related protein). J. Biol. Chem 266, 14011–14017.

Moestrup, S. & Gliemann, J. (1989). Purification of the rat hepatic $\alpha_2$macroglobulin receptor as an approximately 440-kDa single chain protein. J. Biol. Chem. 264, 15574–15577.

Moestrup, S., Holtet, T., Etzerodt, M., Thogersen, H., Nykjaer, A., Andreasen, P., Rasmussen, H., Sottrup-Jensen, L., & Gliemann, J. (1993). α_2Macroglobulin-proteinase complexes, plasminogen activator inhibitor type-1-plasminogen activator complexes, and receptor-associated protein bind to a region of the $\alpha_2$macroglobulin receptor containing a cluster of eight complement-type repeats. J. Biol. Chem. 268, 13691–13696.

Morris, R. E. (1990). Interaction between *Pseudomonas* exotoxin A and mouse LM fibroblast cells. In: *Trafficking of Bacterial Toxins*, Ed. C. B. Saelinger, CRC Press, Boca Raton Fl. pp. 49–71.

Morris, R. E. & Saelinger, C. B. (1983). Diphtheria toxin does not enter resistant cells by receptor-mediated endocytosis. Infect. Immun. 42, 812–817.

Morris, R. E. & Saelinger, C. B. (1986). Reduced temperature alters *Pseudomonas* exotoxin A entry into the mouse LM cell. Infect. Immun. 52, 445–453.

Nykjaer, A., Bengtsson-Olivecrona, G., Lookene, A., Moestrup, S., Petersen, C., Weber, W., Beisiegel, U., & Gliemann, J. (1993). The α2-macroglobulin receptor/low density lipoprotein receptor-related protein binds lipoprotein lipase and β-migrating very low density lipoprotein associated with the lipase. J. Biol. Chem. 268, 15048–15055.

Nykjaer, A., Petersen, C., Moller, B., Jensen, P., Moestrup, S., Holtet, T., Etzerodt, M., Thogersen, H., Munch, M., Andreasen, P., & Gliemann, J. (1992). Purified α2-macroglobulin receptor/low density lipoprotein receptor-related protein binds urokinase-plasminogen activator inhibitor type-1 complex; evidence that the α2-macroglobulin receptor mediates cellular degradation of urokinase receptor-bound complexes. J. Biol. Chem. 267, 14543–14546.

Ogata, M., Chaudhary, V. K., Pastan, I., & FitzGerald, D. J. (1990). Processing of *Pseudomonas* exotoxin by a cellular protease results in the generation of a 37,000-Da toxin fragment that is translocated to the cytosol. J. Biol. Chem. 265, 20678–20685.

Orth, K., Madison, E., Gething, M.-J., Sambrook, J., & Herz, J. (1992). Complexes of tissue-type plasminogen activator and its serpin inhibitor plasminogen-activator inhibitor type 1 are internalized by means of the low density lipoprotein receptor-related protein/α2macroglobulin receptor. Proc. Natl. Acad. Sci. USA 89, 7422–7426.

Pavlovskis, O. R. & Shackelford, A. H. (1974). *Pseudomonas aeruginosa* exotoxin in mice: Localization and effect on protein synthesis. Infect. Immun. 9, 540–546.

Saelinger, C., Snell, K., & Holder, I. (1977). Experimental studies on the pathogenesis of infections due to *Pseudomonas aeruginosa*: direct evidence for toxin production during *Pseudomonas* infection of burned skin tissues. J. Infect. Dis. 136, 555–561.

Seetharam, S., Chaudhary, V. K., FitzGerald, D., & Pastan, I. (1991). Increased cytotoxic activity of *Pseudomonas* exotoxin and two chimeric toxins ending in KDEL. J. Biol. Chem. 266, 17376–17381.

Siegall, C. B., Chaudhary, V. K., FitzGerald, D. J., & Pastan I. (1989). Functional analysis of Domains II, Ib, and III of *Pseudomonas* exotoxin. J. Biol. Chem. 264, 14256–14261.

Snell, K., Holder, I. A., Leppla, S., & Saelinger, C. B. (1978). Role of exotoxin and protease as possible virulence factors in experimental infections with *Pseudomonas aeruginosa*. Infect. Immun. 19, 839–845.

Strickland, D., Ashcom, J., Williams, S., Burgess, W., Migliorini, M., & Argraves, W. (1990). Sequence identity between the α2macroglobulin receptor and low density lipoprotein receptor-related protein suggests that this molecule is a multifunctional receptor. J. Biol. Chem. 265, 17401–17404.

Strickland, D. K., Ashcom, J., Williams, S., Battey, F., Behre, E., McTigue, K., Battey, J., & Argraves, W. (1991). Primary structure of α2macroglobulin receptor-associated protein. J. Biol. Chem. 266, 13364–13369.

Thompson, M., Forristal, J., Kauffmann, P., Madden, T., Kozak, K., Morris, R. E., & Saelinger, C. B. (1991). Isolation and characterization of *Pseudomonas aeruginosa* exotoxin A binding glycoprotein from mouse LM cells. J. Biol. Chem. 266, 2390–2396.

Vasil, M. & Iglewski, B. (1978). Comparative toxicities of diphtherial toxin and *Pseudomonas aeruginosa* exotoxin A: evidence for different cell receptors. J. Gen. Microbiol. 108, 333–337.

Wick, M., Frank, D., Storey, D., & Iglewski, B. H. (1990). Structure, function, and regulation of *Pseudomonas aeruginosa* exotoxin A. Ann. Rev. Microbiol. 44, 335–363.

Williams, S., Ashcom, J., Argraves, W., & Strickland, D. K. (1992). A novel mechanism for controlling the activity of α2macroglobulin receptor/low density lipoprotein receptor-related protein. J. Biol. Chem. 267, 9035–9040.

Willnow, T. E., Goldstein, J. L., Orth, K., Brown, M. S., & Herz, J. (1992). Low density lipoprotein receptor-related protein and gp330 bind similar ligands, including plasminogen activator-inhibitor complexes and lactoferrin, an inhibitor of chylomicron remnant clearance. J. Biol. Chem. 267, 26172–26180.

Subject Index